# Modern Data Access with Entity Framework Core

## Database Programming Techniques for .NET, .NET Core, UWP, and Xamarin with C#

Holger Schwichtenberg

Apress®

## Modern Data Access with Entity Framework Core

Holger Schwichtenberg
Essen, Germany

ISBN-13 (pbk): 978-1-4842-3551-5 ISBN-13 (electronic): 978-1-4842-3552-2
https://doi.org/10.1007/978-1-4842-3552-2

Library of Congress Control Number: 2018947392

Managing Director, Apress Media LLC: Welmoed Spahr
Acquisitions Editor: Joan Murray
Development Editor: Laura Berendson
Coordinating Editor: Jill Balzano

Cover designed by eStudioCalamar

Cover image designed by Freepik (www.freepik.com)

Distributed to the book trade worldwide by Springer Science+Business Media New York, 233 Spring Street, 6th Floor, New York, NY 10013. Phone 1-800-SPRINGER, fax (201) 348-4505, e-mail orders-ny@springer-sbm.com, or visit www.springeronline.com. Apress Media, LLC is a California LLC and the sole member (owner) is Springer Science + Business Media Finance Inc (SSBM Finance Inc). SSBM Finance Inc is a **Delaware** corporation.

For information on translations, please e-mail rights@apress.com, or visit www.apress.com/rights-permissions.

Apress titles may be purchased in bulk for academic, corporate, or promotional use. eBook versions and licenses are also available for most titles. For more information, reference our Print and eBook Bulk Sales web page at www.apress.com/bulk-sales.

Any source code or other supplementary material referenced by the author in this book is available to readers on GitHub via the book's product page, located at www.apress.com/9781484235515. For more detailed information, please visit www.apress.com/source-code.

Printed on acid-free paper

*For Heidi, Felix, and Maja*

# Table of Contents

# About the Author

**Holger Schwichtenberg** is a .NET MVP with more than 20 years of experience as a developer and trainer. He is currently a technical lead with the German company IT-Visions, where he consults and trains at companies throughout Europe. He also serves as a software architect for 5Minds IT-Solutions. Holger is a huge fan of Entity Framework (EF) and Entity Framework Core and regularly speaks about both. He has used EF in projects of all sizes, most recently on a big data project containing billions of records. He is a prolific writer, having published more than 65 books and 1,000 technical articles in well-known IT professional and developer journals, including MSDN. He has presented at events such as TechEd Europe, Microsoft IT Forum, Advanced Developer Conference, Microsoft Launch, MSDN Technical Summit, and others. Holger has a PhD in business informatics.

His company web sites are `www.IT-Visions.de` and `www.5minds.de`, and he regularly blogs at `dotnet-doktor.de`. His office can be reached at `office@IT-Visions.de`.

# Introduction

I have always been a big fan of object-relational mapping (ORM); in fact, I developed my own OR mapper for my software development projects in the early days of .NET. I switched to the ADO.NET Entity Framework when Microsoft introduced it in .NET 3.5 Service Pack 1. Nowadays, I prefer its successor, Entity Framework Core. Interestingly, some of my projects are still running the classic Entity Framework. As Microsoft continues to do incremental releases of Entity Framework Core, many of the challenges and gripes developers had with earlier versions have gone away, so my plan is to switch the management of all my projects to Entity Framework Core.

The book you hold in your hands came from an idea I had to cover all the important database access scenarios. I hadn't found much collective information in one place and felt that a compendium could be of great value to others. In this book, you will be introduced to database access concepts, get hands-on experience installing Entity Framework Core, and learn about reverse engineering and forward engineering for existing or legacy databases. I'll delve into topics such as schema migrations, data reading, and data modification with LINQ, Dynamic LINQ, APIs, SQL, stored procedures and table-valued functions, object relationships, and asynchronous programming. I'll also talk about third-party products such as LINQPad, Entity Developer, Entity Framework Profiler, Entity Framework Plus, and AutoMapper.

I'll discuss how to apply Entity Framework Core through case studies using Universal Windows Platform (UWP) apps, Xamarin, and ASP.NET Core. Of course, no book would be complete without sharing a healthy dose of hard-earned tips and tricks from my experience with Entity Framework and Entity Framework Core over the years.

## Expectations of the Reader

This book is intended for software developers who have experience with .NET and C# as well as some relational database experience and who now want to use Entity Framework Core to create data access code in .NET, .NET Core, UWP apps, and Xamarin. Previous knowledge in predecessor technologies such as ADO.NET and the classic ADO.NET Entity Framework is useful but not necessary to understand this book.

# Programming Language Used in This Book

I chose to use C# in this book because it is by far the most commonly used programming language in .NET. While I still occasionally develop .NET applications in Visual Basic .NET, it doesn't make sense to print all the listings in both languages.

If you are interested, a language converter between C# and Visual Basic .NET is freely available on several web sites, including `http://converter.telerik.com` and `https://www.mindfusion.eu/convert-cs-vb.html`.

# The Use of Case Studies and Fictitious Enterprises

Most of the sample code in this book revolves around the fictitious airline World Wide Wings, abbreviated as WWWings or just WWW (see Figure 1).

***Figure 1.*** *Logo of the fictional airline World Wide Wings*

---

**Note**   You'll see other case studies used in some chapters, such as the task management app MiracleList.

---

The World Wide Wings use case deals with the following entities:

- *Flights* between two places where the places were deliberately not modeled as separate entities but as strings (this simplifies the understanding of many examples).

- *Passengers* flying on a flight.

- *Employees* of the airline, who have supervisors who are also employees.

- *Pilots* as a specialization of employees. A flight has only one pilot. There are no copilots at World Wide Wings.

- *Persons* as a collection of common characteristics for all people in this example. A person is not available on their own, but only in one of three specializations: passenger, employee, and pilot. In the object-oriented sense, therefore, `Person` is an abstract base class that cannot own instances but is used only for inheritance.

The World Wide Wings use case has two data models, explained here:

- The slightly simpler model version 1 (see Figures 2 and 3) is the result of classic relational database design with normalization. The object model is created by reverse engineering.

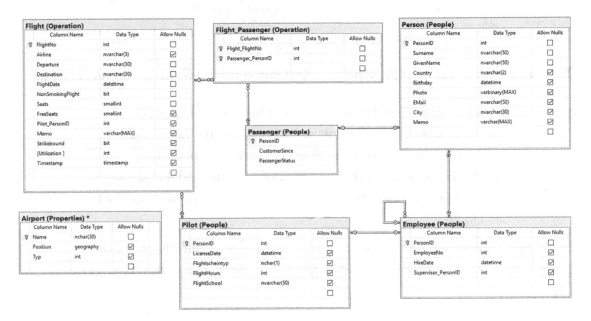

***Figure 2.*** *World Wide Wings data model in the simpler version 1*

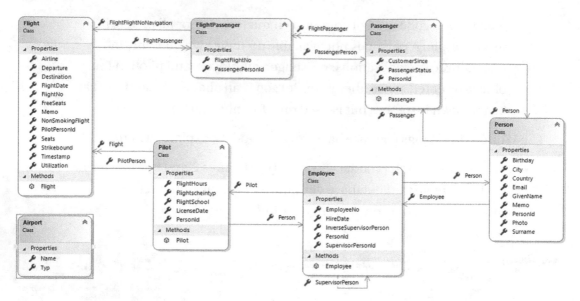

***Figure 3.*** *Object model of the World Wide Wings data model in the simpler version 1*

- Model version 2 (see Figures 4 and 5) is the result of forward engineering with Entity Framework Core from an object model. In addition, there are other entities (Airline, Persondetail, AircraftType, and AircraftTypeDetail) in this model to show further modeling aspects. In this case, there is an optional copilot for each flight.

In model version 1 there is a separate table for people (called Person), staff, pilots, employee, and passengers. This separation corresponds to the classes in the object model.

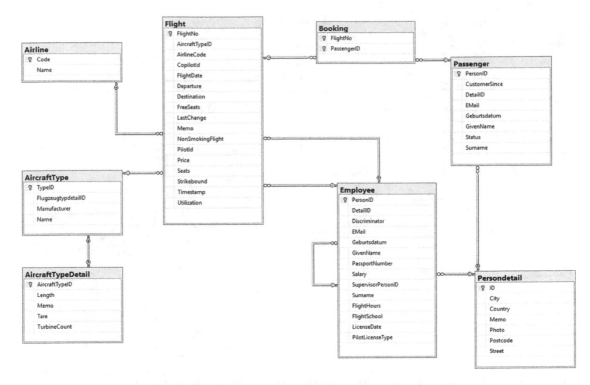

**Figure 4.** *World Wide Wings data model in the more complex version 2*

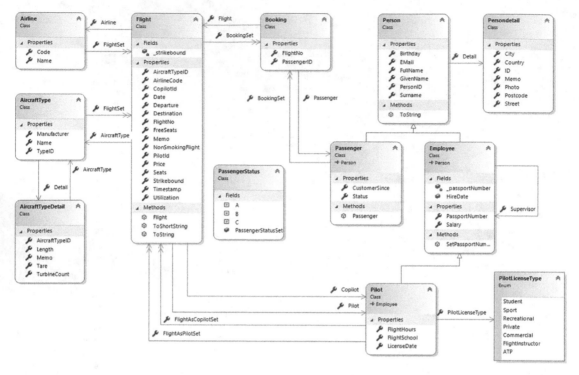

**Figure 5.** *Object model for the World Wide Wings data model in the more complex version 2*

---

**Note** The object models that were created in this book for the data models do not represent an ideal object model because Entity Framework Core does not support some mapping capabilities, such as N:M mapping, yet.

The object model for the data schema of World Wide Wings version 6.1 (Figure 3) was automatically generated by the Entity Framework Core from the database (through reverse engineering); I deliberately did not change it, even if some of the generated names are ugly.

---

In model version 2, there are only the Passenger and Employee tables for these four entities. Entity Framework Core is currently somewhat limited and does not support table per type mapping (a separate table for each class). Therefore, the table Passenger also includes all the characteristics of Person. In addition to the Person properties, the

Employee table includes the properties of the Employee and Pilot entities. In the table, a Discriminator column distinguishes between records that are an employee and those that are a pilot. Entity Framework Core mixes the concepts of table by concrete type (TPC) and table by hierarchy (TPH). The developer has no definite influence on the inheritance mapping in Entity Framework Core 1.*x*/2.0. The classic Entity Framework offers more options here.

The following are the extra dependencies in model version 2:

- A Flight belongs to Airline (there will be only World Wide Wings and its subsidiary Never Come Back Airline in this book).

- There a Copilot entity here, but it is optional.

- A Flight can optionally have an AircraftType object assigned. AircraftType must have an AircraftTypeDetail object.

- Each Person and therefore each Pilot and Passenger must own a Persondetail object.

In this book, both data models are used, partly in modified form, to show certain scenarios (for example, database schema migrations).

# Application Types in This Book

In this book, the examples are for the most time shown via a text-based console in console applications because this allows me to focus on database access. When using graphical user interfaces such as WPF, Windows Forms, ASP.NET Web Forms, or ASP. NET MVC, the representation is decoupled by data binding, which means I would always need to show a second listing so you could understand that the data access was actually delivered. I simulate user inputs in the console examples by writing variables at the beginning of the program code.

I have provided training and consultancy on data access for many years and have learned that console editions are didactically the best tool for teaching because otherwise the listings are large and thus inferior.

Of course, console output is not common practice in 99 percent of software development, but graphical user interfaces are covered in other books, and data binding typically has no impact on the form of data access. Where data access is relevant, this book will also show data binding examples.

# Helper Functions for Console Output

I will show the screen output on the console using the standard method
`Console.WriteLine()` in several places; in addition, I use auxiliary routines that
generate colored screen output. Listing 1 shows these auxiliary routines in the class CUI
from `ITV_DemoUtil.dll` for a better understanding.

*Listing 1.* Class CUI with Subroutines for Screen Output to the Console

```
using System;
using System.Runtime.InteropServices;
using System.Web;
using ITVisions.UI;
using System.Diagnostics;

namespace ITVisions
{
 /// <summary>
 /// Helper utilities for Console UIs
 /// (C) Dr. Holger Schwichtenberg 2002-2018
 /// </summary>
 public static class CUI
 {
  public static bool IsDebug = false;
  public static bool IsVerbose = false;

  #region Print only under certain conditions
 public static void PrintDebug(object s)
  {
    PrintDebug(s, System.Console.ForegroundColor);
  }

  public static void PrintVerbose(object s)
  {
    PrintVerbose(s, System.Console.ForegroundColor);
  }
  #endregion
```

```csharp
#region Print with predefined colors
public static void MainHeadline(string s)
{
  Print(s, ConsoleColor.Black, ConsoleColor.Yellow);

}
public static void Headline(string s)
{
  Print(s, ConsoleColor.Yellow);
}
public static void HeaderFooter(string s)
{
  Console.ForegroundColor = ConsoleColor.Green;
  Console.WriteLine(s);
  Console.ForegroundColor = ConsoleColor.Gray;
}

public static void PrintSuccess(object s)
{
  Print(s, ConsoleColor.Green);
}

public static void PrintStep(object s)
{
  Print(s, ConsoleColor.Cyan);
}

public static void PrintDebugSuccess(object s)
{
  PrintDebug(s, ConsoleColor.Green);
}

public static void PrintVerboseSuccess(object s)
{
  PrintVerbose(s, ConsoleColor.Green);
}
```

```
public static void PrintWarning(object s)
{
 Print(s, ConsoleColor.Cyan);
}

public static void PrintDebugWarning(object s)
{
 PrintDebug(s, ConsoleColor.Cyan);
}

public static void PrintVerboseWarning(object s)
{
 PrintVerbose(s, ConsoleColor.Cyan);
}

public static void PrintError(object s)
{
 Print(s, ConsoleColor.White, ConsoleColor.Red);
}

public static void PrintDebugError(object s)
{
 PrintDebug(s, ConsoleColor.White, ConsoleColor.Red);
}

public static void PrintVerboseError(object s)
{
 Print(s, ConsoleColor.White, ConsoleColor.Red);
}

public static void Print(object s)
{
 PrintInternal(s, null);
}
#endregion

#region Print with selectable color
```

```csharp
public static void Print(object s, ConsoleColor frontcolor, ConsoleColor?
backcolor = null)
{
 PrintInternal(s, frontcolor, backcolor);
}

public static void PrintDebug(object s, ConsoleColor frontcolor,
ConsoleColor? backcolor = null)
{
 if (IsDebug || IsVerbose) PrintDebugOrVerbose(s, frontcolor, backcolor);
}

public static void PrintVerbose(object s, ConsoleColor frontcolor)
{
 if (!IsVerbose) return;
 PrintDebugOrVerbose(s, frontcolor);
}
#endregion

#region Print with additional data

/// <summary>
/// Print with Thread-ID
/// </summary>
public static void PrintWithThreadID(string s, ConsoleColor c =
ConsoleColor.White)
{
 var ausgabe = String.Format("Thread #{0:00} {1:}: {2}", System.Threading.
 Thread.CurrentThread.ManagedThreadId, DateTime.Now.ToLongTimeString(), s);
 CUI.Print(ausgabe, c);
}

/// <summary>
///  Print with time
/// </summary>
public static void PrintWithTime(object s, ConsoleColor c = ConsoleColor.
White)
```

```
  {
    CUI.Print(DateTime.Now.Second + "." + DateTime.Now.Millisecond + ":" + s);
  }

  private static long count;
  /// <summary>
  /// Print with counter
  /// </summary>
  private static void PrintWithCounter(object s, ConsoleColor frontcolor,
  ConsoleColor? backcolor = null)
  {
    count += 1;
    s = $"{count:0000}: {s}";
    CUI.Print(s, frontcolor, backcolor);
  }

  #endregion

  #region internal helper routines
  private static void PrintDebugOrVerbose(object s, ConsoleColor
  frontcolor, ConsoleColor? backcolor = null)
  {
    count += 1;
    s = $"{count:0000}: {s}";
    Print(s, frontcolor, backcolor);
    Debug.WriteLine(s);
    Trace.WriteLine(s);
    Trace.Flush();
  }

  /// <summary>
  /// Output to console, trace and file
  /// </summary>
  /// <param name="s"></param>
  [DebuggerStepThrough()]
```

```
  private static void PrintInternal(object s, ConsoleColor? frontcolor =
  null, ConsoleColor? backcolor = null)
  {
   if (s == null) return;

   if (HttpContext.Current != null)
   {
    try
    {
if (frontcolor != null)
     {
       HttpContext.Current.Response.Write("<span style='color:" +
       frontcolor.Value.DrawingColor().Name + "'>");
     }
     if (!HttpContext.Current.Request.Url.ToString().ToLower().Contains(".
asmx") && !HttpContext.Current.Request.Url.ToString().ToLower().
Contains(".svc") && !HttpContext.Current.Request.Url.ToString().
ToLower().Contains("/api/")) HttpContext.Current.Response.Write(s.
ToString() + "<br>");

     if (frontcolor != null)
     {
      HttpContext.Current.Response.Write("</span>");
     }
    }
    catch (Exception)
    {
    }
   }
   else
   {
    object x = 1;
    lock (x)
    {
```

```
        ConsoleColor altefrontcolor = Console.ForegroundColor;
        ConsoleColor alteHfrontcolor = Console.BackgroundColor;

        if (frontcolor != null) Console.ForegroundColor = frontcolor.Value;
        if (backcolor != null) Console.BackgroundColor = backcolor.Value;

        Console.WriteLine(s);
        Console.ForegroundColor = altefrontcolor;
        Console.BackgroundColor = alteHfrontcolor;
      }
     }
    }
    #endregion

    #region Set the position of the console window
    [DllImport("kernel32.dll", ExactSpelling = true)]
    private static extern IntPtr GetConsoleWindow();
    private static IntPtr MyConsole = GetConsoleWindow();

    [DllImport("user32.dll", EntryPoint = "SetWindowPos")]
    public static extern IntPtr SetWindowPos(IntPtr hWnd, int
    hWndInsertAfter, int x, int Y, int cx, int cy, int wFlags);

    // Set the position of the console window without size
    public static void SetConsolePos(int xpos, int ypos)
    {
     const int SWP_NOSIZE = 0x0001;
     SetWindowPos(MyConsole, 0, xpos, ypos, 0, 0, SWP_NOSIZE);
    }

    // Set the position of the console window with size
    public static void SetConsolePos(int xpos, int ypos, int w, int h)
    {
     SetWindowPos(MyConsole, 0, xpos, ypos, w, h, 0);
    }
    #endregion
   }
  }
```

# Access to Code Examples, Updates, and Figures on GitHub

You can download all the examples for this book as Visual Studio projects via the book's product page, located at www.apress.com/9781484235515. This page will also point you to additional complementary content on Entity Framework Core updates, discussion, and errata.

Microsoft continues to release incremental updates to Entity Framework Core and readily admits the following: "These are things we think we need before we can say EF Core is the recommended version of EF for everyone" (https://github.com/aspnet/EntityFrameworkCore/wiki/roadmap). To help make sense of the changes coming in 2.1, you can find this book's Appendix C, which outlines new features, on the book's GitHub repository.

You can also find information on the book page I maintain at www.efcore.net.

Please note that not every single line of program code in this book is included in the downloadable projects. This book also discusses alternative solutions for individual cases that do not necessarily fit into a total solution.

| Visual Studio Solution | Content |
|---|---|
| EFC_WWWings | Most examples of this book, based on the World Wide Wings scenario. |
| EFC_UWP_SQLite and EFC_Xamarin_SQLite | Sample application MiracleList Light as a UWP app for Windows 10 and cross-platform app for iOS, Android, and Windows 10. The app stores data using Entity Framework Core in SQLite. |

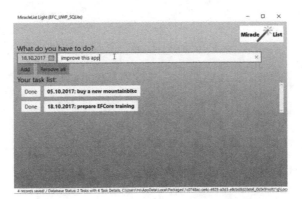

| | |
|---|---|
| EFC_Countries_NMSelf | Country border example from Appendix A. |

The figures that appear in this book are available for download as well, which will allow you to see them in a larger, color format.

Now that you have been introduced to the plan for the book, the mock company I'll be using, and the location for all the code, let's get started on your Entity Framework Core learning journey!

Thank you for reading my book, and I welcome any comments and feedback at EFCoreBookFeedback@dotnet-doctor.net.

# CHAPTER 1

# Introducing Entity Framework Core

In this chapter, you will learn about Entity Framework Core and how it is an OR mapper for .NET (.NET Framework, .NET Core, Mono, and Xamarin). Entity Framework Core is a completely new implementation of the ADO.NET Entity Framework.

Together with .NET Core version 1.0 and ASP.NET Core version 1.0, Entity Framework Core version 1.0 was released on June 27, 2016. Version 2.0 was released on August 14, 2017. Version 2.1 is in progress.

## What Is an Object-Relational Mapper?

In the database world, relational databases are prevalent. The programming world is all about objects. There are significant semantic and syntactic differences between the two worlds, called *impedance mismatch*; see `https://en.wikipedia.org/wiki/Object-relational_impedance_mismatch`.

Working with objects as instances of classes in memory is at the core of object-oriented programming (OOP). Most applications also include the requirement to permanently store data in objects, especially in databases. Basically, there are object-oriented databases (OODBs) that are directly able to store objects, but OODBs have only a small distribution so far. Relational databases are more predominant, but they map the data structures differently than object models.

To make the handling of relational databases more natural in object-oriented systems, the software industry has been relying on object-relational mappers for years. These tools translate concepts from the object-oriented world, such as classes, attributes, or relationships between classes, to corresponding constructs of the relational world, such as tables, columns, and foreign keys (see Figure 1-1). Developers can thus remain

1

H. Schwichtenberg, *Modern Data Access with Entity Framework Core*,
https://doi.org/10.1007/978-1-4842-3552-2_1

in the object-oriented world and instruct the OR mapper to load or store certain objects
that are in the form of records in tables of the relational database. Less interesting tasks
and error-prone ones, such as manually creating INSERT, UPDATE, and DELETE statements,
are also handled by the OR mapper, further reducing the load on the developer.

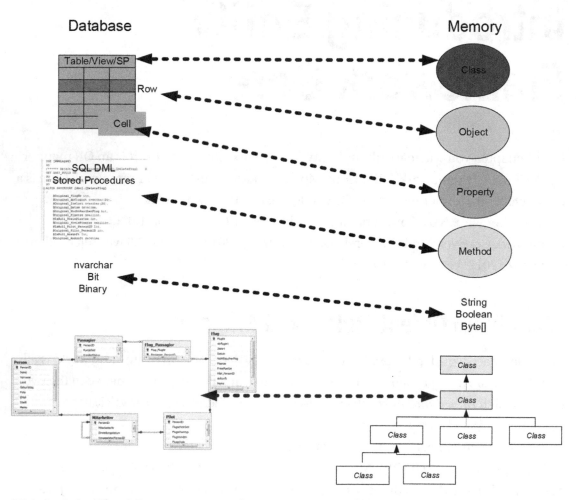

***Figure 1-1.*** *The OR mapper translates constructs of the OOP world to the
relational world*

Two particularly striking differences between the object model and the relation
model are N:M relations and inheritance. While in an object model you can map an
N:M relationship between objects through a reciprocal set of objects, you need an
intermediate table in the relational database. Relational databases do not support
inheritance. There are different ways of replicating, but you'll learn more about that later
in the book.

# OR Mappers in the .NET World

When a .NET developer reads in data from a database with a `DataReader` or `DataSet`, the developer is not doing object-relational mapping at this point. Although `DataReader` and `DataSet` are .NET objects, they only manage table structures. `DataReader` and `DataSet` are untyped, nonspecific containers from the point of view of an object model. Only when the developer defines specific classes for the structures stored in the tables and copies the contents from `DataSet` or `DataReader` into these specific data structures is the developer performing OR mapping. Such "manual" object-relational mapping is time-consuming, tedious, and monotonous programming work for read access (especially for very wide tables). If you then want to save changes in the objects again, the work becomes an intellectual challenge because you have to be able to recognize which objects have been changed. Otherwise, you constantly have to save all the data anew, which is an absurdity in multiuser environments.

While OR mappers have long been established in the Java world, Microsoft failed to bring a suitable product to market for a long time. The first version of .NET did not include an OR mapper but limited itself to direct data access and mapping between XML documents and the relational model. In .NET 3.5, there was an OR mapper named LINQ to SQL, but it was limited to Microsoft SQL Server and had many other restrictions.

Many .NET developers have therefore set about simplifying this work with auxiliary libraries and tools. In addition to the publicly known OR mappers for .NET, you will find many in-house solutions being built.

The following are third-party OR mappers for .NET (some of them open source):

- NHibernate

- Telerik Data Access (aka Open Access)

- Genome

- LLBLGen Pro

- Wilson

- SubSonic

- OBJ.NET

- DataObjects.NET

- Dapper

- PetaPoco

- Massive

- DevExpress XPO

With LINQ to SQL, ADO.NET Entity Framework, and Entity Framework, Microsoft itself now has three ORM products. The software company has meanwhile announced that further development efforts will concentrate solely on Entity Framework Core.

# Version History of Entity Framework Core

Figure 1-2 shows the version history of Entity Framework Core.

| Version | Downloads |
| --- | --- |
| **2.0.1 (current version)** | 128,851 |
| 2.0.0 | 517,081 |
| 2.0.0-preview2-final | 23,623 |
| 2.0.0-preview1-final | 46,953 |
| 1.1.5 | 1,754 |
| 1.1.4 | 18,574 |
| 1.1.3 | 72,553 |
| 1.1.2 | 483,726 |
| 1.1.1 | 410,096 |
| 1.1.0 | 840,208 |
| 1.1.0-preview1-final | 19,623 |
| 1.0.6 | 690 |
| 1.0.5 | 1,379 |
| 1.0.4 | 8,946 |
| 1.0.3 | 91,927 |
| 1.0.2 | 46,577 |
| 1.0.1 | 312,274 |
| 1.0.0 | 435,614 |
| 1.0.0-rc2-final | 72,792 |

***Figure 1-2.*** *Entity Framework Core version history (`https://www.nuget.org/`*
*`packages/Microsoft.EntityFrameworkCore`)*

Major and minor versions (1.0, 1.1, and so on) indicate feature releases from Microsoft, and revision releases (1.0.1, 1.0.2, and so on) indicate bug fix releases. This book mentions the minimum version when discussing a function that requires a particular version.

---

**Note**   The Entity Framework Core tools for Entity Framework Core 1.*x* were released on March 6, 2017, within the framework of Entity Framework Core 1.1.1 and Visual Studio 2017. Previously, there were only "preview" versions of the tools. Since Entity Framework Core 2.0, the tools are always delivered with the new product releases.

---

# Supported Operating Systems

Like the other products in the Core product family, Entity Framework Core (formerly Entity Framework 7.0) is platform independent. The Core version of the established object-relational mapper runs not only on the "full" .NET Framework but also on .NET Core and Mono, including Xamarin. This allows you to use Entity Framework Core on Windows, Windows Phone/Mobile, Linux, macOS, iOS, and Android.

# Supported .NET Versions

Entity Framework Core 1.*x* runs on .NET Core 1.*x*, .NET Framework 4.5.1, Mono 4.6, Xamarin.iOS 10, Xamarin Android 7.0 or higher, and the Universal Windows Platform (UWP).

Entity Framework Core 2.0 is based on .NET Standard 2.0 and therefore requires one of the following .NET implementations (see Figure 1-3):

- .NET Core 2.0 (or higher)

- .NET Framework 4.6.1 (or higher)

- Mono 5.4 (or higher)

- Xamarin.iOS 10.14 (or higher)

- Xamarin.Mac 3.8 (or higher)

- Xamarin.Android 7.5 (or higher)

- UWP 10.0.16299 (or higher)

| .NET Standard | 1.0 | 1.1 | 1.2 | 1.3 | 1.4 | 1.5 | 1.6 | 2.0 |
|---|---|---|---|---|---|---|---|---|
| .NET Core | 1.0 | 1.0 | 1.0 | 1.0 | 1.0 | 1.0 | 1.0 | 2.0 |
| .NET Framework (with .NET Core 1.x SDK) | 4.5 | 4.5 | 4.5.1 | 4.6 | 4.6.1 | 4.6.2 | | |
| .NET Framework (with .NET Core 2.0 SDK) | 4.5 | 4.5 | 4.5.1 | 4.6 | 4.6.1 | 4.6.1 | 4.6.1 | 4.6.1 |
| Mono | 4.6 | 4.6 | 4.6 | 4.6 | 4.6 | 4.6 | 4.6 | 5.4 |
| Xamarin.iOS | 10.0 | 10.0 | 10.0 | 10.0 | 10.0 | 10.0 | 10.0 | 10.14 |
| Xamarin.Mac | 3.0 | 3.0 | 3.0 | 3.0 | 3.0 | 3.0 | 3.0 | 3.8 |
| Xamarin.Android | 7.0 | 7.0 | 7.0 | 7.0 | 7.0 | 7.0 | 7.0 | 8.0 |
| Universal Windows Platform | 10.0 | 10.0 | 10.0 | 10.0 | 10.0 | 10.0.16299 | 10.0.16299 | 10.0.16299 |
| Windows | 8.0 | 8.0 | 8.1 | | | | | |
| Windows Phone | 8.1 | 8.1 | 8.1 | | | | | |
| Windows Phone Silverlight | 8.0 | | | | | | | |

***Figure 1-3.*** *Implementations of .NET Standard (https://docs.microsoft.com/en-us/dotnet/standard/library)*

---

**Note**    Microsoft justifies the restriction to .NET Standard in Entity Framework Core 2.0 at https://github.com/aspnet/Announcements/issues/246. Among other things, it can significantly reduce the size of the NuGet packages.

---

# Supported Visual Studio Versions

For the use of Entity Framework Core 2.0/2.1 you need Visual Studio 2017 Update 3 or higher, even if you are programming with the classic .NET Framework, because Visual Studio only with this update recognizes .NET Standard 2.0 and understands that .NET Framework 4.6.1 and later are implementations of .NET Standard 2.0.

# Supported Databases

Table 1-1 shows the database management systems supported by Entity Framework Core, including ones by Microsoft (SQL Server, SQL Compact, and SQLite) and third-party vendors (PostgreSQL, DB2, Oracle, MySQL, and others).

***Table 1-1.*** *Available Database Drivers for Entity Framework Core*

| Database | Company/Price | URL |
|---|---|---|
| Microsoft SQL Server | Microsoft/free | `www.nuget.org/packages/Microsoft.EntityFrameworkCore.SqlServer` |
| Microsoft SQL Server Compact 3.5 | Microsoft/free | `www.nuget.org/packages/EntityFrameworkCore.SqlServerCompact35` |
| Microsoft SQL Server Compact 4.0 | Microsoft/free | `www.nuget.org/packages/EntityFrameworkCore.SqlServerCompact40` |
| SQLite | Microsoft/free | `www.nuget.org/packages/Microsoft.EntityFrameworkCore.sqlite` |
| In Memory | Microsoft/free | `www.nuget.org/packages/Microsoft.EntityFrameworkCore.InMemory` |
| MySQL | Oracle/free | `www.nuget.org/packages/MySQL.Data.EntityFrameworkCore` |
| PostgreSQL | Open source team npgsql.org/free | `www.nuget.org/packages/Npgsql.EntityFrameworkCore.PostgreSQL` |
| DB2 | IBM/free | `www.nuget.org/packages/EntityFramework.IBMDataServer` |
| MySQL, Oracle, PostgreSQL, SQLite, DB2, Salesforce, Dynamics CRM, SugarCRM, Zoho CRM, QuickBooks, FreshBooks, MailChimp, ExactTarget, Bigcommerce, Magento | Devart/$99 to $299 per driver type | `www.devart.com/purchase.html#dotConnect` |

On mobile devices running Xamarin or Windows 10 UWP apps, Entity Framework Core 1.*x* can address only local databases (SQLite). With the introduction of .NET Standard 2.0, the Microsoft SQL Server client is now also available on Xamarin and the Windows 10 UWP (from the fall of 2017 Creators Update).

The planned support for NoSQL databases such as Redis and Azure Table storage is not yet included in Entity Framework Core version 1.*x*/2.*x*. However, there is an open source development project on GitHub for MongoDB; see `https://github.com/crhairr/EntityFrameworkCore.MongoDb`.

---

**Caution**    Because of breaking changes in the provider interfaces, Entity Framework Core 1.*x* providers are not compatible with Entity Framework Core 2.0. So, you need new providers for version 2.0!

---

# Features of Entity Framework Core

Figure 1-4 visualizes that Entity Framework Core (right areas) contains some new features compared to the previous Entity Framework (left area). Some features are included in Entity Framework 6.*x*, but not in Entity Framework Core 1.*x*/2.0. Microsoft will be upgrading some of these features in the upcoming versions of Entity Framework Core, but will no longer add new features.

---

**Note**    If you download the figures from the book's web site, you'll be able to distinguish the products in the figure based on color.

---

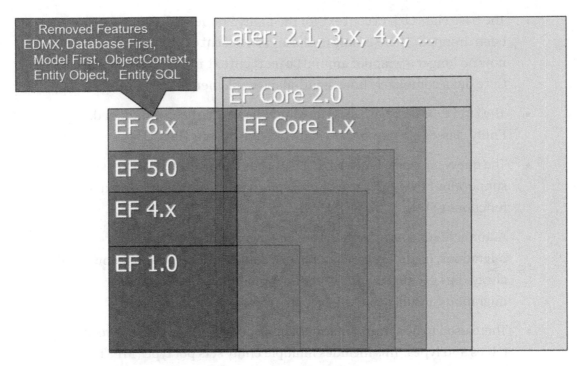

***Figure 1-4.*** *Functional scope of the classic Entity Framework compared to Entity Framework Core. On the left, a balloon shows some features that have been permanently eliminated.*

## Functions That Have Been Eliminated

The following features from the classic Entity Framework have been removed in Entity Framework Core:

- The process models Database First and Model First have been eliminated. In Entity Framework Core, there is only code-based modeling (formerly Code First), with which you can generate both program code from databases (reverse engineering) and databases from program code (forward engineering).

- The Entity Data Model (EDM) and the XML representation (EDMX) have been eliminated. So far, an EDM has also been generated internally in RAM in the Code First model. This overhead has also been eliminated.

- The base class `ObjectContext` for the Entity Framework context has been dropped. There is only the base class `DbContext`. `DbContext` is now no longer a wrapper around `ObjectContext` in Entity Framework Core but a completely new, stand-alone implementation.

- The `EntityObject` base class for entity classes has been eliminated. Entity classes are now always plain old CLR objects (POCOs).

- The query language Entity SQL (ESQL) has been omitted. There is support for LINQ, SQL, stored procedures (SPs), and table-valued functions (TVFs).

- Automatic schema migrations are no longer offered. Schema migrations, including the creation of a database schema, must now always be executed manually at development time. At runtime, a migration can still occur the first time the database is accessed.

- There used to be some scenarios of more complex mapping between tables and types. This includes multiple entity sets per type (MEST, which maps different tables to the same entity), and a combination of the table by hierarchy (TPH), table by type (TPT), and table by concrete type (TPC) strategies in an inheritance hierarchy. All these features have been removed.

# Missing Critical Features

In the Entity Framework Core road map (`https://github.com/aspnet/EntityFramework/wiki/Roadmap`), Microsoft developer Rowan Miller documents what features are missing in Entity Framework Core that will be upgraded soon. This is not backed by a specific timetable, but Microsoft calls some of these features "critical."

- Entity Framework Core supports access only to tables, not to views in the database. You can use a view only if you create the view and the program code manually and treat the view like a table.

- Previously, stored procedures could be used only to query data (`SELECT`) but not to insert (`INSERT`), update (`UPDATE`), and delete (`DELETE`).

- Some LINQ commands are currently executed in RAM, not in the database. This also includes the group by operator, which means that all data sets from the database are read into RAM and grouped there, which leads to catastrophic performance for all tables (except very small ones).

- There is neither automatic lazy loading nor explicit reloading in the Entity Framework Core API. Currently, developers can only load linked data sets directly (eager loading) or reload them with separate commands.

- Direct SQL and stored procedures can be used only if they return entity types. Other types are not supported yet.

- The reverse engineering of existing databases can be started only from the command line or from the NuGet console in Visual Studio. The GUI-based wizard is gone.

- There is also no Update Model from Database command for existing databases; in other words, after reverse engineering a database, the developer must manually add database schema changes in the object model or regenerate the entire object model. This function was also not available in Code First but only in Database First.

- There are no complex types, in other words, classes that do not represent their own entity but are part of another entity.

---

**Preview**   Some of these features will be added in version 2.1; see Appendix C.

---

# High-Priority Features

In a second list, Microsoft calls other features that it does not consider critical "high priority":

- So far, there is no graphic visualization of an object model, as was previously possible with EDMX.

- Some of the previously existing type conversions, such as between XML and strings, do not yet exist.

- The geography and geometry data types of Microsoft SQL Server are not yet supported.

- Entity Framework Core does not support N:M mappings. So far, the developer has to replicate this with two 1:N mappings and an intermediate entity analogous to the intermediate table in the database.

- Table per type is not yet supported as an inheritance strategy. Entity Framework Core uses TPH if there is a `DBSet<T>` for the base class; otherwise, it uses TPC. You cannot explicitly configure TPC.

- It is not possible to populate the database with data as part of the migration (the `seed()` function).

- The Entity Framework 6.0 command interceptors, which allow an Entity Framework software developer to manipulate commands sent to the database before and after execution in the database, do not yet exist.

Some of the items on this high-priority list from Microsoft are also new features that Entity Framework 6.*x* itself does not (yet) master.

- Definition of conditions for data records to be loaded in eager loading (eager loading rules)

- Support for e-tags

- Support for nonrelational data stores (NoSQL) such as Azure Table storage and Redis

This prioritization is from Microsoft's perspective. Based on my practical experience, I would prioritize some points differently; for example, I would upgrade the N:M mapping to critical. A replication of N:M by two 1:N relationships in the object model is possible but makes the program code more complex. Migrating from existing Entity Framework solutions to the Entity Framework Core becomes difficult.

This also applies to the lack of support for table per type inheritance. Again, existing program code has to be extensively changed. And even for new applications with a new database schema and forward engineering, there is a problem: if the inheritance is first realized with TPH or TPC, you have to laboriously rearrange the data in the database schema if you later want to bet on TPH.

Also missing in Microsoft's lists are features such as validating entities, which can save unnecessary round-trips to the database when it's already clear in RAM that the entity does not meet the required conditions.

---

**Preview**   Some of these features will be added in version 2.1; see Appendix C.

---

# New Features in Entity Framework Core

Entity Framework Core can boast the following advantages over its predecessor:

- Entity Framework Core runs not only on Windows, Linux, and macOS but also on mobile devices running Windows 10, iOS, and Android. On mobile devices, of course only access to local databases (such as SQLite) is provided.

- Entity Framework Core provides faster execution speeds, especially when reading data (almost the same performance as manually copying data from a `DataReader` object to a typed .NET object).

- Projections with `Select()` can now be mapped directly to entity classes. The detour via anonymous .NET objects is no longer necessary.

- Batching allows the Entity Framework Core to merge `INSERT`, `DELETE`, and `UPDATE` operations into one database management system round-trip rather than sending each command one at a time.

- Default values for columns in the database are now supported in both reverse engineering and forward engineering.

- In addition to the classic auto-increment values, newer methods such as sequences are now also allowed for key generation.

- The term *shadow properties* in Entity Framework Core refers to the now possible access to columns of the database table for which there are no attributes in the class.

# When to Use Entity Framework Core

Given this long list of missing features, the question arises as to whether and when Entity Framework Core in version 1.$x$/2.0 can be used.

The main application area is on platforms where Entity Framework does not run so far: Windows Phone/Mobile, Android, iOS, Linux, and macOS.

- UWP apps and Xamarin apps can use only Entity Framework Core. The classic Entity Framework is not possible here.

- If you want to develop a new ASP.NET Core web application or web API and you do not want to base it on the full .NET Framework but on .NET Core, there is no way around to do this with Entity Framework Core because the classic Entity Framework 6.$x$ does not work on .NET Core. However, in ASP.NET Core it is also possible to use the .NET Framework 4.6.$x$/4.7.$x$ as a basis so that you use Entity Framework 6.$x$ too.

- Another scenario in which the use of Entity Framework Core on a web server can be recommended is an offline scenario, where there should be a local copy of the server database on the mobile device. In this case, you can use the same data access code on both the client and the server. The client uses Entity Framework Core to access SQLite, and the web server uses the same Entity Framework Core code to access a Microsoft SQL Server.

For projects on other platforms, note the following:

- Migrating existing code from Entity Framework 6.$x$ to Entity Framework Core is expensive. It's important to consider whether the improved features and performance of Entity Framework Core justify the effort.

- However, in new projects, developers can already use Entity Framework Core as a high-performance future technology and, if necessary, use the existing Entity Framework there as an intermediate solution for existing gaps.

# CHAPTER 2

# Installing Entity Framework Core

Entity Framework Core has no `setup.exe`. Entity Framework Core is installed in a project via NuGet packages.

## NuGet Packages

Entity Framework Core, in contrast to the classic Entity Framework, consists of several NuGet packages. Table 2-1 shows only the root packages. Their dependencies, to which NuGet then automatically adds the associated packages, are not listed here. If you're using a project template such as the ASP.NET Core Web Application template, you might have some of the dependencies already included.

15

© Holger Schwichtenberg 2018
H. Schwichtenberg, *Modern Data Access with Entity Framework Core*,
https://doi.org/10.1007/978-1-4842-3552-2_2

***Table 2-1.***  *Main Packages Available on nuget.org for Entity Framework Core*

| Database Management System | NuGet Package Needed at Runtime | NuGet Package Needed at Development Time for Reverse Engineering or Schema Migrations |
|---|---|---|
| Microsoft SQL Server Express, Standard, Enterprise, Developer, LocalDB (from version 2008) | `Microsoft.EntityFrameworkCore.SqlServer` | `Microsoft.EntityFrameworkCore.Tools` `Microsoft.EntityFrameworkCore.SqlServer` (for EF Core 2.0) `Microsoft.EntityFrameworkCore.SQLServer.Design` (for EF Core 1.*x*) |
| Microsoft SQL Server Compact 3.5 | `EntityFrameworkCore.SqlServerCompact35` | Not available |
| Microsoft SQL Server Compact 4.0 | `EntityFrameworkCore.SqlServerCompact40` | Not available |
| SQLite | `Microsoft.EntityFrameworkCore.Sqlite` | `Microsoft.EntityFrameworkCore.Tools` `Microsoft.EntityFrameworkCore.Sqlite` (for EF Core 2.0) `Microsoft.EntityFrameworkCore.Sqlite.Design` (for EF Core 1.*x*) |
| In-Memory | `Microsoft.EntityFrameworkCore.InMemory` | Not available |

*(continued)*

16

***Table 2-1.*** (*continued*)

| Database Management System | NuGet Package Needed at Runtime | NuGet Package Needed at Development Time for Reverse Engineering or Schema Migrations |
|---|---|---|
| PostgreSQL | Npgsql.<br>EntityFrameworkCore.<br>PostgreSQL | Microsoft.<br>EntityFrameworkCore.Tools<br>Npgsql.<br>EntityFrameworkCore.<br>PostgreSQL (for EF Core 2.0)<br>Npgsql.<br>EntityFrameworkCore.<br>PostgreSQL.Design<br>(for EF Core 1.*x*) |
| MySQL | MySQL.Data.<br>EntityFrameworkCore | MySQL.Data.<br>EntityFrameworkCore.<br>Design |
| Oracle (Devart Provider) | Devart.Data.Oracle.<br>EFCore | Microsoft.<br>EntityFrameworkCore.Tools |

In Entity Framework Core version 2.0, Microsoft changed the tailoring of the packages (as it did during the alpha and beta versions), as shown in Figure 2-1. Before, there were two packages for each driver, one of them with *design* in the name. The "design" packages were dissolved and integrated into the actual driver assemblies.

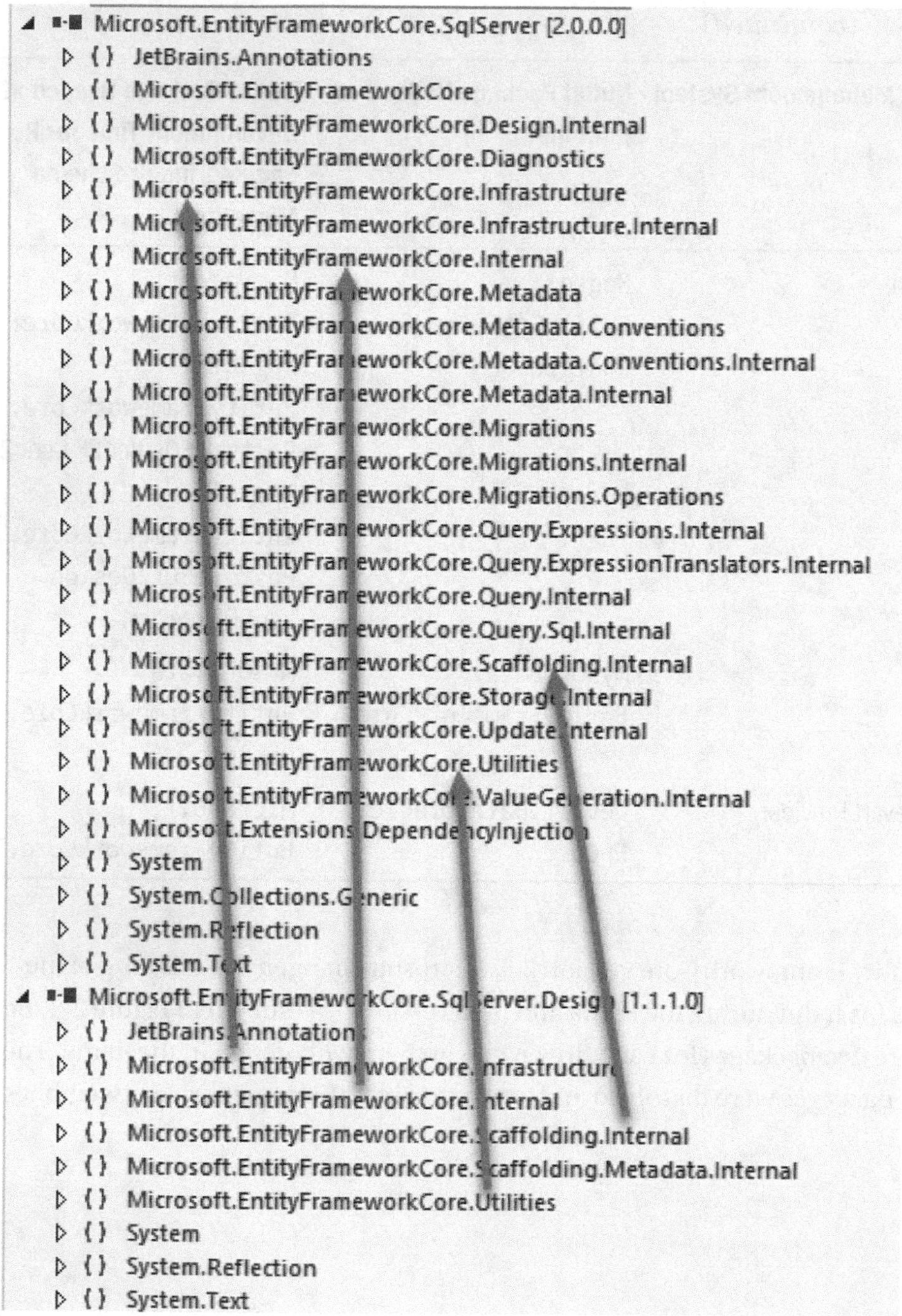

**Figure 2-1.** *In Entity Framework Core 2.0, Microsoft has integrated the classes of Microsoft.Entity FrameworkCore.SQL Server.Design.dll into Microsoft. EntityFrameworkCore.SqlServer.dll.*

# Installing the Packages

You can install the packages with the NuGet Package Manager (Figure 2-2 and 2-3 and 2-4) or the PowerShell cmdlet Install-Package in Visual Studio (Figure 2-5 and 2-6).

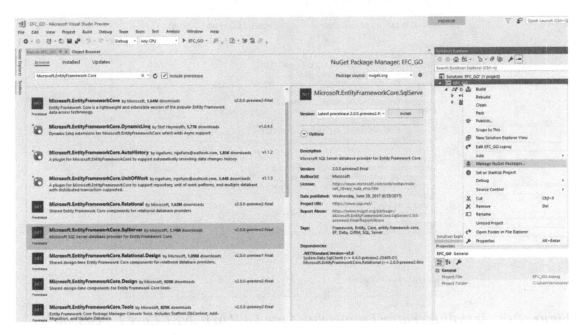

***Figure 2-2.*** *Installing the driver for Microsoft SQL Server with the NuGet Package Manager GUI*

At the command line (select NuGet Package Manager Console ➤ PMC), you can install the current stable version and associated dependencies with the following:

```
Install-Package Microsoft.EntityFrameworkCore.SqlServer
```

You can install the current prerelease version with the following:

```
Install-Package Microsoft.EntityFrameworkCore.SqlServer -Pre
```

You can install a specific version with the following:

```
Install-Package Microsoft.EntityFrameworkCore.SqlServer version 2.0.0
```

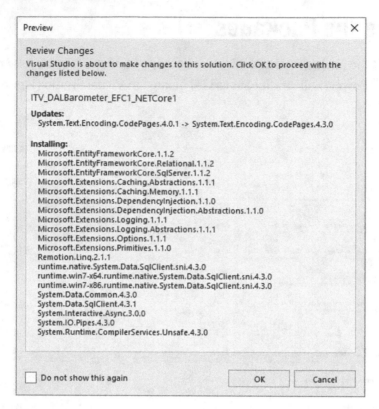

**Figure 2-3.**  *Installing Entity Framework Core 1.1.2 in a .NET Core 1.1 console*

**Figure 2-4.**  *Installing Entity Framework Core 2.0 in a .NET Core 2.0 application includes a different set of dependencies*

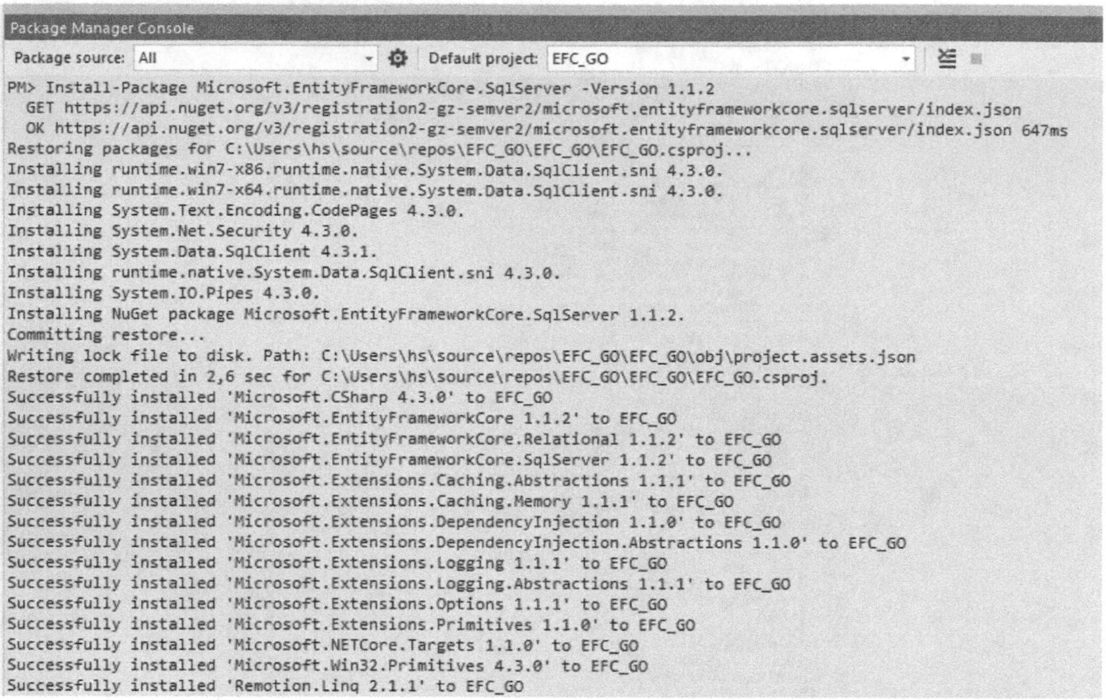

**Figure 2-5.** *Installing the Microsoft SQL Server driver with the NuGet Package Manager Console (shown here in version 1.1.2)*

You can list all the available versions of a package in the NuGet Package Manager Console with the following:

```
(Find-Package Microsoft.Entity FrameworkCore.SqlServer -ExactMatch
-allversions -includeprerelease).Versions | Format-Table Version, Release
```

You can see the versions of a package referenced in the projects of the current solution with the following:

```
(Get-Package Microsoft.EntityFrameworkCore.SqlServer) | Format-Table
Projectname, id, Versions
```

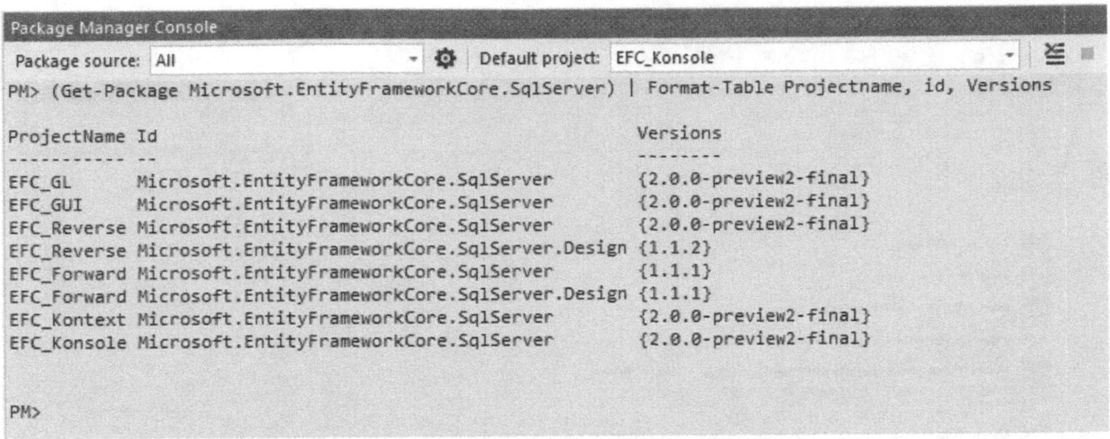

```
Package Manager Console
Package source:  All              ▼  ⚙  Default project:  EFC_Konsole                    ▼  ⅀ ▪
PM> (Get-Package Microsoft.EntityFrameworkCore.SqlServer) | Format-Table Projectname, id, Versions

ProjectName Id                                                     Versions
----------- --                                                     --------
EFC_GL      Microsoft.EntityFrameworkCore.SqlServer                {2.0.0-preview2-final}
EFC_GUI     Microsoft.EntityFrameworkCore.SqlServer                {2.0.0-preview2-final}
EFC_Reverse Microsoft.EntityFrameworkCore.SqlServer                {2.0.0-preview2-final}
EFC_Reverse Microsoft.EntityFrameworkCore.SqlServer.Design         {1.1.2}
EFC_Forward Microsoft.EntityFrameworkCore.SqlServer                {1.1.1}
EFC_Forward Microsoft.EntityFrameworkCore.SqlServer.Design         {1.1.1}
EFC_Kontext Microsoft.EntityFrameworkCore.SqlServer                {2.0.0-preview2-final}
EFC_Konsole Microsoft.EntityFrameworkCore.SqlServer                {2.0.0-preview2-final}


PM>
```

***Figure 2-6.*** *The Get-Package cmdlet shows that some projects have already been upgraded to Entity Framework Core 2.0, others not yet*

## Updating to a New Version

Existing projects are upgraded to a new version of Entity Framework Core with the NuGet Package Manager, either in its graphical version or at the command line.

The NuGet Package Manager GUI indicates that many NuGet packages are to be updated when a new Entity Framework Core version is available.

---

**Tip**    Since the NuGet Package Manager sometimes "gets tangled up" with many updates, you should not update all the packages at once, as Figure 2-7 shows. You should update only the actual root package, in other words, the package with the desired Entity Framework Core driver (for example, `Microsoft.EntityFrameworkCore.SqlServer`, as shown in Figure 2-8). This package update also entails updating its dependencies.

---

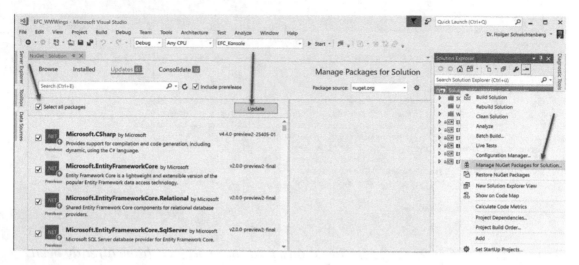

***Figure 2-7.*** *Graphical update of all NuGet packages (not recommended!)*

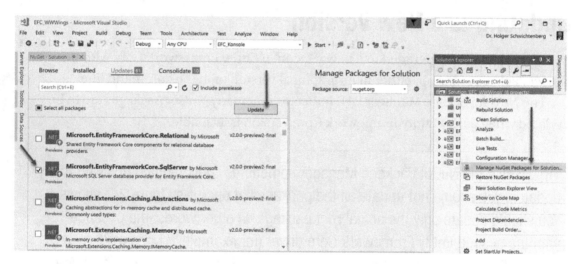

***Figure 2-8.*** *It is better to choose only the root packages, in other words, the package with the database driver*

This corresponds to the procedure on the command line, on which you do not want to type in all packages but only the root package updates, for example, when upgrading to Entity Framework Core 2.0:

```
Update-Package Microsoft.EntityFrameworkCore.SqlServer version 2.0.0
```

**Tip**    If you get the error message "Could not install package 'Microsoft. EntityFrameworkCore.SqlServer 2.0.0'. You are trying to install this package into a project that targets '.NETFramework,Version=v4.x, but the package does not contain any assembly references or content files that are compatible with that framework," this may have the following causes:

- You are using a .NET version prior to 4.6.1 that is not compatible with .NET Standard 2.0 and therefore cannot use Entity Framework Core 2.0.

- However, if the version number in the error message is 4.6.1 or higher (see Figure 2-9), this is because you are using a too-old version of Visual Studio. Entity Framework Core 2.0 can be used only as of Visual Studio 2015 Update 3 with .NET Core installed. (Even if you use the classic .NET Framework, .NET Core must be installed on the development system!)

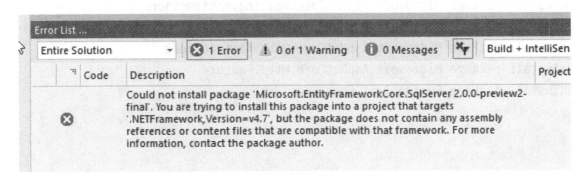

***Figure 2-9.***  *Error message when updating to Entity Framework Core 2.0*

**Tip**   When upgrading from Entity Framework Core 1.*x* to version 2.0, you will need remove the reference to `Microsoft.EntityFrameworkCore.SQLServer.Design` manually.

`uninstall-package Microsoft.EntityFrameworkCore.SqlServer.Design`

If you also have a reference to the package `Microsoft.EntityFrameworkCore.Relational.Design`, then remove this (Figure 2-10):

`uninstall-package Microsoft.EntityFrameworkCore.Relational.Design`

In Entity Framework Core 2.0, Microsoft has moved the content of the NuGet packages with the suffix `.Design` into the eponymous packages without this suffix.

If you still have packages named `Microsoft.AspNetCore...`, even though you are not using an ASP.NET Core–based web application, you can remove them as well. These references are a relic from the first versions of the Entity Framework Core tools:

```
uninstall-package Microsoft.AspNetCore.Hosting.Abstractions
uninstall-package Microsoft.AspNetCore.Hosting.Server.Abstractions
uninstall-package Microsoft.AspNetCore.Http.Abstractions
uninstall-package Microsoft.AspNetCore.Http.Feature
uninstall-package System.Text.Encodings.Web
```

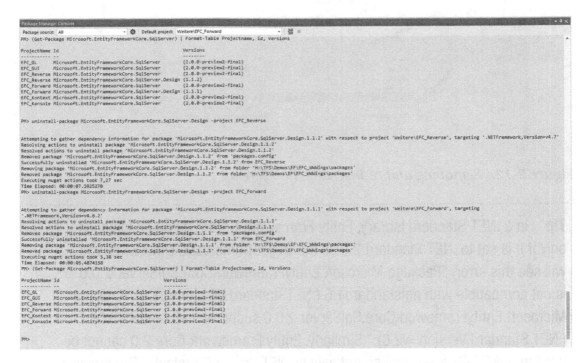

***Figure 2-10.*** *Uninstalling the package Microsoft.EntityFrameworkCore.SqlServer. Design, which is no longer required in Entity Framework Core 2.0*

---

**Tip**   Sometimes Visual Studio will not find the compiled output of other projects in the same solution after an update (see Figure 2-11). In this case, briefly deactivate the project in the Reference Manager (References ➤ Add Reference) and then directly select it again (see Figure 2-12).

---

| Error List | | | | | | | | |
|---|---|---|---|---|---|---|---|---|
| Entire Solution ▼ | | ⊗ 1 Error | ⚠ 0 of 13 Warnings | ❶ 0 Messages | ✖ᵧ | Build + IntelliSense ▼ | | Search E |
| | Code | Description | | | | Project ▲ | File | |
| ⊗ | CS0006 | Metadata file 'H:\TFS\Demos\EF\EFC_WWWings\EFC_Kontext\bin\Debug \EFC_Kontext.dll' could not be found | | | | EFC_Konsole | CSC | |

***Figure 2-11.*** *The project exists, but the compilation is not found*

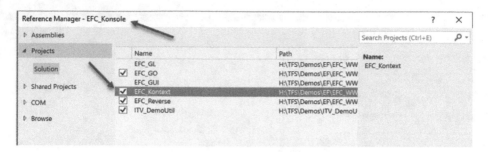

*Figure 2-12.  Removing and re-inserting the reference*

---

**Tip**   In a .NET Standard Library, Entity Framework Core 2.0 can be installed
only if it is set to .NET Standard 2.0 as the target framework. Otherwise, you
will see this error: "Package Microsoft.EntityFrameworkCore.SqlServer 2.0.0
is not compatible with netstandard1.6 (.NETStandard,Version=v1.6). Package
Microsoft.EntityFrameworkCore.SqlServer 2.0.0 supports: netstandard2.0
(.NET Standard,Version=v2.0)." Similarly, Entity Framework Core 2.0 cannot be
used in a .NET Core 1.*x* project, but only in .NET Core 2.0 projects. The projects
may need to be upgraded beforehand (see Figure 2-13 and Figure 2-14).

---

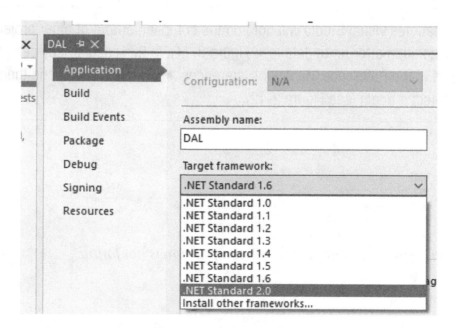

*Figure 2-13.  Updating the target framework to .NET Standard version 2.0 in the*
*project settings*

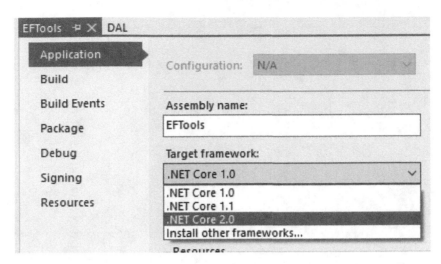

***Figure 2-14.*** *Updating the target framework to .NET Core version 2.0 in the project settings*

# CHAPTER 3

# Concepts of Entity Framework Core

In this chapter, you will learn about the core concepts of Entity Framework Core, broken down according to the procedural models and artifacts of Entity Framework Core.

## Process Models for Entity Framework Core

Entity Framework Core supports the following:

- Reverse engineering of existing databases (an object model is created from an existing database schema)

- Forward engineering of databases (a database schema is generated from an object model).

Reverse engineering (often referred to as *database first*) is useful if you already have a database or if developers choose to create a database in a traditional way. The second option, called *forward engineering*, gives the developer the ability to design an object model. From this, the developer can then generate a database schema.

For the developer, forward engineering is usually better because you can design an object model that you need for programming.

Forward engineering can be used at development time (via so-called schema migrations) or at runtime. A *schema migration* is the creation of the database with an initial schema or a later extension/modification of the schema.

At runtime, this means the database is created (`EnsureCreated()`) or updated (`Migrate()`) when the Entity Framework Core–based application is running.

Reverse engineering always takes place during development.

© Holger Schwichtenberg 2018
H. Schwichtenberg, *Modern Data Access with Entity Framework Core*,
https://doi.org/10.1007/978-1-4842-3552-2_3

The ADO.NET Entity Framework, the predecessor of Entity Framework Core, supported four process models, as shown here:

- Reverse engineering with EDMX files (aka Database First)

- Reverse engineering with Code First

- Forward engineering with EDMX files (aka Model First)

- Forward engineering with Code First

Because there is no EDMX in Entity Framework Core, two of the models have been eliminated. Reverse engineering and forward engineering in Entity Framework Core are the successors of the corresponding Code First practices. However, Microsoft no longer speaks of Code First because this name suggests to many developers forward engineering. Microsoft generally references *code-based modeling*. Figure 3-1 illustrates forward engineering and reverse engineering.

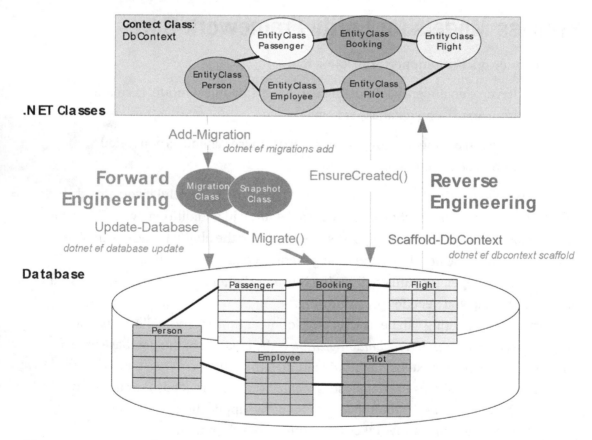

***Figure 3-1.*** *Forward engineering versus reverse engineering for Entity Framework Core*

Table 3-1 compares the features of the two process models in Entity Framework Core.

***Table 3-1.***  *Forward Engineering vs. Reverse Engineering in Entity Framework Core*

| Feature | Reverse Engineering | Forward Engineering |
|---|---|---|
| Import existing database | ✓ | ✖ |
| Changes and extensions of the database schema | ✖ (Microsoft)<br>✓ (with third-party tool Entity Developer) | ✓ (migrations) |
| Graphically model | ✖ (Microsoft)<br>✓ (with third-party tool Entity Developer) | ✖ (Microsoft)<br>✓ (with third-party tool Entity Developer) |
| Stored procedures | • Can be used manually (Microsoft) ✓<br>  Mapping code generation with third-party tool Entity Developer | • Can be used manually (Microsoft) |
| Table-valued functions | • Can be used manually (Microsoft) ✓<br>  Mapping code generation with third-party tool Entity Developer | • Can be used manually (Microsoft) |
| Views | • Can be used manually (Microsoft) ✓<br>  Mapping code generation with third-party tool Entity Developer | ✖ |
| Own metadata/ annotations in the object model | ✓ Possible but not easy<br>✓✓ Easier with third-party tool Entity Developer | ✓✓ Very easy! |
| Control over the object design | ✖ | ✓ |
| Clarity | ✖ | ✓ |

# Components of Entity Framework Core

Figure 3-2 illustrates the key components of an Entity Framework Core project and their relationship to traditional database objects.

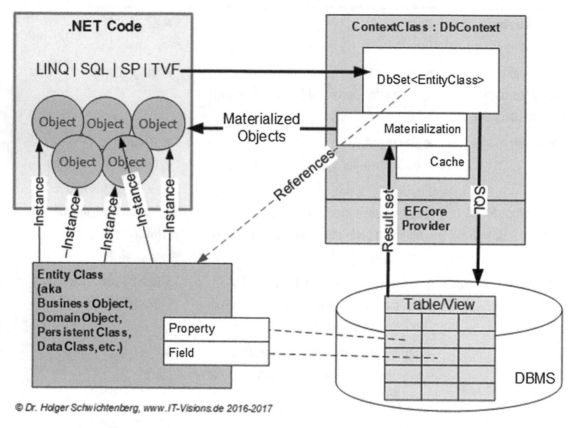

**Figure 3-2.** *The central artifacts in Entity Framework Core and their context*

The database management system (DBMS) contains a database with tables and views.

**Note**   Currently only tables with primary keys or with views that include a primary key can be used. In Entity Framework Core version 2.1, there will be the option to read (but not change) tables without a primary key (see Appendix C).

Entity classes (aka domain object classes, business object classes, data classes, or persistent classes) are representations of tables and views. They contain properties or fields that are mapped to columns of the tables/views. Entity classes can be plain old CLR objects (POCO classes); in other words, they need no base class and no interface. However, you cannot access the database with only these objects.

---

**Best Practice**    Although the use of fields is possible, you should work only with properties because a lot of other libraries and frameworks require properties.

---

A *context class* is a class always derived from the DbContext base class. It has properties of type DbSet<EntityClass> for each of the entity classes. The context class or DbSet properties take the commands of the self-created program code in the form of LINQ commands, SQL commands, stored procedure and table-valued function (TVF) calls, or special API calls for append, modify, and delete. The context class sends the commands to the DBMS-specific provider, which sends the commands to the database via DbCommand objects and receives result sets in a DataReader from the database. The context class transforms the contents of the DataReader object into instances of the entity class. This process is called *materialization*.

# CHAPTER 4

# Reverse Engineering of Existing Databases (Database First Development)

This chapter discusses the reverse engineering of existing databases. Reverse engineering is when an object model is created from an existing database schema.

This chapter covers the simpler version 1 of the World Wide Wings database schema. You can install this database schema with the SQL script `WWWings66.sql`, which also provides the data (10,000 flights and 200 pilots).

## Using Reverse Engineering Tools

Out of the box there are no visual tools for this process, but future releases from Microsoft might include options. In Chapter 19, I will introduce a few additional tools that might assist you with this process, as described here:

- PowerShell cmdlets for the NuGet Package Manager Console (PMC) within the Visual Studio development environment. These commands can be used not only in .NET core projects but also in "full" .NET Framework projects.

- The command-line .NET Core tool (called `dotnet.exe` in Windows), which can also be used independently of Visual Studio and Windows. However, this is available only for .NET Core or ASP.NET Core–based projects.

37

© Holger Schwichtenberg 2018
H. Schwichtenberg, *Modern Data Access with Entity Framework Core*,
https://doi.org/10.1007/978-1-4842-3552-2_4

# Reverse Engineering with PowerShell Cmdlets

For reverse engineering, two NuGet packages are relevant in Entity Framework Core 2.0.

- The package `Microsoft.EntityFrameworkCore.Tools` is needed at development time in the current startup project of Visual Studio.

- The package for each Entity Framework Core database driver (for example, `Microsoft.EntityFrameworkCore.SqlServer` or `Microsoft.EntityFrameworkCore.Sqlite`) is needed in the project where the program code was generated and in the project with the tools.

---

**Tip**    While it is theoretically possible to work with only one project and include both packages, you should in practice create your own project for the Entity Framework Core tools. At the moment, only the Entity Framework Core tools are used. The startup project is made but otherwise remains unused. Alternatively, it is also possible to uninstall the Entity Framework Core tools after the program code generation. This keeps your projects cleaner and more focused.

---

For the example in this chapter, go ahead and create these two projects:

- `EFC_Tools.csproj` exists only for the tools. This project will be a .NET Framework console application (EXE).

- The program code is generated in `EFC_WWWWingsV1_Reverse.csproj`. This project is a .NET Standard 2.0 library and thus can be used in the .NET Framework as well as .NET Core, Mono, and Xamarin.

In this case, you have to install the packages first, so do the following in `EFC_Tools.csproj`:

```
Install-Package Microsoft.EntityFrameworkCore.Tools Install-package
Microsoft.EntityFrameworkCore.SqlServer
```

Do the following in `EFC_WWWingsV1_Reverse.csproj`:

```
Install-package Microsoft.EntityFrameworkCore.SqlServer
```

Or do the following for SQLite:

```
Install-package Microsoft.EntityFrameworkCore.Sqlite
```

> **Note**    In Entity Framework Core 1.*x*, you must also include the package
> `Microsoft.EntityFrameworkCore.SqlServer.Design` or `Microsoft.`
> `EntityFrameworkCore.Sqlite.Design` in the tool project. These packages
> are no longer needed in Entity Framework Core 2.0.

When you run the package installation commands for the tools, a total of 33
assembly references will be added to a .NET 4.7 project (see Figure 4-1). In Entity
Framework Core 1.*x*, there were even more, including ASP.NET Core assemblies, even
though you were not in an ASP.NET Core project at all.

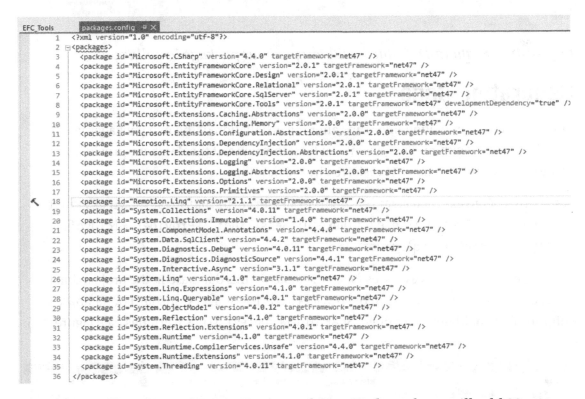

***Figure 4-1.*** *The Microsoft.EntityFrameworkCore.Tools package will add 33 more*
*packages!*

If a code generation command was executed without the previous package
installation, the developer sees an error in the Package Manager Console (see Figure 4-2).

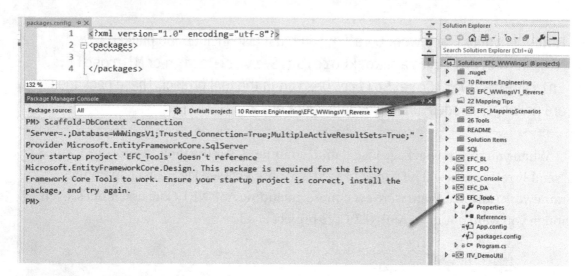

***Figure 4-2.*** *Scaffold-DbContext without previous package installation*

# Generating Code

The actual code generation then runs after the installation of the two packages via the
Scaffold-DbContext cmdlet, to which at least the name of the database provider and a
connection string have to be transferred by the developer.

```
Scaffold-DbContext -Connection "Server=DBServer02;Database=WWWings;Trust
ed_Connection=True;MultipleActiveResultSets=True;" -Provider Microsoft.
EntityFrameworkCore.SqlServer
```

This command creates classes for all the tables in this database in the project that is
set as the current target project in the NuGet Package Manager Console. For database
columns that could not be mapped, Scaffold-DbContext issues warnings (see Figure 4-3).

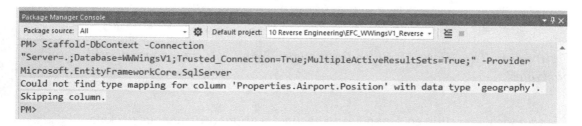

***Figure 4-3.*** *Scaffold-DbContext warns that a column of type Geography has been
ignored*

Alternatively, you can use a schema or tables to limit the generation to specific database schema names or table names. For both parameters, you can specify several names separated by semicolons.

```
Scaffold-DbContext -Connection "Server=DBServer02;Database=WWWWingsV1;Tru
sted_Connection=True;MultipleActiveResultSets=True;" -Provider Microsoft.
EntityFrameworkCore.SqlServer -Tables Flight,Person,Pilot,Passenger,Airport,
Employee,Flight_Passenger -force
```

You can use the table names with or without schema names (in other words, you can use Flight or operation.Flight). But beware that if a table with the same name exists in multiple schemas, then a specification without a schema name will generate entity classes for all tables of that name from all schemas.

By default, the code is generated in the project currently selected in the NuGet Package Manager Console in its root directory using the default namespace of that project. With the parameters -Project and -OutputDir, the developer can influence the project and the output folder. Unfortunately, with the existing parameters, it is not possible to direct the code generation of the entity class and the context class into different projects.

With respect to the data model shown in Figure 4-4, the Scaffold-DbContext cmdlet now generates the following outputs:

- One entity class each in POCO style is generated for each of the six tables, including the N:M intermediate table Flight_Passenger, which has always eliminated the classic Entity Framework in the object model. Unfortunately, Entity Framework Core version 2.0 does not yet support N:M relationships; it only replicates them with two 1:N relationships, as does the relational model.

- A context class derived by the base class Microsoft.Entity FrameworkCore.DbContext is derived. Unlike before, this class is no longer a wrapper around the ObjectContext class but a whole new, stand-alone implementation. The name of this class can be influenced by the developer with the command-line parameter -Context. Unfortunately, specifying a namespace is not possible here. Using points in the parameter value is acknowledged by Visual Studio with "The context class name passed in is not a valid C# identifier."

- If code generation is not possible for individual columns, there will be a yellow alert output in the Package Manager Console. This happens for the Geometry and Geography data types, for example, because Entity Framework Core is not yet supported.

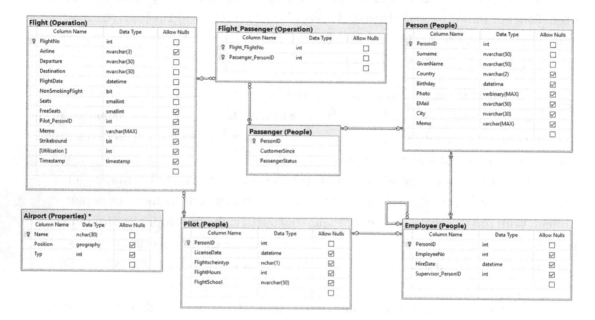

***Figure 4-4.***  *Example database for the Word Wide Wings airline (version 1)*

Figure 4-5 shows the generated classes for the sample database from Figure 4-4, and Figure 4-6 shows the object model.

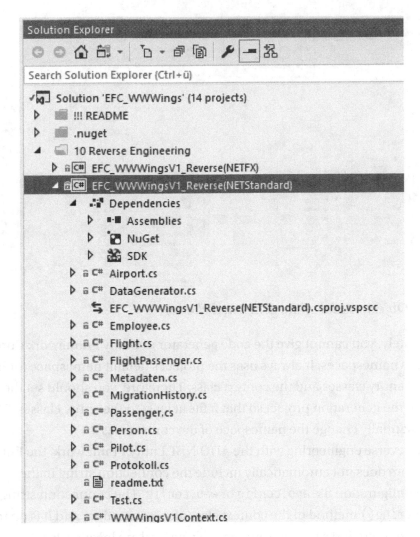

**Figure 4-5.** *Project with the generated classes for the sample database from Figure 4-4*

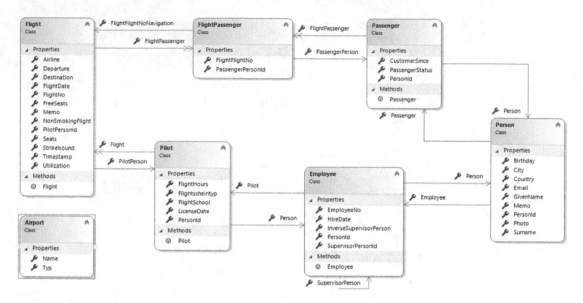

**Figure 4-6.** *Object model of the generated classes*

Unfortunately, you cannot give the code generator of Entity Framework Core any settings for the namespaces; it always uses the project's default namespace for both the generated entity classes and the context class. Therefore, you should set the default namespace in the generation project so that it fits at least to the entity classes. Then you just have to manually change the namespace of the context class.

Unlike in reverse engineering with the ADO.NET Entity Framework, the Entity Framework Core does not automatically include the connection string in the application configuration file `app.config` or `web.config`. The connection string is in the `OnConfiguring()` method of the context class after generation, and it is up to the software developer to find a suitable and possibly backed-up location for it.

A *pluralization* (in other words, changing the table names to plural in the class names) does not take place. So far, there is no option in Entity Framework Core to do this; however, this is not a big loss because the pluralization service works only for English table names.

# Looking at the Generated Program Code

The following listings show the program code generated by `Scaffold-DbContext` for the context class and, by way of example, for the entity classes `Flight` and `Passenger`.

The definition of how the object model is mapped to the database schema continues as in the classic Entity Framework in these three ways:

- Conventions that are applied by Entity Framework Core automatically

- Data annotations that are applied by the entity classes and their members

- Fluent API used in the OnModelCreating() method of the DbContext class

The code generated by the Entity Framework Core tools focuses on the third way. The OnModelCreating() method in the context class is accordingly full of Fluent API calls. But conventions continue to work, for example, that the properties of the class in the standard have the same names as the columns in the table.

So far, the assistant for the classic Entity Framework in Visual Studio also used data annotations, which are now no longer used in the generated program code. If you want to get the old behavior back, you can use the parameter -DataAnnotations in Scaffold-DbContext.

The Fluent API contains the following definitions:

- It defines the table names in the database if they differ or if they have a schema name different from dbo, using ToTable().

- It defines names for primary key columns and indexes, using HasName().

- It defines column types and column properties if the .NET type names are not unique to a data type in the database management system, using HasColumnType(), IsRequired(), HasMaxLength().

- It defines default values for columns, using HasDefaultValueSql() or HasDefaultValue().

- It defines the cardinalities between tables and their foreign keys, using HasOne(), HasMany(), WithOne(), WithMany(), HasForeignKey(), and HasConstraintName().

- It defines the indices, using HasIndex().

- It determines whether a column's content must be reread after inserting or modifying an Entity Framework Core record because it is generated by the database management system, using `ValueGeneratedOnAddOrUpdate()` and `ValueGeneratedOnAdd()` and `ValueGeneratedNever()`.

- It sets cascading delete settings, using `OnDelete()`.

In the source code, the Fluent API configuration is organized by classes, as shown here:

```
modelBuilder.Entity<Person>(entity => {...});
```

Within these method calls you will find the configuration for the individual columns of these tables.

```
entity.Property(e => e.PersonId)...
```

Compared to the previous ADO.NET Entity Framework, there are some syntactical changes and also improvements. So, the index configuration is now much more concise. The following features are just like the classic Entity Framework:

- The reverse engineering code generator does not create inheritance relationships between entity classes, even though this would be possible as in the case of Passenger ➤ Person, Employee ➤ Person, and Pilot ➤ Employee. Instead, the code generator always generates associations and associated navigation properties. Such inheritance relationships must be defined by the developer later and then remove the navigation properties.

    - For example, the navigation properties in the entity classes are declared `virtual`, even though the Entity Framework Core lazy loading does not yet support what `virtual` is necessary for.

    - For example, navigation properties for sets are declared with `ICollection<T>` and are then filled in the constructor with the new `HashSet<T>()`.

- For each entity class, there is a `DbSet<T>` property in the context class.

The generated source code can be changed (see Listings 4-1, Listing 4-2, and Listing 4-3), such as if you want to have property names other than column names in the database in the object model. You would do this with the Fluent API method `HasColumnName("column name")` or the data annotation `Column("column name")`.

***Listing 4-1.*** Generated Context Class

```
using System;
using Microsoft.EntityFrameworkCore;
using Microsoft.EntityFrameworkCore.Metadata;

namespace EFC_WWWingsV1_Reverse
{
    public partial class WWWingsV1Context : DbContext
    {
        public virtual DbSet<Airport> Airport { get; set; }
        public virtual DbSet<Employee> Employee { get; set; }
        public virtual DbSet<Flight> Flight { get; set; }
        public virtual DbSet<FlightPassenger> FlightPassenger { get; set; }
        public virtual DbSet<Metadaten> Metadaten { get; set; }
        public virtual DbSet<MigrationHistory> MigrationHistory { get; set; }
        public virtual DbSet<Passenger> Passenger { get; set; }
        public virtual DbSet<Person> Person { get; set; }
        public virtual DbSet<Pilot> Pilot { get; set; }
        public virtual DbSet<Protokoll> Protokoll { get; set; }
        public virtual DbSet<Test> Test { get; set; }

        protected override void OnConfiguring(DbContextOptionsBuilder
        optionsBuilder)
        {
            if (!optionsBuilder.IsConfigured)
            {
#warning To protect potentially sensitive information in your connection
string, you should move it out of source code. See http://go.microsoft.com/
fwlink/?LinkId=723263 for guidance on storing connection strings.
```

```
            optionsBuilder.UseSqlServer(@"Server=.;Database=WWWingsV1;
            Trusted_Connection=True;MultipleActiveResultSets=True;");
        }
    }

    protected override void OnModelCreating(ModelBuilder modelBuilder)
    {
        modelBuilder.Entity<Airport>(entity =>
        {
            entity.HasKey(e => e.Name);

            entity.ToTable("Airport", "Properties");

            entity.Property(e => e.Name)
                .HasColumnType("nchar(30)")
                .ValueGeneratedNever();
        });

        modelBuilder.Entity<Employee>(entity =>
        {
            entity.HasKey(e => e.PersonId);

            entity.ToTable("Employee", "People");

            entity.Property(e => e.PersonId)
                .HasColumnName("PersonID")
                .ValueGeneratedNever();

            entity.Property(e => e.HireDate).HasColumnType("datetime");

            entity.Property(e => e.SupervisorPersonId).
            HasColumnName("Supervisor_PersonID");

            entity.HasOne(d => d.Person)
                .WithOne(p => p.Employee)
                .HasForeignKey<Employee>(d => d.PersonId)
                .OnDelete(DeleteBehavior.ClientSetNull)
                .HasConstraintName("FK_MI_Employee_PE_Person");
```

```
        entity.HasOne(d => d.SupervisorPerson)
            .WithMany(p => p.InverseSupervisorPerson)
            .HasForeignKey(d => d.SupervisorPersonId)
            .HasConstraintName("FK_Employee_Employee");
    });

    modelBuilder.Entity<Flight>(entity =>
    {
        entity.HasKey(e => e.FlightNo);

        entity.ToTable("Flight", "Operation");

        entity.Property(e => e.FlightNo).ValueGeneratedNever();

        entity.Property(e => e.Airline).HasMaxLength(3);

        entity.Property(e => e.Departure)
            .IsRequired()
            .HasMaxLength(30);

        entity.Property(e => e.Destination)
            .IsRequired()
            .HasMaxLength(30);

        entity.Property(e => e.FlightDate).HasColumnType("datetime");

        entity.Property(e => e.Memo).IsUnicode(false);

        entity.Property(e => e.PilotPersonId).HasColumnName("Pilot_
        PersonID");

        entity.Property(e => e.Timestamp).IsRowVersion();

        entity.Property(e => e.Utilization).
        HasColumnName("Utilization ");

        entity.HasOne(d => d.PilotPerson)
            .WithMany(p => p.Flight)
```

```
                    .HasForeignKey(d => d.PilotPersonId)
                    .HasConstraintName("FK_FL_Flight_PI_Pilot");
            });

            modelBuilder.Entity<FlightPassenger>(entity =>
            {
                entity.HasKey(e => new { e.FlightFlightNo,
                e.PassengerPersonId })
                    .ForSqlServerIsClustered(false);

                entity.ToTable("Flight_Passenger", "Operation");

                entity.Property(e => e.FlightFlightNo).
                HasColumnName("Flight_FlightNo");

                entity.Property(e => e.PassengerPersonId).
                HasColumnName("Passenger_PersonID");

                entity.HasOne(d => d.FlightFlightNoNavigation)
                    .WithMany(p => p.FlightPassenger)
                    .HasForeignKey(d => d.FlightFlightNo)
                    .OnDelete(DeleteBehavior.ClientSetNull)
                    .HasConstraintName("FK_Flight_Passenger_Flight");

                entity.HasOne(d => d.PassengerPerson)
                    .WithMany(p => p.FlightPassenger)
                    .HasForeignKey(d => d.PassengerPersonId)
                    .OnDelete(DeleteBehavior.ClientSetNull)
                    .HasConstraintName("FK_Flight_Passenger_Passenger");
            });

            modelBuilder.Entity<Passenger>(entity =>
            {
                entity.HasKey(e => e.PersonId);

                entity.ToTable("Passenger", "People");

                entity.Property(e => e.PersonId)
```

```
            .HasColumnName("PersonID")
            .ValueGeneratedNever();

    entity.Property(e => e.CustomerSince).
    HasColumnType("datetime");

    entity.Property(e => e.PassengerStatus).
    HasColumnType("nchar(1)");

    entity.HasOne(d => d.Person)
        .WithOne(p => p.Passenger)
        .HasForeignKey<Passenger>(d => d.PersonId)
        .OnDelete(DeleteBehavior.ClientSetNull)
        .HasConstraintName("FK_PS_Passenger_PE_Person");
});

modelBuilder.Entity<Person>(entity =>
{
    entity.ToTable("Person", "People");

    entity.Property(e => e.PersonId).HasColumnName("PersonID");

    entity.Property(e => e.Birthday).HasColumnType("datetime");

    entity.Property(e => e.City).HasMaxLength(30);

    entity.Property(e => e.Country).HasMaxLength(2);

    entity.Property(e => e.Email)
        .HasColumnName("EMail")
        .HasMaxLength(50);

    entity.Property(e => e.GivenName)
        .IsRequired()
        .HasMaxLength(50);

    entity.Property(e => e.Memo).IsUnicode(false);
```

```
        entity.Property(e => e.Surname)
            .IsRequired()
            .HasMaxLength(50);
    });

    modelBuilder.Entity<Pilot>(entity =>
    {
        entity.HasKey(e => e.PersonId);

        entity.ToTable("Pilot", "People");

        entity.Property(e => e.PersonId)
            .HasColumnName("PersonID")
            .ValueGeneratedNever();

        entity.Property(e => e.FlightSchool).HasMaxLength(50);

        entity.Property(e => e.Flightscheintyp).
        HasColumnType("nchar(1)");

        entity.Property(e => e.LicenseDate).
        HasColumnType("datetime");

        entity.HasOne(d => d.Person)
            .WithOne(p => p.Pilot)
            .HasForeignKey<Pilot>(d => d.PersonId)
            .OnDelete(DeleteBehavior.ClientSetNull)
            .HasConstraintName("FK_PI_Pilot_MI_Employee");
    });

    }
    }
}
```

***Listing 4-2.*** Generated Entity Class Flight

```
using System;
using System.Collections.Generic;

namespace EFC_WWWingsV1_Reverse
{
    public partial class Flight
    {
        public Flight()
        {
            FlightPassenger = new HashSet<FlightPassenger>();
        }

        public int FlightNo { get; set; }
        public string Airline { get; set; }
        public string Departure { get; set; }
        public string Destination { get; set; }
        public DateTime FlightDate { get; set; }
        public bool NonSmokingFlight { get; set; }
        public short Seats { get; set; }
        public short? FreeSeats { get; set; }
        public int? PilotPersonId { get; set; }
        public string Memo { get; set; }
        public bool? Strikebound { get; set; }
        public int? Utilization { get; set; }
        public byte[] Timestamp { get; set; }

        public Pilot PilotPerson { get; set; }
        public ICollection<FlightPassenger> FlightPassenger { get; set; }
    }
}
```

***Listing 4-3.*** Generated Entity Class Passenger

```
using System;
using System.Collections.Generic;

namespace EFC_WWWingsV1_Reverse
{
    public partial class Passenger
    {
        public Passenger()
        {
            FlightPassenger = new HashSet<FlightPassenger>();
        }

        public int PersonId { get; set; }
        public DateTime? CustomerSince { get; set; }
        public string PassengerStatus { get; set; }

        public Person { get; set; }
        public ICollection<FlightPassenger> FlightPassenger { get; set; }
    }
}
```

# Seeing an Example Client

The program shown in Listing 4-4 uses the generated Entity Framework context class and the entity class Passenger.

The method illustrated creates a new passenger, attaches the passenger to the DbSet<Passenger>, and then stores the new passenger in the database using the SaveChanges() method.

Then all passengers are loaded for control, and their numbers are printed. Listing 4-4 shows a version of all passengers with the name Schwichtenberg. This filtering then takes place in RAM with LINQ to Objects over the previously loaded passengers. Figure 4-7 shows the output on the screen.

**Note**   The commands used in this example are of course described in more detail in later chapters in this book. However, this listing is didactically necessary here to prove the functionality of the Entity Framework Core context class created.

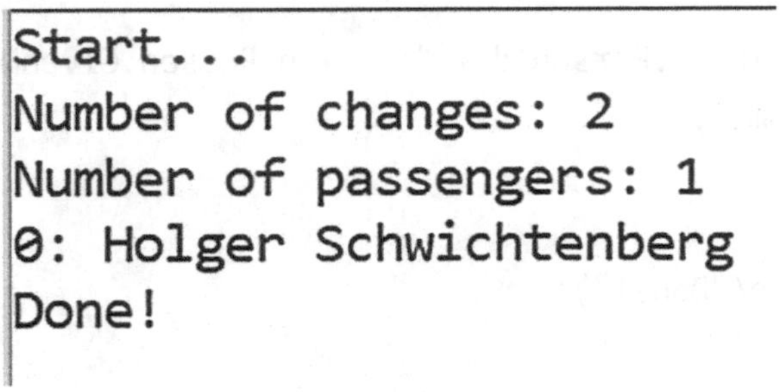

```
Start...
Number of changes: 2
Number of passengers: 1
0: Holger Schwichtenberg
Done!
```

***Figure 4-7.***  *Output of the sample client*

***Listing 4-4.***  Program Code That Uses the Created Entity Framework Core Model

```
public static void Run()
  {
   Console.WriteLine("Start...");
   using (var ctx = new WWWingsV1Context())
   {
    // Create Person object
    var newPerson = new Person();
    newPerson.GivenName = "Holger";
    newPerson.Surname = "Schwichtenberg";
    // Create Passenger object
    var newPassenger = new Passenger();
    newPassenger.PassengerStatus = "A";
    newPassenger.Person = newPerson;
    // Add Passenger to Context
    ctx.Passenger.Add(newPassenger);
    // Save objects
    var count = ctx.SaveChanges();
    Console.WriteLine("Number of changes: " + count);
    // Get all passengers from the database
```

```
    var passengerSet = ctx.Passenger.Include(x => x.Person).ToList();
    Console.WriteLine("Number of passengers: " + passengerSet.Count);
    // Filter with LINQ-to-Objects
    foreach (var p in passengerSet.Where(x=>x.Person.Surname ==
    "Schwichtenberg").ToList())
    {
     Console.WriteLine(p.PersonId + ": " + p.Person.GivenName + " " +
     p.Person.Surname);
    }
   }
  Console.WriteLine("Done!");
  Console.ReadLine();
 }
```

# Using the .NET Core Tool dotnet

When developing .NET Core projects, the command-line tool dotnet (also known as
the .NET Core command-line interface [CLI]) from the .NET Core SDK can be used as
an alternative to the PowerShell cmdlets (https://www.microsoft.com/net/download/
core). Unlike the PowerShell cmdlets, dotnet is available not only for Windows but also
for Linux and macOS.

This form of generation works for the following:

- .NET Core console applications

- ASP.NET Core projects based on .NET Core or the .NET Framework
  4.6.2 and later

First, the package Microsoft.EntityFrameworkCore.Tools.DotNet has to be
installed, which does not work via a command-line tool but only through a manual entry
in the XML-based .csproj project file (see Figure 4-8):

```
<ItemGroup>
    <DotNetCliToolReference Include="Microsoft.EntityFrameworkCore.Tools.DotNet"
      Version="2.0.1" />
</ItemGroup>
```

```
EF_NETCoreConsole.csproj*  ⊕ ✕  EF_NETCoreConsole*       NuGet: EF_NETCoreConsole        Program.cs        Object Browser
    1  ⊟<Project Sdk="Microsoft.NET.Sdk">
    2  ⊟  <PropertyGroup Label="Globals">
    3        <SccProjectName>SAK</SccProjectName>
    4        <SccProvider>SAK</SccProvider>
    5        <SccAuxPath>SAK</SccAuxPath>
    6        <SccLocalPath>SAK</SccLocalPath>
    7     </PropertyGroup>
    8  ⊟  <PropertyGroup>
    9        <OutputType>Exe</OutputType>
   10        <TargetFramework>netcoreapp2.0</TargetFramework>
   11     </PropertyGroup>
   12  ⊟  <ItemGroup>
   13        <DotNetCliToolReference Include="Microsoft.EntityFrameworkCore.Tools.DotNet" Version="1.0.0" />
   14     </ItemGroup>
   15  </Project>
```

***Figure 4-8.*** *Manual extension of the .csproj file*

Then you have to add the following package:

```
<ItemGroup>
    <PackageReference Include="Microsoft.EntityFrameworkCore.Design"
Version="2.0.1" />
</ItemGroup>
```

However, this is also possible via the command line in the project directory, as shown here:

```
dotnet add package Microsoft.EntityFrameworkCore.design
```

Now add the desired Entity Framework Core provider, shown here:

```
dotnet add package Microsoft.EntityFrameworkCore.SqlServer
```

The following package was also necessary in Entity Framework Core 1.*x* but is not needed anymore in Entity Framework Core 2.0:

```
dotnet add package Microsoft.EntityFrameworkCore.SQL Server.design
```

Then you can do the code generation (see Figure 4-9).

```
dotnet ef dbcontext scaffold "server =.; Database = WWWings66; Trusted_
Connection = True; MultipleActiveResultSets = True; "Microsoft.
EntityFrameworkCore.SqlServer --output-dir model
```

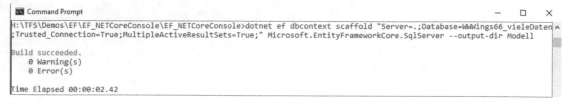

**Figure 4-9.**  *Reverse engineering with dotnet.exe*

---

**Note**    Microsoft did not release the final version of `dotnet.exe` 1.0 until March 6, 2017, as part of Entity Framework Core 1.1.1 and Visual Studio 2017. Previously, there were only "preview" versions. These preview versions used a `project.json` file. If you still use this outdated format, you do not have to make the entries in the `.csproj` file; you have to do it in the `project.json` file!

---

# Understanding the Weaknesses of Reverse Engineering

As in the classic Entity Framework, you can create entity types only for tables with primary keys. Composite primary keys, however, are not a problem for Entity Framework Core.

---

**Note**    Microsoft will introduce the mapping of a table without a primary key in version 2.1; see Appendix C for more information.

---

In the case of the temporal tables (called *system-versioned* tables) added in SQL Server 2016, the history tables cannot be mapped using Entity Framework Core. However, this is already possible for the actual table, whereby querying the historical values is possible only via SQL and so far not via LINQ.

For database views and stored procedures, in contrast to the classic Entity Framework, classes and functions cannot be generated.

Once the object model is generated using the Entity Framework Core command-line tools, you cannot update it. The Update Model from Database command available for the Database First approach is currently not implemented. You can only restart the generation. If the classes to be generated already exist, the cmdlet Scaffold-DbContext complains about it. With the additional parameter -force, the cmdlet will overwrite existing files. However, any manually made changes to the source code files will be lost.

If in a new Scaffold-DbContext command you let not all previously generated tables be generated, but only a few selected ones, then in the context class the DbSet<T> declarations and the Fluent API configurations are missing for all the tables that are now no longer generated. Once again, this is a reason to generate a project from which you can then copy the generated parts you need to another project. However, Microsoft has announced (at https://github.com/aspnet/EntityFramework/wiki/Roadmap) that it plans to improve the tools and provide an Update Model from Database feature.

Until then, it's the best way to at least limit the code generation that occurs in changes to new tables; the changes for new, changed, or deleted columns are better made manually in the source code. Or, after a reverse engineering of a database, you can switch to forward engineering; in that case, changes will be recorded in the object model and used to generate DDL commands for changing the database schema.

# CHAPTER 5

# Forward Engineering for New Databases

Although Entity Framework Core supports the reverse engineering of existing database models, the ideal process model is forward engineering, where the database model is generated from an object model. This is because the developer can design the object model according to the needs of the business case.

Forward engineering is available in the classic Entity Framework in two variants: Model First and Code First. In Model First, you graphically create an entity data model (EDM) to generate the database schema and .NET classes. In Code First, you write classes directly, from which the database schema is created. The EDM is invisible. In the redesigned Entity Framework Core, there is only the second approach, which however is not called Code First but *code-based modeling* and no longer uses an invisible EDM.

## Two Types of Classes

Code-based modeling in Entity Framework Core happens through these two types of classes:

- You create entity classes, which store the data in RAM. You create navigation properties in the entity classes that represent the relationships between the entity classes. These are typically plain old CRL objects (POCOs) with properties for each database column.

- You write a context class (derived from `DbContext`) that represents the database model, with each of the entities listed as a `DBSet`. This will be used for all queries and other operations.

© Holger Schwichtenberg 2018
H. Schwichtenberg, *Modern Data Access with Entity Framework Core*,
https://doi.org/10.1007/978-1-4842-3552-2_5

Ideally, these two types of classes are implemented in different projects (DLL assemblies) because entity classes are often used in several or even all layers of the software architecture, while the context class is part of the data access layer and should be used only by the layer above it.

# Examples in This Chapter

This chapter shows how to create a preliminary edition of the World Wide Wings object model in version 2. Initially, you will consider only the entities Person, Employee, Pilot, Passenger, Flight, and Booking. You will set only the minimum options needed to create a database schema from the object model. You will expand and refine the object model in the following chapters. You can find the program code in the solution EFC_WWWings. The entity classes are in a DLL project named EFC_BO_Step1 (for business objects), and the context class is in a DLL project named EFC_DA_Step1 (for data access). The startup application is a console application (EFC_Konsole). This includes the data access code and screen output. See Figure 5-1.

---

**Note**    To keep the examples simple and focused on using the Entity Framework Core API, I am not doing any further delineation of business logic or creating a dedicated data access layer above the context class. This is not an architectural example; those types of examples will follow later in this book.

---

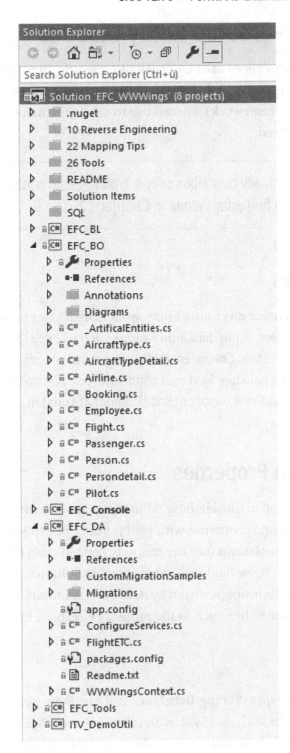

***Figure 5-1.*** *Solution for the example in this and the following chapters*

# Rules for Self-Created Entity Classes

As mentioned, the entity classes are POCOs. In other words, they do not have to inherit from a base class or implement an interface. However, there must be a parameterless constructor that Entity Framework Core can use to create the instances when database table rows are materialized.

---

**Note**    This chapter initially describes only a typical basic configuration for the entity classes. You will find adaptations in Chapter 12.

---

## NuGet Packages

You do not need to reference any Entity Framework Core NuGet packages to implement the entity classes. However, using data annotations such as [Key] and [StringLength] requires a reference to System.ComponentModel.Annotations.dll in the classic .NET Framework or the NuGet package System.ComponentModel.Annotations (https://www.nuget.org/packages/System.ComponentModel.Annotations) in .NET Core and .NET Standard.

## Data Annotation Properties

Each column to be created in the database table must be represented by a property. The properties can be automatic properties with {get; set;} or explicit properties with getter and setter implementations (see the property Memo in class Flight). A class may also own fields; however, these fields are not mapped to columns. In other words, the information in these fields is not persisted by default. In addition, properties that have no setter are not persisted either, such as the property Fullname in the class Person.

## Data Types

The .NET primitive data types (String, DateTime, Boolean, Byte, Byte[], Int16, Int32, Int64, Single, Double, Decimal, and System.Guid) are allowed. Nullable<T> may indicate that the corresponding column in the database table may be left empty (NULL). Enumeration types are also allowed; for example, see PilotLicenseType in class Pilot.

The data types DbGeometry and DbGeography, which have been supported in the classic Entity Framework since version 5.0, unfortunately do not exist in Entity Framework Core for now.

## Relationships (Master-Detail)

There may also be properties that are declared as different entity types. These are called *navigation properties,* and they express the relationships between two entity classes. Entity Framework Core supports the following:

- *1:0/1 relationships*: Here, the property is declared to a single object of the related type (see Pilot and Copilot in the Flight class).

---

**Important**    For a single object, it is semantically wrong to assign an instance of the related type in the navigation property declaration or in the constructor because an OR mapper, such as the Entity Framework Core, sees a new entity object. This instantiation makes sense only if a new top object always requires a new subobject. In the case of Flight and Pilot, this is not so since a new pilot is not set for each flight.

---

- *1:0 / N relations*: Here, the property is declared to a collection type of a related type (see List<Flight> for FlightAsPilotSet and FlightAsCopilotSet in the Pilot class). It is allowed to declare the navigation property as ICollection or any other interface based on it (like IList) or as an ICollection<T> class (like List<T> or HashSet<T>).

---

**Important**    It is usually a good idea to assign concrete set types directly in the declaration or in the constructor so that the calling program code does not have to do this. Entity Framework Core handles set instantiation within relationship fixup. Since only an empty list is created here, Entity Framework Core will not want to persist anything here, as long as the list is not filled.

---

---

**Note**    Entity Framework Core does not currently support N:M relationships. More specifically, Entity Framework Core does not support the abstraction of representing two 1:N relationships with an intermediate table in the object model as N:M. Neither does the relational database model support N:M relationships; it needs an intermediate table for two 1:Nrelationships, where the intermediate table contains a composite primary from the primary keys of the entities to be joined. In Entity Framework Core, you have an intermediate class for the intermediate table in the object model and, just like in the relational model, two 1:N relationships instead of an N:M relationship. Such an intermediate entity class can be seen in the class Booking. It is referenced by a booking set in class `Flight` and class `Passenger`.

---

The navigation properties must be labeled `virtual` in the classic Entity Framework Core for lazy loading to work. Since there is no lazy loading in Entity Framework Core 1.*x*/2.0, this edition is no longer required. Lazy loading will come in a later version in Entity Framework Core; so far, it is still unclear whether Microsoft will require this again as a label. But it does not hurt to declare the navigation properties as `virtual` today.

Navigation properties can be bidirectional, as the entity classes `Flight` and `Pilot` show. The 0/1 pages of a navigation property may have (but not need) explicit foreign key properties (see `PilotId` and `CopilotId`).

## Inheritance

Entity classes can inherit from each other. This can be seen in the classes `Pilot` and `Passenger`, which inherit from the class `Person`. In this case, `Person` is an abstract class. But it could also be a class with instances.

## Primary Key

Another prerequisite for Entity Framework Core is that each entity class must have a primary key (PK) made up of one or more simple properties. In the simplest case, according to a convention, you create a property with the name ID or Id or `classnameID` or `classnameId`. The case sensitivity of the ID and the class name is not relevant here, even if the still incomplete documentation of Entity Framework Core at `www.efproject.net/en/latest/modeling/keys.html` suggests differently.

---

**Note**    If you want to rename the primary key or define a composite primary key, then you must explicitly configure it, as you'll learn in Chapter 12.

---

# Examples

Listings 5-1, 5-2, 5-3, 5-4, 5-5, and 5-62 reflect part of the World Wide Wings instance classes.

***Listing 5-1.*** Class Flight

```
using System;
using System.Collections.Generic;
using System.ComponentModel.DataAnnotations;
using System.ComponentModel.DataAnnotations.Schema;

namespace BO
{

 [Serializable]
 public class Flight
 {
  public Flight()
  {  }

  #region Key
  [Key]
  [DatabaseGenerated(DatabaseGeneratedOption.None)] // No identity column!
  public int FlightNo { get; set; }
  #endregion

  #region Primitive Properties
  [StringLength(50), MinLength(3)]
  public string Departure { get; set; }
  [StringLength(50), MinLength(3)]
  public string Destination { get; set; }
```

```csharp
[Column("FlightDate", Order = 1)]
public DateTime Date { get; set; }
public bool? NonSmokingFlight { get; set; }

[Required]
public short? Seats { get; set; }

public short? FreeSeats { get; set; }


public decimal? Price { get; set; }


public string Memo { get; set; }
#endregion

#region Related Objects
public Airline { get; set; }
public ICollection<Booking> BookingSet { get; set; }
public Pilot { get; set; }
public Pilot Copilot { get; set; }

// Explicit foreign key properties for the navigation properties
public string AirlineCode { get; set; } // mandatory!
public int PilotId { get; set; } // mandatory!
public int? CopilotId { get; set; } // optional
public byte? AircraftTypeID { get; set; } // optional
#endregion

public override string ToString()
{
 return String.Format($"Flight #{this.FlightNo}: from {this.Departure} to
 {this.Destination} on {this.Date:dd.MM.yy HH:mm}: {this.FreeSeats} free
 Seats.");
}

public string ToShortString()
{
```

```
  return String.Format($"Flight #{this.FlightNo}: {this.Departure}->{this.
  Destination} {this.Date:dd.MM.yy HH:mm}: {this.FreeSeats} free Seats.");
 }
 }
}
```

*Listing 5-2.* Class Person

```
using System;

namespace BO
{
 public class Person
 {
  #region Primitive properties
  // --- Primary Key
  public int PersonID { get; set; }
  // --- Additional properties
  public string Surname { get; set; }
  public string GivenName { get; set; }
  public Nullable<DateTime> Birthday { get; set; }
  public virtual string EMail { get; set; }
  // --- Relations
  public Persondetail Detail { get; set; } = new Persondetail();
  // mandatory (no FK property!)
  #endregion

  // Calculated property (in RAM only)
  public string FullName => this.GivenName + " " + this.Surname;

  public override string ToString()
  {
   return "#" + this.PersonID + ": " + this.FullName;
  }
 }
}
```

***Listing 5-3.*** Class Employee

```csharp
using System;

namespace BO
{
 public class Employee : Person
 {
  public DateTime? HireDate;
  public float Salary { get; set; }
  public Employee Supervisor { get; set; }

  public string PassportNumber => this._passportNumber;
  private string _passportNumber;

  public void SetPassportNumber(string passportNumber)
  {
   this._passportNumber = passportNumber;
  }
 }
}
```

***Listing 5-4.*** Class Pilot

```csharp
using System;
using System.Collections.Generic;
using System.ComponentModel.DataAnnotations;

namespace BO
{
 public enum PilotLicenseType
 {
  // https://en.wikipedia.org/wiki/Pilot_licensing_and_certification
  Student, Sport, Recreational, Private, Commercial, FlightInstructor, ATP
 }

 [Serializable]
 public partial class Pilot : Employee
 {
```

```csharp
// PK ist inherited from Employee

#region Primitive
public virtual DateTime LicenseDate { get; set; }
public virtual Nullable<int> FlightHours { get; set; }

public virtual PilotLicenseType
{
 get;
 set;
}
[StringLength(50)]
public virtual string FlightSchool
{
 get;
 set;
}
#endregion

#region Related Objects
public virtual ICollection<Flight> FlightAsPilotSet { get; set; }
public virtual ICollection<Flight> FlightAsCopilotSet { get; set; }
#endregion
 }
}
```

***Listing 5-5.*** Class Passenger

```csharp
using System;
using System.Collections.Generic;
using System.ComponentModel.DataAnnotations;

namespace BO
{
 public class PassengerStatus
 {
  public const string A = "A";
  public const string B = "B";
```

```csharp
  public const string C = "C";
  public static string[] PassengerStatusSet = { PassengerStatus.A,
  PassengerStatus.B, PassengerStatus.C };
 }

 [Serializable]
 public partial class Passenger : Person
 {

  public Passenger()
  {
   this.BookingSet = new List<Booking>();
  }

  // Primary key is inherited!
  #region Primitive Properties
  public virtual Nullable<DateTime> CustomerSince { get; set; }

  [StringLength(1), MinLength(1), RegularExpression("[ABC]")]
  public virtual string Status { get; set; }
  #endregion

  #region Relations
  public virtual ICollection<Booking> BookingSet { get; set; }
  #endregion
 }
}
```

*Listing 5-6.* Class Booking

```csharp
namespace BO
{
 /// <summary>
 /// Join class for join table
 /// </summary>
 public class Booking
 {
  // Composite Key: [Key] not possible, see Fluent API!
  public int FlightNo { get; set; }
```

```
 // Composite Key: [Key] not possible, see Fluent API!
 public int PassengerID { get; set; }

 public Flight { get; set; }
 public Passenger { get; set; }
 }
}
```

# Rules for the Self-Created Context Class

The context class is the linchpin for Entity Framework Core programming, and there are a few rules to follow when implementing it.

---

**Note**    This chapter only describes a typical basic configuration for the context class. You will find adaptations in Chapter 12.

---

## Installing the NuGet Packages

For the realization of the context class, you need a NuGet package for your respective database management system (see Table 5-1). For example, enter the following in the NuGet Package Manager Console:

```
Install-Package Microsoft.EntityFrameworkCore.SqlServer
```

For SQLite, enter the following:

```
Install-Package Microsoft.EntityFrameworkCore.Sqlite
```

For Oracle, enter the following:

```
Install-Package Devart.Data.Oracle.EFCore
```

While in the classical Entity Framework just two assemblies had to be referenced (and these references had to be created manually), the new NuGet package (in the sense of the modularization of the core products) entails a jumble of 32 references (see the project DAL), which you do not want to have to create manually. For the project BO, no reference to an Entity Framework DLL is necessary!

***Table 5-1.*** *The Entity Framework Core Providers Available on nuget.org*

| Database Management System | NuGet Package |
| --- | --- |
| Microsoft SQL Server Express, Standard, Enterprise, Developer, LocalDB 2008+ | `Microsoft.EntityFrameworkCore.SqlServer` |
| Microsoft SQL Server Compact 3.5 | `EntityFrameworkCore.SqlServerCompact35` |
| Microsoft SQL Server Compact 4.0 | `EntityFrameworkCore.SqlServerCompact40` |
| SQLite | `Microsoft.EntityFrameworkCore.sqlite` |
| PostgreSQL | `Npgsql.EntityFrameworkCore.PostgreSQL` |
| In memory (for unit tests) | `Microsoft.EntityFrameworkCore.InMemory` |
| MySQL | `MySQL.Data.EntityFrameworkCore` |
| Oracle (DevArt) | `Devart.Data.Oracle.EFCore` |

# Base Class

The context class is not a POCO class. It must inherit from the base class `Microsoft.EntityFrameworkCore.DbContext`. The alternative base class `ObjectContext`, which existed in the classic Entity Framework, no longer exists.

# Constructor

The context class must have a parameterless constructor to use the schema migration tools in Visual Studio or the command line because these tools must instantiate the context class at design time. It does not require a parameterless constructor if the database schema is to be generated exclusively at application startup. Then the developer has the opportunity to call the context class with constructor parameters.

**Note**    Without an explicit constructor, C# automatically has a parameterless constructor.

## References to Entity Classes

The developer must create a property of type DbSet<EntityType> for each entity class, like so:

```
public DbSet<Flight> FlightSet {get; set; }
public DbSet<Pilot> PilotSet {get; set; }
```

**Caution**    By default, Entity Framework Core uses the property names shown here for the table names in the database schema. You will learn later how to change this behavior.

## Provider and Connection String

The connection string for the database to be addressed had to be passed in the classic Entity Framework via the constructor to the local implementation of the base class DbContext. Entity Framework Core has a different approach, namely, a new method called OnConfiguring(), which has to be overridden. This method is called by Entity Framework Core for the first instantiation of the context in a process. The method OnConfiguring() receives as a parameter an instance of DbContextOptionsBuilder. In OnConfiguring(), you then invoke an extension method on this instance of DbContextOptionsBuilder, which determines the database provider and the connection string. The extension method to be invoked is provided by the Entity Framework Core database provider. In the case of Microsoft SQL Server, it is named UseSqlServer() and expects the connection string as a parameter. It is up to you to move the connection string to a suitable location (for example, the configuration file) and to load from there.

**Note**    While having a connection string in the code is a bad practice for real projects, it is the best solution here to make the example clear. Therefore, many listings in this book save the connection string within the code. In real projects, you should read the connection string from a configuration file.

The ability to outsource configuration data is highly dependent on the type of project, and a solution like the one shown here would not run in any other type of project. The treatment of various configuration systems and associated APIs is not part of this book. Please refer to basic documentation on .NET, .NET Core, UWP, and Xamarin.

The connection string must contain `MultipleActiveResultSets = True` because otherwise Entity Framework Core may not work correctly in some cases; you'll get the following error message: "There is already an open DataReader associated with this Command which must be closed first."

```
builder.UseSqlServer(@"Server=MyServer;Database=MyDatabase;Trusted_Connection=
True;MultipleActiveResultSets=True");
```

**Attention**    If no `UseXY()` method is called in `OnConfiguring()`, then the following runtime error appears: "No database provider has been configured for this DbContext. A provider can be configured by overriding the DbContext. OnConfiguring method or by using AddDbContext on the application service provider. If AddDbContext is used, then also ensure that your DbContext type accepts a DbContextOptions<TContext> object in its constructor and passes it to the base constructor for DbContext."

# Seeing an Example

Listing 5-7 shows the context class of the World Wide Wings sample in a basic configuration.

***Listing 5-7.*** Context Class

```
using BO;
using Microsoft.EntityFrameworkCore;

namespace DA
{
 /// <summary>
 /// EFCore context class for World Wings Wings database schema version 7.0
 /// </summary>
 public class WWWingsContext : DbContext
 {

  #region Tables
  public DbSet<Flight> FlightSet { get; set; }
  public DbSet<Pilot> PilotSet { get; set; }
  public DbSet<Passenger> PassengerSet { get; set; }
  public DbSet<Booking> BookingSet { get; set; }
  #endregion

  public static string ConnectionString { get; set; } =
  @"Server=.;Database=WWWingsV2_EN_Step1;Trusted_Connection=True;Multiple
  ActiveResultSets=True;App=Entityframework";

  public WWWingsContext() { }

  protected override void OnConfiguring(DbContextOptionsBuilder builder)
  {
   builder.UseSqlServer(ConnectionString);
  }
 }
}
```

# Your Own Connections

UseSqlServer() and other drivers can also receive a connection object (an instance
of the class DbConnection) instead of the connection string. The connection does
not necessarily have to be opened beforehand. It can be open, and then the existing
connection is used. The Entity Framework Core context does not close it either. If it is not
open, the Entity Framework Core context opens and closes the connection as needed.

---

**Best Practice**   Basically, you should keep away from Entity Framework Core connection management! Only in exceptional cases, where this is absolutely necessary (for example, transactions across multiple context instances), should you open the connection yourself!

---

## Thread Safety

The DbContext class is not thread-safe, meaning that the self-created context class inherited from DbContext must under no circumstances be used in several different threads. Each thread needs its own instance of the context class! Those who disregard this risk unpredictable behavior and strange runtime errors in Entity Framework Core! For those using dependency injection, DbContext should be scoped as a Transient object.

---

**Note**   This, of course, also applies to DbContext classes generated using reverse engineering.

---

# Rules for Database Schema Generation

Entity Framework Code then generates a database schema from the entity classes and the context class capable of storing all instances of the entity classes. The structure of the database schema is based on conventions and configurations. The principle of convention before configuration applies here.

There are numerous conventions. The following are the most important:

- From each entity class, for which there is a DbSet<T> in the context class, a table is created. In the classic Entity Framework, the class name of the entity class was plural in the standard system. In Entity Framework Core, the standard now uses the name of the DbSet<T> property in the context class.

- Each elementary property in an entity class becomes a column in the table.

- Properties named ID or classes named ID automatically become primary keys with auto-increment values.

- For every 1/0 side of a navigation property, an additional foreign key column is created, even if there is no explicit foreign key property.

- Properties that are named as navigation properties plus the suffix ID represent automatically generated foreign key columns.

- Enumerated types become int columns in the database.

---

**Note**  While in many cases these conventions suffice to create a database schema from an object model, the conventions in this case are not sufficient. Unfortunately, you will not detect this at compile time; you will detect it only when executing program code that uses the context class.

---

# Looking at an Example Client

The program in Listing 5-8 now uses the Entity Framework context class created and the entity class Passenger. First, by calling the EnsureCreated() method the program makes sure that the database is created, if it does not already exist. The database initialization classes known from the classic Entity Framework no longer exist in Entity Framework Core.

Thereafter, the program creates a new passenger, attaches the passenger to the DbSet<Passenger>, and then stores the new passenger in the database using the SaveChanges() method.

Then all passengers are loaded, and their numbers are printed. Finally, a version of all passengers with the name Schwichtenberg follows. This filtering then takes place in RAM with LINQ to Objects over the previously loaded passengers.

---

**Note**  The commands used in this example are of course described in more detail in later chapters in this book. However, this discussion is didactically necessary to prove the functionality of the Entity Framework Core context class created.

The example is unfortunately not yet running error-free. In the next chapter, you will learn why this is so and how to fix the problems.

---

***Listing 5-8.*** Program Code That Uses the Created Entity Framework Core Model

```
using DA;
using BO;
using System;
using System.Linq;

namespace EFC_Console
{
 class SampleClientForward
 {
  public static void Run()
  {
   Console.WriteLine("Start...");
   using (var ctx = new WWWingsContext())
   {
    // Create database at runtime, if not available!
    var e = ctx.Database.EnsureCreated();
    if (e) Console.WriteLine("Database has been created!");
    // Create passenger object
    var newPassenger = new Passenger();
    newPassenger.GivenName = "Holger";
    newPassenger.Surname = "Schwichtenberg";
    // Append Passenger to EFC context
    ctx.PassengerSet.Add(newPassenger);
    // Save object
    var count = ctx.SaveChanges();
    Console.WriteLine("Number of changes: " + count);
    // Read all passengers from the database
    var passengerSet = ctx.PassengerSet.ToList();
    Console.WriteLine("Number of passengers: " + passengerSet.Count);
    // Filter with LINQ-to-Objects
    foreach (var p in passengerSet.Where(x => x.Surname ==
    "Schwichtenberg").ToList())
    {
```

```
    Console.WriteLine(p.PersonID + ": " + p.GivenName + " " + p.Surname);
   }
  }
  Console.WriteLine("Done!");
  Console.ReadLine();
 }
}
```

# Adaptation by Fluent API (OnModelCreating())

When you start the program code in Listing 5-8, the EnsureCreated() method first encounters the following runtime error: "Unable to determine the relationship represented by navigation property Flight.Pilot' of type 'Pilot'. Either manually configure the relationship, or ignore this property using the '[NotMapped]' attribute or by using 'EntityTypeBuilder.Ignore' in 'OnModelCreating'."

By this the Entity Framework Core tells you that in the case of the two-way relationship between Flight and Pilot (with the properties Pilot and Copilot), it does not know which of the two navigation properties on the Pilot side (FlightsAsPilotSet and FlightAsCopilotSet) correspond to the navigation properties Pilot and Copilot on the Flight side.

To clarify this, there is the so-called Fluent API in Entity Framework Core, which was available in Code First in the classic Entity Framework. The Fluent API consists of the method protected override void OnModelCreating(ModelBuilder modelBuilder), which is to be overridden in the context class. On the modelBuilder object, you then make the configuration in call chains of methods.

```
protected override void OnModelCreating(ModelBuilder builder)
{
...
}
```

In the case of the two-way relationship between `Pilot` and `Flight`, the following two chains of methods are to be entered in `OnModelCreating()`, where the pilot clearly associates with `FlightAsPilot` and where `Copilot` associates with `FlightAsCopilot`:

```
modelBuilder.Entity<Pilot>().HasMany(p => p.FlightAsPilotSet).WithOne(p =>
p.Pilot).HasForeignKey(f => f.PilotId).OnDelete(DeleteBehavior.Restrict);
modelBuilder.Entity<Pilot>().HasMany(p => p.FlightAsCopilotSet).WithOne
(p => p.Copilot).HasForeignKey(f => f.CopilotId).OnDelete(DeleteBehavior.
Restrict);
```

With `.OnDelete(DeleteBehavior.Restrict)`, you turn off the cascading deletion, which makes no sense in this case.

If you start the program again afterward, you get the runtime error "The entity type 'BO.Booking' requires a key to be defined." Entity Framework Core does not know what the primary key is in the intermediate class booking because there is no property there by convention that can be the primary key. There should be a composite primary key. Therefore, you have to add the following in the Fluent API:

```
modelBuilder.Entity<Booking>().HasKey(b => new { b.FlightNo, b.PassengerID });
```

Entity Framework is still not satisfied and complains at the next start of the program about a missing primary key in the classes `Flight`, `Passenger`, and `staff`. In `Flight` this is clear because `FlightNr` does not correspond to the convention (that would be `FlightID`). So, add the following:

```
modelBuilder.Entity<Flight>().HasKey(x => x.FlightNo);
```

Passengers and employees, however, inherit the primary key `PersonID` of the base class `Person`. Unfortunately, Entity Framework Core is not smart enough to notice. Therefore, it should also be added.

```
modelBuilder.Entity<Employee>().HasKey(x => x.PersonID);
modelBuilder.Entity<Passenger>().HasKey(x => x.PersonID);
```

Thus, the program code is finally executable!

Now the question arises why Entity Framework Core does not complain that there is no primary key for `Pilot`. This is because of the way Entity Framework Core maps inheritance in the database. The pilots are not stored in a separate table but are in the same table with the employees. Therefore, Entity Framework Core does not complain for the pilots.

Listing 5-9 shows the improved version of the context class. With this version, the program is now executing as expected. Figure 5-2 shows the output.

***Listing 5-9.*** Improved Version of the Context Class

```
using BO;
using Microsoft.EntityFrameworkCore;

namespace DA
{
 /// <summary>
 /// EFCore context class for World Wings Wings database schema version 7.0
 /// </summary>
 public class WWWingsContext : DbContext
 {

  #region Tables
  public DbSet<Flight> FlightSet { get; set; }
  public DbSet<Passenger> PassengerSet { get; set; }
  public DbSet<Pilot> PilotSet { get; set; }
  public DbSet<Booking> BookingSet { get; set; }
  #endregion

  public static string ConnectionString { get; set; } =
  @"Server=.;Database=WWWingsV2_EN_Step1;Trusted_Connection=True;Multiple
  ActiveResultSets=True;App=Entityframework";

  public WWWingsContext() { }

  protected override void OnConfiguring(DbContextOptionsBuilder builder)
  {
   builder.UseSqlServer(ConnectionString);
  }

  protected override void OnModelCreating(ModelBuilder modelBuilder)
  {
   #region Configure the double relation between Flight and Pilot
```

```
    // fix for problem:   "Unable to determine the relationship represented
    by navigation property Flight.Pilot' of type 'Pilot'. Either manually
    configure the relationship, or ignore this property using the
    '[NotMapped]' attribute or by using 'EntityTypeBuilder.Ignore' in
    'OnModelCreating'."
    modelBuilder.Entity<Pilot>().HasMany(p => p.FlightAsPilotSet).WithOne
    (p => p.Pilot).HasForeignKey(f => f.PilotId).OnDelete(DeleteBehavior.
    Restrict);
    modelBuilder.Entity<Pilot>().HasMany(p => p.FlightAsCopilotSet).WithOne
    (p => p.Copilot).HasForeignKey(f => f.CopilotId).OnDelete(DeleteBehavior.
    Restrict);
    #endregion

    #region Composite key for BookingSet
    // fix for problem: 'The entity type 'Booking' requires a primary key to
    be defined.'
    modelBuilder.Entity<Booking>().HasKey(b => new { FlightNo = b.FlightNo,
    PassengerID = b.PassengerID });
    #endregion

    #region Other Primary Keys
    // fix for problem: 'The entity type 'Employee' requires a primary key
    to be defined.'
    modelBuilder.Entity<Employee>().HasKey(x => x.PersonID);

    // fix for problem: 'The entity type 'Flight' requires a primary key to
    be defined.'
    modelBuilder.Entity<Flight>().HasKey(x => x.FlightNo);

    // fix for problem: 'The entity type 'Passenger' requires a primary key
    to be defined.'
    modelBuilder.Entity<Passenger>().HasKey(x => x.PersonID);
    #endregion

    base.OnModelCreating(modelBuilder);
  }
 }
}
```

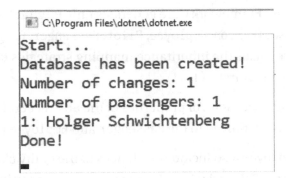

**Figure 5-2.** *Output of the sample client using the improved context class (in a .NET Core console app)*

# Viewing the Generated Database Schema

Figure 5-3 shows the resulting database.

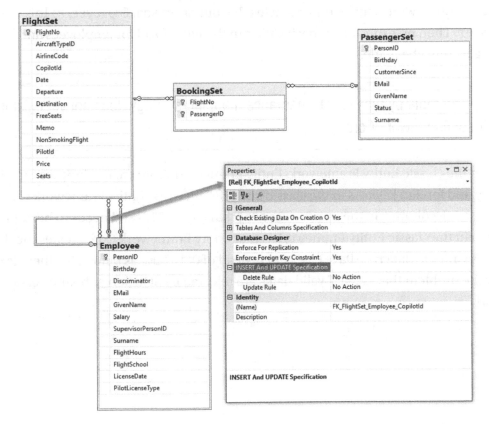

**Figure 5-3.** *The resulting database model*

As you can see, Entity Framework Core generates only four tables in the database from the six entity classes (`Person`, `Employee`, `Pilot`, `Passenger`, `Flight`, and `Booking`). Entity Framework Core mixes the inheritance mapping strategies table per concrete type (TPC) and table per hierarchy (TPH).

- There is no table `Person`. All properties of the abstract entity class `Person` have been moved to the `Passenger` and `Employee` tables.

- The table `Employee` also includes instances of the entity class `Pilot` according to the TPH principle. There is a column `Discriminator` that automatically populates Entity Framework Core with the value `Pilot` or `Employee`.

So far, developers using Entity Framework Core have very little influence on the inheritance mapping strategy. You can only indirectly influence the decision between TPC and TPH. If there were a `DbSet<Person>` and `DbSet<Employee>` in the context class, Entity Framework Core would have completely applied the table by hierarchy (TPH) strategy, in other words, made only a single table out of `Person`, `Employee`, `Pilot`, and `Passenger`. Then no explicit key specification in the Fluent API for employees and passengers must be made!

---

**Note**   The table by type (TPT) inheritance mapping strategy does not yet exist in the Entity Framework Core.

---

As in the classic Entity Framework, Entity Framework Core creates indexes for all primary and foreign keys. As in the classic Entity Framework, Entity Framework Core sets the string columns in the standard as `nvarchar(max)`. This still needs to be adjusted. Better than the classic Entity Framework is that Entity Framework Core creates the date columns with the data type `DateTime2(7)` instead of `DateTime` as before. So, there is no longer the problem that in .NET valid dates are rejected by SQL Server before version 1.1.1601.

# CHAPTER 6

# Customizing the Database Schema

In many cases, Entity Framework Core can create a database schema in forward engineering based solely on conventions from an object model. However, the previous chapter has already shown that conventions are not always enough to create a valid database schema. The computer needed tutoring by the software developer to create composite primary keys, to create primary keys for inheriting classes, and to deactivate the cascading deletion; otherwise, the database results in circular deletion operations.

In other cases, although Entity Framework Core can create a database schema, the result is unsatisfactory. Both of these cases were shown in the previous chapter (see the table names and lengths of the character string columns).

In this chapter, you will learn how to override or supplement the conventions by explicitly configuring via data annotations in the entity classes or via the Fluent API in the OnModelCreating() method.

## Examples in This Chapter

While the previous chapter used a preliminary stage of the World Wide Wings object model version 2, the book will now cover the complete object model version 2. You will find the program code in the project folder EFC_WWWings in the projects EFC_GO, EFC_DA, and EFC_Console (see Figure 6-1).

© Holger Schwichtenberg 2018
H. Schwichtenberg, *Modern Data Access with Entity Framework Core*,
https://doi.org/10.1007/978-1-4842-3552-2_6

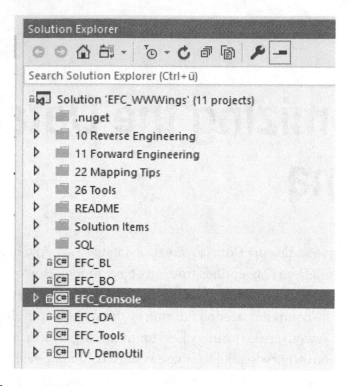

***Figure 6-1.*** *Solution EFC_WWWings*

# Convention vs. Configuration

There are two ways to configure the database schema in Entity Framework Core.

- With data annotations in the entity classes

- With the Fluent API in the `OnModelCreating()` method in the context class

There are three basic rules, listed here:

- Configuration via data annotations or the Fluent API weighs heavier than the conventions. In other words, configuration overrides conventions for individual cases. Microsoft speaks about Entity Framework Core using "convention before configuration." However, this means that the goal is to make the explicit configuration as far as possible superfluous by conventions.

- When there are conflicting data annotations and Fluent API calls, the Fluent API call always wins.

- You can express all configuration options via the Fluent API. A subset of this is also possible via data annotations.

# Persistent vs. Transient Classes

A *persistent class* is a class from the object model whose instances are stored in the database. Within Entity Framework Core, persistent classes are also called *entity classes*. By contrast, transient classes have only volatile instances that are purely in main memory.

Basically, every .NET class is transient. Entity Framework Core makes a class persistent if any of the following are true:

- There is a `DbSet<EntityClass>` in the context class

- There is a call to `modelBuilder.Entity<EntityClass>()` in the Fluent API

- Another persistent class refers to this class by navigation property

Using the second option usually makes no sense because without a `DbSet<EntityClass>` or a navigation property, the data access class through Entity Framework Core is not available.

A developer may sometimes want to define deviations from the third rule if a persistent class is to have a relation to a transient class. In this case, the developer must annotate the related class that is to remain transient with `[NotMapped]` or `modelBuilder. Ignore<Class>()` in the Fluent API.

`[NotMapped]` can also be used at the level of properties of the class if you do not want to persist individual properties of a persistent class in the database because by default Entity Framework Core persists all properties of an entity class that have a getter and a setter. The Fluent API uses the `Ignore()` method for this, but this time it does so after calling `Entity<T>()`: `modelBuilder.Entity<EntityClass>().Ignore(x => x.Property)`.

An entity class property, in particular, must be ignored if the property has a more complex .NET data type that Entity Framework Core cannot map. This applies, for example, to the class `system.Xml.XmlDocument`. Entity Framework Core fails

89

to generate a database schema and gives the following error: "The key {'TempId'} contains properties in shadow state and is referenced by a relationship from 'XmlSchemaCompilationSettings' to 'XmlSchemaSet.CompilationSettings'. Configure a non-shadow principal key for this relationship." Although there is an XML data type in Microsoft SQL Server and other database management systems, in Entity Framework Core a mapping to the .NET class `system.Xml.XmlDocument` is not realized yet.

# Names in the Database Schema

By convention, Entity Framework Core assigns the following:

- Each table gets the name of the property used in the context class for the `DbSet<EntityClass>`.

- For each entity class for which there is no `DbSet <entity class>`, Entity Framework Core uses the class name as the table name.

- Each column gets the name of the property in the entity class.

To change this, you can use the options described in Table 6-1.

***Table 6-1.*** *Changing Conventionally Specified Table and Column Names in the Database Schema*

| | Data Annotation | Fluent API |
| --- | --- | --- |
| **Table name** | In front of a class: `[Table("TableName")]` or with additional specification of the schema name: `[Table("TableName", schema = "SchemaName")]` Without the schema name, the table always ends up in the default schema, which is dbo. | `modelBuilder.Entity<EntityClass>().ToTable( "TableName");` or `modelBuilder.Entity<EntityClass>().ToTable ("TableName", schema: "SchemaName");` |
| **Column name** | In front of a property: `[Column("Column Name")]` | `modelBuilder.Entity<EntityClass>().Property(b => b.Property).HasColumnName("Column Name");` |

# Order of Columns in a Table

Entity Framework Core sorts the columns in a table as follows:

1. First the primary key columns appear in alphabetical order.

2. Then all other columns appear in alphabetical order.

3. Columns added later are not sorted in the order but added at the back.

Unlike the classic Entity Framework, Entity Framework Core does not follow the order of the properties in the source code. Microsoft explains this at `https://github.com/aspnet/EntityFramework/issues/2272` with the following: "In EF6 we tried having column order match the order of properties in the class. The issue is that reflection can return a different order on different architectures."

In the classic Entity Framework, the order could be configured by the annotation `[Column(Order = Number)]`. However, this affects only the first time the table is created, not columns that are added later, because sorting new columns between existing columns in many database management systems requires a table rebuild. According to Microsoft, "There isn't any way to do this since SQL Server requires a table rebuild (rename existing table, create new table, copy data, delete old table) to re-order the columns" (`https://github.com/aspnet/EntityFramework/issues/2272`). Microsoft has therefore decided not to respect the `Order` property of the annotation `[Column]` in Entity Framework Core.

# Column Types/Data Types

The database type used in the database schema for a .NET type is not determined by Entity Framework Core but by the database provider. For example, Table 6-2 shows what is selected by default in Microsoft SQL Server, SQLite, and the DevArt Oracle provider.

---

**Note**    While there are fixed mappings of column types to .NET data types in Entity Framework Core 2.0, Microsoft will introduce value converters in Entity Framework Core 2.1; see Appendix C. A value converter allows property values to be converted when reading from or writing to the database.

---

***Table 6-2.*** *Mapping .NET Data Types to Column Types*

| .NET Data Type | Microsoft SQL Server Column Type | SQLite Column Type | Oracle Column Type |
|---|---|---|---|
| Byte | Tinyint | INTEGER | NUMBER(5, 0) |
| Short | Smalintl | INTEGER | NUMBER(5, 0) |
| Int32 | Int | INTEGER | NUMBER(10, 0) |
| Int64 | Bitint | INTEGER | NUMBER(19, 0) |
| DateTime | DateTime2 | TEXT | TIMESTAMP(7) |
| DateTimeOffset | datetimeoffset | TEXT | TIMESTAMP(7) WITH TIME ZONE |
| TimeSpan | time | TEXT | INTERVAL DAY(2) TO SECOND(6) |
| String | nvarchar(MAX) | TEXT | NCLOB |
| String limited length | nvarchar(x) | TEXT | NVARCHAR2(x) |
| Guid | Uniqueidentifier | BLOB | RAW(16) |
| Float | Real | REAL | BINARY_FLOAT |
| Double | Float | REAL | BINARY_DOUBLE |
| Decimal | decimal(18,2) | TEXT | NUMBER |
| Byte[] | varbinary(MAX) | BLOB | BLOB |
| [Timestamp] Byte[] | Rowversion | BLOB | BLOB |
| Byte | Tinyint | INTEGER | NUMBER(5, 0) |

*(continued)*

***Table 6-2.*** (*continued*)

| .NET Data Type | Microsoft SQL Server Column Type | SQLite Column Type | Oracle Column Type |
|---|---|---|---|
| Other array types, such as short[], int[], and string[] | Mapping is not yet supported by Entity Framework Core. You will get the following error:<br>"The property 'xy' could not be mapped, because it is of type 'Int16[]' which is not a supported primitive type or a valid entity type. Either explicitly map this property, or ignore it using the '[NotMapped]' attribute or by using 'EntityTypeBuilder.Ignore' in 'OnModelCreating'." | | |
| Char | Mapping is not yet supported by Entity Framework Core. You will get the following error:<br>"The property 'xy' is of type 'char' which is not supported by current database provider. Either change the property CLR type or ignore the property using the '[NotMapped]' attribute or by using 'EntityTypeBuilder.Ignore' in 'OnModelCreating'." | INTEGER | Mapping is not yet supported by Entity Framework Core. You will get the following error:<br>"The property 'xy' is of type 'char' which is not supported by current database provider. Either change the property CLR type or ignore the property using the '[NotMapped]' attribute or by using 'EntityTypeBuilder.Ignore' in 'OnModelCreating'." |
| XmlDocument | Mapping is not yet supported by Entity Framework Core. You will get the following error:<br>"The entity type 'XmlSchemaCompilationSettings' requires a primary key to be defined." | | |

If you do not agree with this data type equivalent, you must use the data annotation [Column] or use HasColumnType() in the Fluent API.

Here's an example:

```
[Column(TypeName = "varchar(200)")]
modelBuilder.Entity<Entitätsklasse>()
.Property(x => x.Destination).HasColumnType("varchar(200)")
```

---

**Caution**    The DbGeometry and DbGeography classes, which have been supported by the classic Entity Framework since version 5.0, cannot yet be used in Entity Framework Core. So far, there is no mapping for the SQL Server Geometry and Geography column types.

---

# Mandatory Fields and Optional Fields

The convention states that only those columns in the database are to be created as "nullable", in which the .NET type in the object model can also tolerate a null (or nothing in Visual Basic .NET). In other words, it can take string, byte[], and the explicit nullable value types Nullable<int>, int?, Nullable<DateTime>, DateTime?, and so on.

With the annotation [Required] or with modelBuilder.Entity<EntityClass>(). Property(x => x. Propertyname).IsRequired(), you determine that a property should not be nullable in the database, even if the property actually allows a null or nothing in your code. It is not possible to enforce a nullable column using an annotation or the Fluent API; if the database allows null values in a column but the corresponding property your code is not nullable, you get runtime errors.

---

**Note**    Because of the behavior of .NET value types, it may be necessary to declare a column as int? with a [Required] attribute to ensure that the value is actually provided and not just set to the .NET default of 0.

---

# Field Lengths

A significant weakness in the generated database schema in Chapter 5 was that all character string columns with the long data type `nvarchar(max)` were generated. Entity Framework Core limits a string column that is a primary key to 450 characters by default.

You can define a length limitation with the annotation `[MaxLength(number)]` or `[StringLength(number)]` or with `modelBuilder.Entity<EntityClass>().Property (x => x.PropertyName).HasMaxLength(number)`.

# Primary Keys

By convention, the primary key of a table is a property named `ID` or `Id` or `ClassNameID` or `ClassNameId`. The case of these names is not relevant. Unfortunately, if you use more than one of these variants in a class (all four are possible in C# , but only two are possible in Visual Basic .NET because the language is case-insensitive), Entity Framework Core takes the first matching property in that order as it stands in the program code. All other properties corresponding to this convention become normal columns of the table.

If another property is to become a primary key, you must annotate it with `[Key]` or write it in the Fluent API: `modelBuilder.Entity<EntityClass>().HasKey(x => x.Property)`. As opposed to the classic Entity Framework, composite primary keys can no longer be specified in Entity Framework Core via data annotations; they can be specified only via the Fluent API, as in `builder.Entity<Booking>().HasKey(x => new { x.FlightNo, x.Passenger ID })`.

For primary keys that are integers (`byte`, `short`, `int`, `long`), Entity Framework Core creates default identity columns (also known as *auto-increment columns*) in the database schema. The documentation says this: "By convention, primary keys that are of an integer or GUID data type will be set up to have values generated on add" (`https:// docs.microsoft.com/en-us/ef/core/modeling/generated-properties`). *Integer* is a misleading generic term (because the sentence is also true for `byte`, `short`, and `long`). If you do not want auto-increment columns, use the annotation `[DatabaseGenerate d(DatabaseGeneratedOption.None)]` in front of the property or use `modelBuilder. Entity<class>().Property(x => x.PropertyName).ValueGeneratedNever()` in `OnModelCreating()`.

# Relationships and Foreign Keys

Entity Framework Core automatically treats properties that reference one or more instances of another entity class as navigation properties. This allows you to create relationships between entities (master-detail relationships such as 1:1 and 1:N).

For sets, the developer can use ICollection or any other interface based on it (such as IList), as well as any ICollection<T>-implementing class (such as List<T> or HashSet<T>). Entity Framework Core automatically creates a foreign key column in the database schema in the table in these spots:

- On the N side of a 1:N relationship

- On one side of a 1:0/1 relationship

The foreign key column contains the name of the navigation property plus the name of the primary key of the related entity class. For each foreign key column, Entity Framework Core automatically generates an index in the database.

For the program code in Listing 6-1 (which introduces the entity type AircraftType into the World Wide Wings sample), the foreign key column AircraftTypeTypeId is created in the Flight table (Type occurs twice because it belongs to both the class name and the name of the primary key). To get Entity Framework Core to use a simpler name, you use the annotation [ForeignKey("AircraftTypeId")] on the navigation property. In the Fluent API, this is a bit more complicated because you have to explicitly formulate the cardinality with HasOne(), HasMany(), WithOne(), and WithMany() before calling the method HasForeignKey() to get the name of the Set foreign key column.

```
builder.Entity<Flight>().HasOne(f => f. AircraftType).WithMany(t=>t.
FlightSet).HasForeignKey("AircraftTypeId");
```

You have the choice to determine from which direction you want to formulate the relationship. Therefore, the following command line is equivalent to the previous one. Having both instructions in program code is not an error but unnecessary.

```
builder.Entity<AircraftType>().HasMany(t => t.FlightSet).WithOne(t => t.
AircraftType).HasForeignKey("AircraftTypeTypeId");
```

This foreign key column can also be explicitly mapped in the object model by a foreign key property (see public byte AircraftTypeID {get; set;} in Listing 6-1). However, this explicit mapping is not mandatory.

**Tip**    The advantage of displaying the foreign key column by a property in the object model is that relationships can be established via the foreign key without having to load the complete, related object. By convention, Entity Framework Core automatically treats a property as a foreign key property if it matches the name Entity Framework Core chooses by default for the foreign key column.

# Optional Relationships and Mandatory Relationships

Listing 6-1 introduces the entity types `AircraftType` and `AircraftTypeDetail`. An Aircraft has exactly one `AircraftType`, and one `AircraftType` has exactly one `AircraftTypeDetail`.

***Listing 6-1.*** New Entity Classes AircraftType and AircraftTypeDetail with the Relevant Cutouts from the Related Flight and Pilot Classes

```
using System.Collections.Generic;
using System.ComponentModel.DataAnnotations;

namespace BO
{
 /// <summary>
 /// AircraftType has a dependent object  AircraftTypeDetail (1:1)
 /// AircraftTypeDetail uses the same primary key as AircraftType
 /// </summary>
 public class AircraftType
 {
  [Key]
  public byte TypeID { get; set; }
  public string Manufacturer { get; set; }
  public string Name { get; set; }
  // Navigation Property 1:N
  public List<Flight> FlightSet { get; set; }
```

```csharp
  // Navigation Property 1:1, unidirectional, no FK Property
  public AircraftTypeDetail Detail { get; set; } }
}

using System.ComponentModel.DataAnnotations;

namespace BO
{
 /// <summary>
 /// AircraftTypeDetail is a dependent object (1:1) of AircraftType
 /// AircraftTypeDetail uses the same primary key as AircraftType
 /// </summary>
 public class AircraftTypeDetail
 {
  [Key]
  public byte AircraftTypeID { get; set; }
  public byte? TurbineCount { get; set; }
  public float? Length { get; set; }
  public short? Tare { get; set; }
  public string Memo { get; set; }

  public AircraftType { get; set; }
 }
}
using System;
using System.Collections.Generic;
using System.ComponentModel.DataAnnotations;
using System.ComponentModel.DataAnnotations.Schema;
using EFCExtensions;

namespace BO
{

public class Flight
 {

  #region Key
  public int FlightNo { get; set; }
  #endregion
```

...

```csharp
  #region Related Objects
  public ICollection<Booking> BookingSet { get; set; }
  public Pilot { get; set; }
  public Pilot Copilot { get; set; }
  [ForeignKey("AircraftTypeID")]
  public AircraftType AircraftType { get; set; }

  // Explicit foreign key properties for the navigation properties
  public int PilotId { get; set; } // mandatory!
  public int? CopilotId { get; set; } // optional
  public byte? AircraftTypeID { get; set; } // optional
  #endregion
}
}

using System;
using System.Collections.Generic;
using System.ComponentModel.DataAnnotations;

namespace BO
{

 [Serializable]
 public partial class Pilot : Employee
 {
  // PK is inherited from Employee

...

  #region Related Objects
  public virtual ICollection<Flight> FlightAsPilotSet { get; set; }
  public virtual ICollection<Flight> FlightAsCopilotSet { get; set; }
  #endregion
 }
}
```

In Listing 6-1, the relationship between `Flight` and `AircraftType` is a mandatory relationship, i.e. each flight must be assigned exactly one `AircraftType`, because the foreign key property `AircraftTypeID` must be assigned a value. See Figure 6-2.

To make this relationship optional, i.e. permit `Flight` objects that have no assigned `AircraftType`, the property has to allow zero or nothing for the foreign key column. In this case, there would have to be: `public byte? Aircraft TypeNr {get; set; }`. You can also create a mandatory relationship with `IsRequired()` in the Fluent API, even if the foreign key column allows null or nothing, as shown here:

```
builder.Entity<Flight>()
.HasOne(f => f.AircraftType)
.WithMany(t => t.FlightSet)
.IsRequired()
.HasForeignKey("AircraftTypeID");
```

---

**Note**   If there is no explicit foreign key property, the relationship is optional by default. Again, you need to call the method `IsRequired()` to force a mandatory relationship.

---

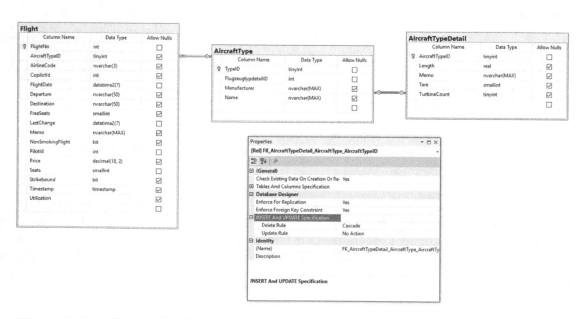

***Figure 6-2.*** *Relationship between Flight and AircraftType and AircraftTypeDetail*

# Unidirectional and Bidirectional Relationships

Relations between two entity classes can be bidirectional in the object model; that is, there are two-way navigation properties, both from Flight to AircraftType (via the AircraftType property in class Flight) and from AircraftType to Flight (via the FlightSet property in class AircraftType). Alternatively, unidirectional relationships are allowed because of simply omitting navigation in one of the two classes (see the relationship between AircraftTypeDetail and AircraftTypeDetail, which is unidirectional). Listing 6-1 showed that AircraftType has a navigation property named Detail that refers to an AircraftTypeDetail object. However, in the implementation of AircraftTypeDetail, there is no navigation property to AircraftType. Bidirectional relationships, however, usually make sense, so the object model is easier to use, especially since they consume no additional space in the database and only minimal space in the main memory.

In bidirectional relationships, Entity Framework Core uses convention to find the two matching navigation properties and their cardinalities. So, if Flight has a navigation property of type AircraftType and AircraftType has a navigation property of type List<Flight>, then Entity Framework Core automatically assumes a 1:N relationship.

However, this convention-based mechanism is not possible for the Flight <-> Pilot relationship because there are two navigation property of type Pilot (named Pilot and Copilot) in the Flight class and two navigation properties of type List<Flight> (FlightAsPilotSet and FlightAsCopilotSet) in the class Pilot. At this point, you must give the Entity Framework Core pertinent hints about what belongs together. This is done either via the data annotation [InverseProperty("FlightAsPilotSet")] or [InverseProperty("FlightAsCopilotSet")] or via the Fluent API, as shown here:

```
builder.Entity<Pilot>().HasMany(p => p.FlightAsCopilotSet)
.WithOne(p => p.Copilot).HasForeignKey(f => f.CopilotId);
builder.Entity<Pilot>().HasMany(p => p.FlightAsPilotSet)
.WithOne(p => p.Pilot).HasForeignKey(f => f.PilotId);
```

In the World Wide Wings example, the relationship between Flight and Pilot over the navigation property Pilot is a mandatory relationship; Copilot is optional.

Abolishing the copilot and landing the plane by the flight attendant in an emergency (as in the movie *Turbulence* from 1997) incidentally was a real proposal of Michael O'Leary, the boss of the Irish airline Ryanair in the year 2010 (see www.dailymail.co.uk/news/article-1308852/Let-stewardesses-land-plane-crisis-says-Ryanair-boss-Airline-wants-ditch-pilots.html).

# 1:1 Relationships

Listing 6-1 also shows the 1:1 relationship between `AircraftType` and `AircraftTypeDetail`. This is a mandatory relationship; that is, there must be exactly one `AircraftTypeDetail` object for each `AircraftType` object because the relationship between the classes is not backed by a foreign key column. `AircraftType` and `AircraftTypeDetail` have a primary key property that is identical in name and type. The relationship thus arises from `AircraftType.TypeID` to `AircraftTypeDetail.AircraftTypeID`.

`AircraftType.TypeID` is created as an auto-increment value. Entity Framework Core is smart enough to also create `AircraftTypeDetail.AircraftTypeID` as an auto-increment value because these two numbers must indeed correspond so that the relationship works.

If `AircraftType.TypeID` is not an auto-increment value, Entity Framework Core would make one for `AircraftTypeDetail.AircraftTypeID`, which leads to problems. `AircraftType.TypeNr` has no auto-increment value, but Entity Framework Core still does not store the value that was explicitly assigned in the source code. Entity Framework Core then uses for `AircraftType.TypeID` the auto-increment value, which `AircraftTypeDetail.AircraftTypeID` has specified. Only if both `AircraftType.TypeID` and `AircraftTypeDetail.AircraftTypeID` are set to `ValueGeneratedNever()` you can set the values freely. Here you would have to help a bit and configure `AircraftTypeDetail.AircraftTypeID` as having no auto-increment value.

```
builder.Entity<AircraftTypeDetail>().Property(x => x. AircraftTypeID).
ValueGeneratedNever().
```

If the entity class `AircraftTypeDetail` had a different primary key name (for example, `No`), Entity Framework Core would have created this primary key column as the auto-increment value column and added a foreign key column in the `AircraftType` table (named `DetailNo`). This relationship would then be a 1:0/1 relationship, so there could be `AircraftType` objects without an `AircraftTypeDetail` object. Then you could not easily see the relationship from the data; for example, `AircraftType` object #456 may be associated with `AircraftTypeDetail` object #72.

A `DbSet<AircraftTypeDetail>` does not have to exist in the context class. In addition, the relationship between `AircraftType` and `AircraftTypeDetail` is a unidirectional relationship since there is only one navigation type from `AircraftType` to `AircraftTypeDetail`, but not from `AircraftTypeDetail` to `AircraftType`. From the point of view of Entity Framework Core, this is fine, and in this case it is technically quite appropriate that `AircraftTypeDetail` exists as a purely `AircraftType`-dependent object.

# Indexes

Entity Framework Core automatically assigns an index to all foreign key columns. In addition, you can assign arbitrary indices with the method `HasIndex()` in the Fluent API (possibly with the addition of `IsUnique()` and `ForSqlServerIsClustered()`). The syntax is simpler than the classic Entity Framework. However, unlike the classic Entity Framework, you cannot use data annotations in Entity Framework Core for indexes.

Here are some examples:

```
// Index with one column
modelBuilder.Entity<Flight>().HasIndex(x => x.FreeSeats).
// Index with two columns
modelBuilder.Entity<Flight>().HasIndex(f => new {f.Departure, f.Destination});
// Unique Index: Then there could be only one Flight on each Flight route ...
modelBuilder.Entity<Flight>().HasIndex(f => new {f.Departure,
f.Destination).IsUnique();
// Unique Index and Clustered Index: there can only be one CI per table
(usually PK)
modelBuilder.Entity<Flight>().HasIndex (f => new {f.Departure,
f.Destination).IsUnique().ForSqlServerIsClustered();
```

Entity Framework Core names the indexes in the database with the prefix `IX_`.

---

**Tip**   With `HasName()`, you can influence the name of an index in the database, as in `modelBuilder.Entity<Flight>().HasIndex(x=>x.FreeSeats).HasName("Index_FreeSeats");`.

---

In Figure 6-3, there are three foreign key relationship indexes, with one based on the primary key. The remaining two are created manually.

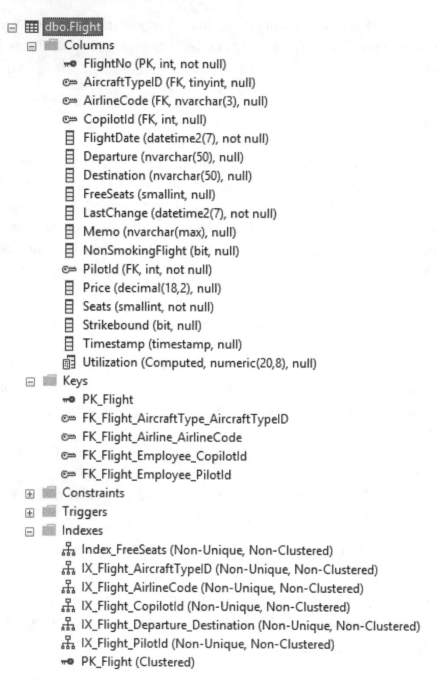

*Figure 6-3.* *Indexes in SQL Server Management Studio*

# Syntax Options for the Fluent API

For larger object models, the Fluent API configuration in the OnModelCreating() method in the Entity Framework Core context class can become very large. Entity Framework Core therefore offers various options for structuring the content differently instead of the sequential calls shown up to now.

## Sequential Configuration

The starting point for the statements is the sequential configuration of the entity class Flight shown in Listing 6-2.

***Listing 6-2.***  Fluent API Calls for the Entity Class Flight Without Structuring

```
modelBuilder.Entity<Flight>().HasKey(f => f.FlightNo);
modelBuilder.Entity<Flight>().Property(b => b.FlightNo).
ValueGeneratedNever();

// ----------- Length and null values
modelBuilder.Entity<Flight>().Property(f => f.Memo).HasMaxLength(5000);
modelBuilder.Entity<Flight>().Property(f => f.Seats).IsRequired();

// ----------- Calculated column
modelBuilder.Entity<Flight>().Property(p => p.Utilization)
        .HasComputedColumnSql("100.0-(([FreeSeats]*1.0)/[Seats])*100.0");

// ----------- Default values
modelBuilder.Entity<Flight>().Property(x => x.Price).
HasDefaultValue(123.45m);
modelBuilder.Entity<Flight>().Property(x => x.Departure).
HasDefaultValue("(not set)");
modelBuilder.Entity<Flight>().Property(x => x.Destination).
HasDefaultValue("(not set)");
modelBuilder.Entity<Flight>().Property(x => x.Date).HasDefaultValueSql
("getdate()");
```

```
//// ----------- Indexes
//// Index over one column
modelBuilder.Entity<Flight>().HasIndex(x => x.FreeSeats).HasName("Index_
FreeSeats");
//// Index over two columns
modelBuilder.Entity<Flight>().HasIndex(f => new { f.Departure,
f.Destination });
```

# Structuring by Lambda Statement

This structuring form eliminates the constant repetition of modelBuilder.
Entity<Flight>() by entering a lambda expression with a command sequence in the
method Entity(); see Listing 6-3.

***Listing 6-3.*** Fluent API Calls Structured by Lambda Statement

```
modelBuilder.Entity<Flight>(f =>
{
 // ----------- PK
 f.HasKey(x => x.FlightNo);
 f.Property(x => x.FlightNo).ValueGeneratedNever();
 //// ----------- Length and null values
 f.Property(x => x.Memo).HasMaxLength(5000);
 f.Property(x => x.Seats).IsRequired();
 // ----------- Calculated column
 f.Property(x => x.Utilization)
 .HasComputedColumnSql("100.0-(([FreeSeats]*1.0)/[Seats])*100.0");

 // ----------- Default values
 f.Property(x => x.Price).HasDefaultValue(123.45m);
 f.Property(x => x.Departure).HasDefaultValue("(not set)");
 f.Property(x => x.Destination).HasDefaultValue("(not set)");
 f.Property(x => x.Date).HasDefaultValueSql("getdate()");

 // ----------- Indexes
 // Index with one column
 f.HasIndex(x => x.FreeSeats).HasName("Index_FreeSeats");
```

```
// Index with two columns
f.HasIndex(x => new { x.Departure, x.Destination });
});
```

## Structuring by Subroutines

In the structured form shown in Listing 6-4, the configuration for an entity class is stored in a subroutine.

***Listing 6-4.*** Fluent API Calls Structured by Subroutine

```
modelBuilder.Entity<Flight>(ConfigureFlight);

private void ConfigureFlight(EntityTypeBuilder<Flight> f)
  {
    // ----------- PK
    f.HasKey(x => x.FlightNo);
    f.Property(x => x.FlightNo).ValueGeneratedNever();
    //// ----------- Length and null values
    f.Property(x => x.Memo).HasMaxLength(5000);
    f.Property(x => x.Seats).IsRequired();
    // ----------- Calculated column
    f.Property(x => x.Utilization)
     .HasComputedColumnSql("100.0-(([FreeSeats]*1.0)/[Seats])*100.0");

    // ----------- Default values
    f.Property(x => x.Price).HasDefaultValue(123.45m);
    f.Property(x => x.Departure).HasDefaultValue("(not set)");
    f.Property(x => x.Destination).HasDefaultValue("(not set)");
    f.Property(x => x.Date).HasDefaultValueSql("getdate()");

    // ----------- Indexes
    // Index with one column
    f.HasIndex(x => x.FreeSeats).HasName("Index_FreeSeats");
    // Index with two columns
    f.HasIndex(x => new { x.Departure, x.Destination });
  }
```

# Structuring Through Configuration Classes

With Entity Framework Core 2.0, Microsoft has introduced another structuring option. Following the outsourcing of configuration in the EntityTypeConfiguration<T> inheriting classes that exist in the classic Entity Framework, Entity Framework Core now provides the IEntityTypeConfiguration <EntityType> interface with which you can implement a separate configuration class for an entity type; see Listing 6-5.

***Listing 6-5.*** Fluent API Calls Structured by IEntityTypeConfiguration

```
using BO;
using Microsoft.EntityFrameworkCore;
using Microsoft.EntityFrameworkCore.Metadata.Builders;

namespace DA
{
 /// <summary>
 /// Configuration Class for Entity Class Flight
 /// EFCore >= 2.0
 /// </summary>
 class FlightETC : IEntityTypeConfiguration<Flight>
 {
  public void Configure(EntityTypeBuilder<Flight> f)
  {
   // ----------- PK
   f.HasKey(x => x.FlightNo);
   f.Property(x => x.FlightNo).ValueGeneratedNever();
   //// ----------- Length and null values
   f.Property(x => x.Memo).HasMaxLength(5000);
   f.Property(x => x.Seats).IsRequired();
   // ----------- Calculated column
   f.Property(x => x.Utilization)
    .HasComputedColumnSql("100.0-(([FreeSeats]*1.0)/[Seats])*100.0");

   // ----------- Default values
   f.Property(x => x.Price).HasDefaultValue(123.45m);
   f.Property(x => x.Departure).HasDefaultValue("(not set)");
```

```
    f.Property(x => x.Destination).HasDefaultValue("(not set)");
    f.Property(x => x.Date).HasDefaultValueSql("getdate()");

    // ----------- Indexes
    // Index with one column
    f.HasIndex(x => x.FreeSeats).HasName("Index_FreeSeats");
    // Index with two columns
    f.HasIndex(x => new { x.Departure, x.Destination });
  }
 }
}
```

You can use this configuration class by calling modelBuilder.ApplyConfiguration
<EntityClass>(ConfigurationObject) in OnModelCreating(), as shown here:

```
modelBuilder.ApplyConfiguration<Flight>(new FlightETC());
```

# Bulk Configuration with the Fluent API

Another option in the Fluent API is not to configure each individual entity classes but
instead configure several at once. The subobject Model of the passed ModelBuilder
object provides a list of all entity classes in the form of objects with the interface
IMutableEntityType via GetEntityTypes(). This interface provides access to all
configuration options for an entity class. The example in Listing 6-6 shows the following:

- It prevents the convention that all table names are named as the
  property name of DbSet<EntityClass> in the context class. With
  entity.Relational().TableName = entity.DisplayName(), all
  tables are named as the entity classes. Exceptions are only those
  classes that have a [Table ] annotation, so you have the opportunity
  to set individual deviations from the rule.

- You ensure that properties ending in the letters NO automatically
  become the primary key and there are no auto-increment values for
  these primary keys.

***Listing 6-6.*** Bulk Configuration in the Fluent API

```
protected override void OnModelCreating (ModelBuilder builder)
{
...
    #region Bulk configuration via model class for all table names
    foreach (IMutableEntityType entity in modelBuilder.Model.GetEntityTypes())
    {
     // All table names = class names (~ EF 6.x),
     // except the classes that have a [Table] annotation
     var annotation = entity.ClrType.GetCustomAttribute<TableAttribute>();
     if (annotation == null)
     {
      entity.Relational().TableName = entity.DisplayName();
     }
    }
...

    #region Bulk configuration via model class for primary key
    foreach (IMutableEntityType entity in modelBuilder.Model.GetEntityTypes())
    {
     // properties ending in the letters "NO" automatically become the primary
     key and there are no auto increment values for these primary keys.
     var propNr = entity.GetProperties().FirstOrDefault(x => x.Name.
     EndsWith("No"));
     if (propNr != null)
     {
      entity.SetPrimaryKey(propNr);
      propNr.ValueGenerated = ValueGenerated.Never;
     }
    }
}
```

# CHAPTER 7

# Database Schema Migrations

Entity Framework Core contains tools to create a database from an object model at the time of development or runtime of the application and to change the schema of an existing database (in simple cases without data loss).

By default, Entity Framework Core assumes at startup that the database to be addressed exists and is in the correct schema version. There is no check to see whether this is really true. For example, if a table or column is missing or a relationship does not exist as intended, then you will get a runtime error (for example, "Invalid object name 'AircraftType'") when you access the object in the database.

## Creating the Database at Runtime

At program start, you can call the EnsureCreated() method in the Database subobject of the context class (see Listing 7-1); this method creates the complete database in the sense of forward engineering if it does not exist and creates the tables with the associated keys and indexes.

However, if the database already exists, EnsureCreated() leaves it as is. The method EnsureCreated() does not then check whether the database schema is correct, meaning whether it corresponds to the current object model. Instead, EnsureCreated() uses the following command:

```
IF EXISTS (SELECT * FROM INFORMATION_SCHEMA.TABLES WHERE TABLE_TYPE = 'BASE
TABLE') SELECT 1 ELSE SELECT 0
```

© Holger Schwichtenberg 2018
H. Schwichtenberg, *Modern Data Access with Entity Framework Core*,
https://doi.org/10.1007/978-1-4842-3552-2_7

This checks whether there is any table in the database at all. If there is no table, all tables are created. However, as soon as there is any table in the database, nothing happens, and the program fails at runtime. You get more "intelligence" with the schema migrations described in the next section.

***Listing 7-1.*** Using EnsureCreated()

```
using DA;
using ITVisions;
using Microsoft.EntityFrameworkCore;

namespace EFC_Console
{
 class CreateDatabaseAtRuntime
 {
  public static void Create()
  {
   CUI.MainHeadline("----------- Create Database at runtime");
   using (var ctx = new WWWingsContext())
   {
    // GetDbConnection() requires using Microsoft.EntityFrameworkCore !
    CUI.Print("Database: " + ctx.Database.GetDbConnection().ConnectionString);
    var e = ctx.Database.EnsureCreated();
    if (e)
    {
     CUI.Print("Database has been created");
    }
    else
    {
     CUI.Print("Database exists!");
    }
   }
  }
 }
}
```

# Schema Migrations at the Time of Development

In the 4.3 version of the classic Entity Framework, Microsoft introduced schema migrations. These schema migrations are now (with slight differences) also available in Entity Framework Core.

Schema migrations allow you to do the following:

- Change the database schema at a later date while retaining the existing data

- Reverse the changes, if necessary

- Run the migration either at development time or at the start of the application

# Commands for the Schema Migrations

As in the classic Entity Framework, there is no graphical user interface (GUI) for performing migrations. Rather, you execute all actions within a command-line interface, either via a PowerShell cmdlet within the Package Manager Console of Visual Studio or via the external command-line tool dotnet.exe (or dotnet on other operating systems).

To use these commands, install a NuGet package.

```
Install-Package Microsoft.EntityFrameworkCore.Tools
```

Unfortunately, this package brings a lot of new assembly references into the project, which will not be needed at runtime later. However, since the NuGet packages are needed only in the startup project of the application and not in other projects, a simple solution is available to not inflate the actual startup project. Follow these steps:

1. Create a new console application project, such as with the name EFC_Tools.

2. Install the Entity Framework Core tools package (Microsoft. EntityFrameworkCore.Tools).

3. From the EFC_Tools project, refer to the project in which the context class is located.

4.  Make this EFC_Tools project the startup project.

5.  Execute the required migration commands here.

6.  Change the startup project again.

You can refrain from changing the startup project if you specify what the startup project should be in the cmdlets using the parameter -startupproject.

---

**Note**   The project EFC_Tools does not need to be deployed to users later.

---

Microsoft has changed some of the details of the procedure for creating and using schema migrations in Entity Framework Core compared to the classic Entity Framework. You do not need to run the Enable-Migrations command at the beginning; you can start the project directly with Add-Migration. The Command-Enable-Migrations command still exists, but it only returns the following message: "Enable-Migrations is obsolete. Use Add-Migration to start using Migrations." Automatic migrations, without calling Add-Migration, are no longer available in Entity Framework Core. Updating the database is done as before with Update-Database. If you prefer a SQL script for your own execution, you will now receive it with Script-Migration instead of with the Update-Database script (see Figure 7-1).

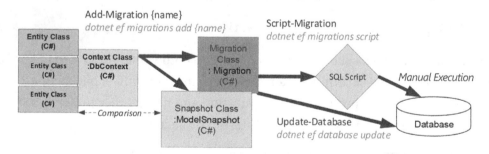

***Figure 7-1.***   *Flow of schema migration in the Entity Framework Core*

# ef.exe

Internally, the PowerShell cmdlets use a classic command-line utility called ef.exe (the Entity Framework Core command-line tools), which is part of the NuGet package Microsoft.EntityFrameworkCore.Tools and is located in the Tools folder. Figure 7-2 shows help for the command.

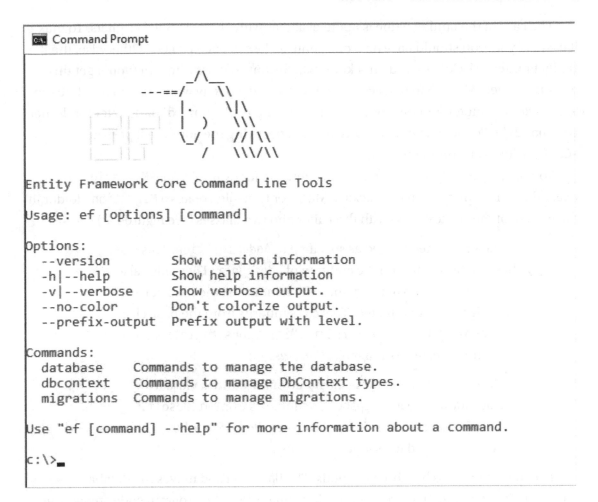

*Figure 7-2.*  *Help for ef.exe*

# Add-Migration

For Entity Framework Core, you start a schema migration (even the first one) within Visual Studio with the PowerShell cmdlet `Add-Migration`. This is done via the NuGet Package Manager Console (PMC).

1.  The Entity Framework Core tools are actually installed in the current startup project.

2.  That this project is selected as the default project in which the context class is located.

3.  That all projects can be compiled in the solution.

The case of the cmdlet name is not relevant, as with all PowerShell cmdlets. To make it easy for you, you should have only one context class in each project. Otherwise, the Entity Framework Core tools do not know which context class is meant (you'll get this error message: "More than one DbContext was found. Specify which one to use. Use the '-Context' parameter for PowerShell commands and the '--context' parameter for dotnet commands."). You always have to solve this by specifying the additional parameter -Context for each command.

You have to specify a freely selectable name for, for example, Add-Migration v1. Executing this command in the Package Manager Console creates a Migrations folder in the project of the context class with three files and two classes (see Figure 7-3).

- A class is created named as specified in Add-Migration. This class has two files, one with the addition of .designer. These files also carry a timestamp in the name, which indicates the time at which the migration was created. This class inherits from the base class Microsoft.EntityFrameworkCore.Migrations.Migration. It is hereafter referred to as a *migration class*.

- A class is created that takes the name of the context class with the addition of ModelSnapshot and inherits from Microsoft. EntityFrameworkCore.Infrastructure.ModelSnapshot. This class is hereafter referred to as a *snapshot class*.

The migration class has three methods. The Up() method moves the database schema to its new state (if there is no previous migration, the program code creates the database in its default state), and the Down() method undoes the changes. The method BuildTargetModel() returns the state of the object model at the time the migration was created. BuildTargetModel() uses an instance of ModelBuilder passed from Entity Framework Core, just like the OnModelCreating() method in the context class.

In the classic Entity Framework, Microsoft stored the current state of the object model in an XML resource file (.resx) with a binary representation of the current state in an embedded BLOB. However, such a binary representation was not suitable for comparisons in source control systems and therefore presented challenges when multiple developers created schema migrations. Entity Framework Core can still be a source of conflict in team environments, but these conflicts can now be solved more easily through the source control system because snapshots are now kept in C# (or Visual Basic .NET).

**Figure 7-3.** *File created by Add-Migration*

The Snapshot class contains a BuildModel() method that contains the same program code as BuildTargetModel() on the first migration. The Snapshot class always reflects the last state of the object model, while BuildTargetModel() refers to the time the migration was created. Common to both methods is that they express the entire object model in Fluent API syntax, not just the content of OnModelCreating(); they also formulate the conventions and the data annotations via the Fluent API. Here you can see that the Fluent API really offers all the configuration options of Entity Framework Core (see www.n-tv.de/mediathek/videos/wirtschaft/Ryanair-will-Co-Piloten-abschaffen-article1428656.html).

Developers can extend the Up() and Down() methods themselves and perform their own steps here. In addition to CreateTable(), DropTable(), AddColumn(), and DropColumn(), further operations are available such as CreateIndex(), AddPrimaryKey(), AddForeignKey(), DropTable(), DropIndex(), DropPrimaryKey(), DropForeignKey(), RenameColumn(), RenameTable(), MoveTable(), and Sql(). With the latter operation, you can execute any SQL command, for example, to update values or create records. A Seed() method for populating the database tables as in the classic Entity Framework does not exist in Entity Framework Core. See Figure 7-4.

***Figure 7-4.*** *Content of BuildModel() vs. BuildTargetModel() on first migration*

Add-Migration does not create a database and does not read the database. Add-Migration solely decides what to do based on the current snapshot class. Thus, in Entity Framework Core you can create multiple migrations one after the other without actually having to update the database in between.

In the classic Entity Framework this was different. Here, Add-Migration always looked first in the database to see whether it was up-to-date. If it was not, the error "Unable to generate an explicit migration because of the following explicit migrations are pending" appeared. Unfortunately, this meant you could not create multiple schema migrations one after the other without updating your own database in between. Although it is advisable to create schema migrations in small steps, you don't want to be forced to update the database every time.

**Attention**   You might see the following error message: "No DbContext was found in assembly. Ensure that you're using the correct assembly and that the type is neither abstract nor generic." This means you have chosen the wrong assembly to run Add-Migration. It also can mean that there is a (small) inconsistency in the version numbers. If, for example, in the context class project (EFC_DA) Entity Framework Core 2.0 is used but the version 2.0.1 of the tools is installed in the EFC_Tools project, this misleading error message occurs.

Listing 7-2 shows the migration v2, which was created after v1 and after I added the property Plz, forgotten in the class Persondetail. The Up() method adds the column AddColumn(), and Down() clears it with DropColumn().

***Listing 7-2.*** The Migration v2 Complements the Column Plz in the Table Persondetail

```
using Microsoft.EntityFrameworkCore.Migrations;
using System;
using System.Collections.Generic;

namespace DA.Migrations
{
    public partial class v2 : Migration
    {
        protected override void Up(MigrationBuilder migrationBuilder)
        {
            migrationBuilder.AddColumn<string>(
                name: "Postcode",
                table: "Persondetail",
                maxLength: 8,
                nullable: true);
        }
```

```
        protected override void Down(MigrationBuilder migrationBuilder)
        {
            migrationBuilder.DropColumn(
                name: "Postcode",
                table: "Persondetail");
        }
    }
}
```

# Update-Database

The command Update-Database then brings the database into the state described by the migration steps any time you want. It is important that at this time, in the OnConfiguring() method of the context class, you pass the correct connection string to the desired database to the Entity Framework Core database provider via UseSqlServer(ConnectionString). Update-Database will instantiate the context class at development time, and OnConfiguring() will execute it. Update-Database will create the database (if it's not there yet), as well as all tables, according to the Up() methods of all the not yet executed schema migrations. Figure 7-5 shows that two schema migrations (v1 and v2) are executed.

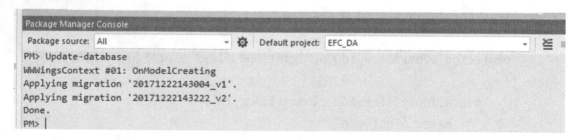

***Figure 7-5.*** *Execution of Update-Database*

Update-Database also creates an additional __EFMigrationsHistory table in the database with the MigrationId and ProductVersion columns. MigrationId corresponds to the file name of the migration class without the file name extension (e.g., 20171222143004_v1), and ProductVersion is the version number of Entity Framework Core (for example, 2.0.1-rtm-125). In the classic Entity Framework, the table was named __MigrationHistory and also contained a BLOB of the object model state. See Figure 7-6.

| | MigrationId | ProductVersion |
|---|---|---|
| 1 | 20171219200937_v1 | 2.0.1-rtm-125 |
| 2 | 20171222091949_v2 | 2.0.1-rtm-125 |

***Figure 7-6.*** *Content of the Table __EFMigrationsHistory*

If the database already exists, Entity Framework Core looks to see whether the __EFMigrationsHistory table already exists. If there is the table and there are all migration steps recorded, nothing happens. Repeated execution of Update-Database does not generate an error (execution is idempotent). Entity Framework Core does not check whether the actual database schema fits. So, if someone has deleted a table (via SQL Management Studio or similar), the problem arises only when the program is running and Entity Framework Core wants to access that table.

If the __EFMigrationsHistory table does not exist, Entity Framework Core creates it, but at the same time it assumes that the database schema does not yet exist and performs all the migration steps. However, if there are already tables with those names, Update-Database fails (with the error message "There is already an object named 'xy' in the database."). So, if someone deletes the __EFMigrationsHistory table because they think it is superfluous, it destroys the ability to load more schema migrations.

## Script-Migration

While forward engineering with Entity Framework Core is adequate for executing the PowerShell commandlet Update-Database or the equivalent command-line command dotnet ef database update for importing schema changes directly to a development system, you will need other mechanisms when distributing the applications. For most companies, it will be necessary to use a SQL script that database administrators (after careful scrutiny) install on the well-shielded database server. Such a SQL script is obtained through the Script-Migration command line or via the command line with dotnet ef migrations script.

Script-Migration creates a SQL Data Definition Language (DDL) script with the migration actions. Script-Migration does not look in the database and therefore does not know what its status is. The cmdlet always creates a SQL script for all migration steps after the first one, without further specification of parameters. If you only want individual migration steps as an SQL script, you must specify this with -from and -to. Here's an example:

```
Script-Migration -from 20170905085855_v2 -to 20170905090511_v3
```

There are two "difficulties" built into this cmdlet.

- You can't use the self-assigned name (e.g., v2); you have to use the complete migration name including the timestamp given by Entity Framework Core. The value 0 in the parameter -from is a fixed name for the initial state.

- The migration step specified in the parameter -from is also executed. So, the previous command does not create a SQL script with the differences between v2 and v3, but a SQL script with the differences between v1 and v3.

# Further Migration Steps

Even after importing one or more migration steps, you can create additional migration steps at any time. Update-Database detects whether a migration step has not been recorded yet and then performs it.

# Migration Scenarios

The question arises for which types of schema changes Entity Framework Core can automatically generate appropriate schema migrations. Schema changes that add tables or columns are not critical. Columns are always added at the end of the table, regardless of the alphabetical order used (otherwise the entire table would have to be deleted and re-created, which would require that the data be saved in a temporary table beforehand).

When you create migration steps that delete tables or columns, Add-Migration warns you with this message: "An operation was scaffolded that may result in the loss of data. Please review the migration for accuracy."

Sometimes you want to do something else besides adding tables and columns. When renaming a table or column, for example, you must intervene manually. Here, Add-Migration generates some program code that contains the old table or column deletes, and a new table or column is created because there is no feature of a .NET class or property that is preserved when renamed. The data was lost during this migration. Now, here in the migration class, you have to change by your own a DropTable() or a CreateTable() to a RenameTable() as well as a DropColumn() and a CreateColumn() to a RenameColumn() methods.

---

**Note**    Since Entity Framework Core version 2.0, Entity Framework Core regards deleting a property and adding a property of the same data type and length as a rename operation and thus creates a RenameColumn() method in the migration class. That may be correct; you might have wanted to delete a column and create a new one instead. Again, you must carefully examine the generated migration classes.

---

When changing a data type (for example, reducing an nvarchar column from eight to five characters), Migration aborts Update-Database if there are longer strings in the database (you get this error message: "String or binary data would be truncated."). In this case, you must first prune the data. For example, you can shorten the Postcode column from eight to five characters by adding this:

```
migrationBuilder.Sql("update Persondetail set Postcode = left(Postcode, 5)")
```

to the Up() method before the execution of this:

```
migrationBuilder.AlterColumn<string>(name: "Postcode", table:
"Persondetail", maxLength: 5, nullable: true).
```

Schema migrations that change cardinalities are difficult. For example, say you must suddenly build a 1:N relationship from the 1:0/1 relationship between Passenger and Persondetail because the requirement has changed so that each Passenger now may have multiple addresses. With this cardinality change, Entity Framework Core can no longer maintain a consistent data state. While there was previously a DetailID column in the Passenger table that referred to a record in the Persondetail table, there must be a `PersonalID` column after the schema construction in `Persondetail` that references `Passenger`. Although Entity Framework Core deletes one column and creates the other new one, it does not fill the new column with the appropriate values. Here, you must manually copy the values using the `Sql()` method in the migration class.

Unfortunately, the Entity Framework Core tools also generate migration code that first deletes the `DetailID` column and then re-creates `PersonID` in the `PersonDetails` table. Obtaining the data of course will not work here. Listing 7-3 shows the correct solution with a different order and uses the `Sql()` method to copy the keys.

***Listing 7-3.*** Up() Migration Class Method for a Cardinality Change from 1:0/1 to 1:N

```
namespace EFC_DA.Migrations
{
 public partial class Kardinaliaet11wird1N : Migration
 {
  protected override void Up(MigrationBuilder migrationBuilder)
  {
   // First create a new column on the N-side
   migrationBuilder.AddColumn<int>(
       name: "PassengerPersonID",
       table: "Persondetail",
       nullable: true);

   // Now copy the values from the 1-side
   migrationBuilder.Sql("update Persondetail set PassengerPersonID =
   Passenger.PersonID FROM Passenger INNER JOIN Persondetail ON Passenger.
   DetailID = Persondetail.ID");

   // Then delete the column on the 1-side first
   migrationBuilder.DropForeignKey(
       name: "FK_Passenger_Persondetail_DetailID",
       table: "Passenger");

   migrationBuilder.DropIndex(
       name: "IX_Passenger_DetailID",
       table: "Passenger");
```

```
migrationBuilder.DropColumn(
    name: "DetailID",
    table: "Passenger");

// Then create index and FK for new column
migrationBuilder.CreateIndex(
    name: "IX_Persondetail_PassengerPersonID",
    table: "Persondetail",
    column: "PassengerPersonID");

migrationBuilder.AddForeignKey(
    name: "FK_Persondetail_Passenger_PassengerPersonID",
    table: "Persondetail",
    column: "PassengerPersonID",
    principalTable: "Passenger",
    principalColumn: "PersonID",
    onDelete: ReferentialAction.Restrict);
  }
}
```

## More Options

With `Update-Database`, you can also return to a previous state of the database schema. For example, you can return to version 2 after importing version 3 with the following command:

```
Update-Database-Migration v2
```

`Update-Database` uses the `Down()` method of the migration class. With `Script-Migration`, you can also create a script for the "down" case. Here's an example:

```
Script-Migration -from v3 -to v2
```

`Remove-Migration` allows you to remove the migration class from Visual Studio for the most recent migration step.

---

**Important**   You should not delete the migration class manually because the snapshot class will no longer be up-to-date. As a result, the next time you create a migration, the manually deleted migration steps would be disregarded. If you manually delete the migration class, you must also manually adjust the snapshot class.

---

`Remove-Migration` checks whether the last migration step has already been applied in the database. If so, deleting the migration class and changing the snapshot class will not occur. The error message is as follows: "The migration has already been applied to the database. Unapply it and try again. If the migration has been applied to other databases, consider reverting its changes using a new migration." You can bypass this check with the parameter `-force`. Without manual intervention in the database schema, you may then no longer be able to create new migration steps in the database because there is a risk that these steps will try to re-create previously created tables or columns.

`Add-Migration`, `Remove-Migration`, `Update-Database`, and `Script-Migration` each have three common parameters, listed here:

- `-StartupProject`: Sets the Visual Studio project containing the Entity Framework Core tool packages if you do not want to change the startup project

- `-Project`: Specifies the Visual Studio project in which the context class resides

- `-Context`: Sets the context class (with namespace) if there are multiple context classes in the Visual Studio project

Here's an example:

```
Update-Database v2 -StartupProject EFC_Tools -Project EFC_DA -Context
WWWingsContext
```

To avoid the repeated use of these parameters in cmdlets, you can set these values with the cmdlet `Use-DbContext` and thus ensure that all subsequent cmdlet calls use these values.

```
Use-dbContext -StartupProject EFC_Tools -Project EFC_DA -context
WWWingsContext
```

# Problems with Schema Migration in Connection with TFS

In combination with the version management system of Team Foundation Server (TFS)—at least in the classic variant with Server Workspace, which works with write protection to files—the tools of Entity Framework Core have difficulties. It reports that it is not possible to modify the files in the migration folder. You'll get the following error: "Access to the path ...WWWingsContextModelSnapshot.cs' is denied." In this case, before running the command, you should unlock the migration folder with Check Out for Edit.

Another issue is that `Remove-Migration` deletes files on disk, but not from TFS versioning. You have to manually select the Visual Studio-Command "Undo" in the Pending Changes window.

# Schema Migrations at Runtime

On rare occasions, a small application (e.g., a console application) will be issued that imports the schema changes to the target database. These cases include the following:

- Providing a tool for inexperienced database administrators or customer service reps who are unfamiliar with SQL

- Installing or updating databases as part of automated integration testing in release pipelines

- Installing or updating local databases on end-user systems (if it is a mobile app, you should install the schema migration directly into the actual app at startup)

For cases where a program execution schema migration is appropriate, Entity Framework Core provides the methods `ctx.Database.GetMigrations()`, `ctx.Database.GetAppliedMigrations()`, and `ctx.Database.Migrate()`. These methods can save developers from having to write a tool that determines which schema migrations are pending and then inject the appropriate SQL scripts.

Unlike the classic Entity Framework, Entity Framework Core does not check whether the schema is up-to-date during the first database access of an application. The program may run on an error (for example, "Invalid column name 'Postcode'"). By calling the method `Migrate()` in the `Database` object of the context class, you can ensure at startup that the database schema is up-to-date (see Listing 7-4). `Migrate()` may execute missing migration steps, which is possible because the migration classes are part of the compilation of the project, which contains the context class.

---

**Note**    For a schema migration at runtime, the NuGet package `Microsoft.EntityFrameworkCore.Tools` is not needed.

---

**Listing 7-4.**  Running a Schema Migration at Runtime Using the Migrate() Method

```
using (var ctx = new WWWingsContext())
{
  ctx.Database.Migrate();
}
```

It is forbidden to use `EnsureCreated()` and `Migrate()` together. Even the tooltip of method `Migrate()` warns against it. If you still want to try it, you'll be rewarded with the unremarkable runtime error "An item with the same key has already been added."

---

**Note**    If the schema is already in the necessary state, the startup costs of `Migrate()` are very low. A schema migration might take some seconds; however, without the migration, your code will likely fail.

---

# CHAPTER 8

# Reading Data with LINQ

Like the classic Entity Framework, Entity Framework Core allows you to write database queries with Language Integrated Query (LINQ).

LINQ is a common query language for different data stores introduced in 2007 in .NET Framework 3.5; it also exists in .NET Core as well as Mono and Xamarin. Microsoft has used LINQ since the beginning in the classic Entity Framework, where it's implemented as LINQ to Entities. Microsoft no longer uses this term in Entity Framework Core; it's just called LINQ. There are some positive and negative differences in LINQ execution between the classic Entity Framework and Entity Framework Core.

## Context Class

The starting point for all LINQ queries in Entity Framework Core is the context class that you create either during the reverse engineering of an existing database or manually while forward engineering. The context class in Entity Framework Core always inherits from the base class `Microsoft.EntityFrameworkCore.DbContext`. The alternative base class of `ObjectContext`, which exists in the classic Entity Framework, has been deleted from Entity Framework Core. Accordingly, you have to use `DbContext` for all LINQ operations. But even the base class `DbContext` has changed a bit in Entity Framework Core.

The `DbContext` class implements the `IDisposable` interface. As part of the `Dispose()` method, `DbContext` frees all allocated resources, including references to all objects loaded with change tracking.

---

**Tip**  Therefore, it is important that the context class user always calls `Dispose()` once the work is done. It's best to use a `using(){ ... }` block!

---

© Holger Schwichtenberg 2018
H. Schwichtenberg, *Modern Data Access with Entity Framework Core*,
https://doi.org/10.1007/978-1-4842-3552-2_8

# LINQ Queries

After instantiating the context class, you can formulate a LINQ query. This query is not necessarily executed immediately; it is initially in the form of an object with the interface IQueryable<T>. In the sense of so-called deferred execution, the LINQ query is executed when the result is actually used (for example, in a foreach loop) or when converted to another collection type. You can force the execution of the query with a LINQ conversion operator, in other words, ToList(), ToArray(), ToLookup(), ToDictionary(), Single(), SingleOrDefault(), First(), FirstOrDefault(), or an aggregate operator such as Count(), Min(), Max(), or Sum().

Since IQueryable<T> is a subtype of IEnumerable<T>, you can start a foreach loop over an object with IQueryable<T>. This causes the query to run immediately. In addition, Entity Framework Core keeps the database connection open until the last object has been fetched, which can lead to unwanted side effects. You should therefore always explicitly use one of the previous conversions or aggregate operators before using the data in RAM because in this case Entity Framework Core will close the database connection. However, because Entity Framework Core is based on ADO.NET, the database connection is not actually closed immediately but returned to the ADO.NET connection pool. Also, data binding to an object with interface IQueryable<T> triggers a retrieval of the data.

Figure 8-1 shows the internal processing for a LINQ query. First, the LINQ query is converted into an expression tree. The expression tree creates a SQL command that Entity Framework Core sends to the database management system using a Command object from ADO.NET. Entity Framework Core has a cache for the SQL commands to prevent the overhead of converting from LINQ to SQL to the same commands twice.

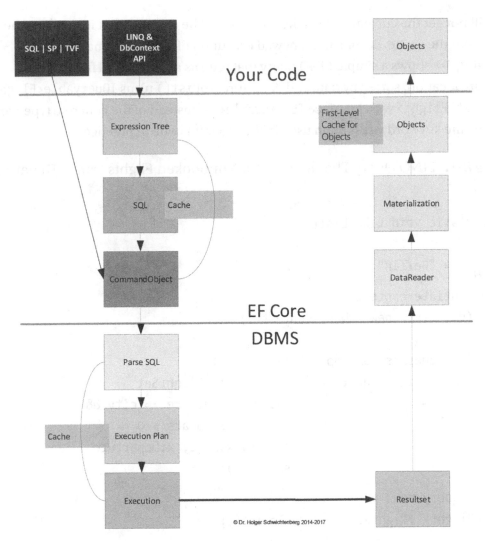

***Figure 8-1.*** *Internals for running a LINQ command through Entity Framework Core*

The database management system analyzes the query and checks whether there is already a suitable execution plan for it. If none exists in the cache, it will be created. Thereafter, the database management system executes the query and delivers a result set to Entity Framework Core. Then Entity Framework Core uses a DataReader object to read the result set, but the user code does not see it because Entity Framework Core materializes the DataReader lines into objects. Except in no-tracking mode, Entity Framework Core looks to see whether the objects to be materialized are already in the first-level cache of the Entity Framework Core context. If the objects are there, the materialization is eliminated. However, this also means that if an object is in RAM, the user will not get the current state of the record from the database but instead get the object from the cache, despite the renewed execution of a SQL command.

Listing 8-1 shows a simple LINQ query that returns all unbooked flights from a departure location, sorted by date and departure. ToList() turns IQueryable<Flight> into a List<Flight> with interface IEnumerable<T>. Instead of the concrete type name, however, the keyword var is often used in the program code in practice.

***Listing 8-1.*** LINQ Query That Returns All Nonbooked Flights from a Departure Location

```
public static void LINQ_List()
{
 var city = "Berlin";

 // Instantiate context
 using (var ctx = new WWWingsContext())
 {
  // Define queries, but do not execute yet
  IQueryable<Flight> query = (from x in ctx.FlightSet
                                 where x.Departure == city &&
                                    x.FreeSeats > 0
                                 orderby x.Date, x.Departure
                                 select x);

  // Run query now
  List<Flight> flightSet = query.ToList();

  // Count loaded objects
  var count = flightSet.Count;
  Console.WriteLine("Number of loaded flights: " + count);

  // Print results
  foreach (var f in flightSet)
  {
   Console.WriteLine($"Flight Nr {f.FlightNo} from {f.Departure} to
   {f.Destination} has {f.FreeSeats} free seats!");
  }
 } // End using-Block -> Dispose() will be called
```

A more concise formulation in the method syntax of LINQ is possible, as shown here:

```
var query2 = ctx.FlightSet.Where(x => x.Departure == city && x.FreeSeats > 0)
    .OrderBy(x => x.Date).ThenBy(x => x.Departure);
```

Of course, it is also possible to combine the call to ToList() with the definition of the LINQ query and thus execute the LINQ query immediately.

```
var flightSet2 = (from x in ctx.FlightSet
                  where x.Departure == city &&
                        x.FreeSeats > 0
                  orderby x.Date, x.Departure
                  select x).ToList();
```

However, the advantage of split notation is that you can attach more operations before executing the query.

The following is the executed SQL query:

```
SELECT [x].[FlightNo], [x].[AircraftTypeID], [x].[AirlineCode], [x].
[CopilotId], [x].[FlightDate], [x].[Departure], [x].[Destination], [x].
[FreeSeats], [x].[LastChange], [x].[Memo], [x].[NonSmokingFlight], [x].
[PilotId], [x].[Price], [x].[Seats], [x].[Strikebound], [x].[Timestamp],
[x].[Utilization]
FROM [Flight] AS [x]
WHERE ([x].[Departure] = @__city_0) AND ([x].[FreeSeats] > 0)
ORDER BY [x].[FlightDate], [x].[Departure]
```

---

**Note**   Theoretically, you can call the method query.Count() instead of the property flightSet.Count. However, this produces a new database query that provides the number of records. This is superfluous because the objects are already materialized and can be counted quickly in RAM. Accessing the DBMS with query.Count()makes sense only if you want to determine whether the number of records in the database has changed.

---

# Step-by-Step Composition of LINQ Queries

Listing 8-2 shows how to attach an extra condition on the departure or destination to the basic query FreeSeats > 0 on a case-by-case basis if the variables for the values do not contain zero or an empty string. This is the typical scenario for user-set filters. If users do not enter anything in a filter field, then they do not want to see those records where the value is empty and want this filter to be ignored during the query.

***Listing 8-2.*** LINQ Query That Returns All Unbooked Flights on a Route, with Both Departure and Destination Optional

```
public static void LINQ_Composition()
  {
   var departure = "";
   var destination = "Rome";

   // Create context instance
   using (var ctx = new WWWingsContext())
   {
    // Define query, but do not execute yet
    IQueryable<Flight> query = from x in ctx.FlightSet
                where x.FreeSeats > 0
                select x;

    // Conditional addition of further conditions
    if (!String.IsNullOrEmpty(departure)) query = query.Where(x =>
    x.Departure == departure);
    if (!String.IsNullOrEmpty(destination)) query = query.Where(x =>
    x.Destination == destination);

    // now use sorting, otherwise there will be problems with variable
    query type (IQueryable<Flight> vs. IOrderedQueryable<Flight>)
    var querySorted = from x in query // IOrderedQueryable<Flight>
                    orderby x.Date, x.Departure
                    select x;

    // Execute query now
    List<Flight> flightSet = querySorted.ToList();
```

```
// Count loaded objects
long c = flightSet.Count;
Console.WriteLine("Number of loaded flights: " + c);

// Print result
foreach (var f in flightSet)
{
 Console.WriteLine($"Flight Nr {f.FlightNo} from {f.Departure} to
 {f.Destination} has {f.FreeSeats} free seats!");
}
} // End using-Block -> Dispose()
```

# Use of var

In practice, instead of the specific type names such as IQueryable<Flight>, the keyword var is used when using LINQ. There are still a lot of arguments among developers regarding the keyword var (specifically when using Dim without a type in Visual Basic .NET). In conjunction with LINQ, var often simplifies the coding. With LINQ, using some operators such as orderby changes the return type. Without orderby, you get an object that implements IQueryable<Flight>. With orderby, it is an IOrderedQueryable<Flight>. So, you often have to change the variable type when changing LINQ queries. This is not necessary when using the keyword var.

# Repository Pattern

Except in small applications, you should not keep the data access code in the user interface. The repository pattern has been established to encapsulate the data access code for one or more (connected) tables. A repository class provides methods that return individual objects or sets of objects or return methods that allow you to insert, delete, and modify records.

An IQueryable<T> can also be used as the return value of a method so that the caller of the method can also extend the query. However, this makes sense only if the context instance still survives after the end of the method and thus can execute the query later. Therefore, you must hold the context instance as an attribute of the class and provide

the IDisposable interface for later destruction of the context instance when calling
Dispose() (see Listing 8-3 with class FlightManager in repository style). The caller can
then expand the query and should use the class FlightManager with a using() block to
ensure the call to Dispose(). See Listing 8-4.

---

**Note**    You can see the repository pattern in action in Appendix A of this book.
There you will also see how to use a common base class for all repository classes,
reducing the code in the repository classes.

---

*Listing 8-3.* Repository Class That Returns an IQuerable <Flight>

```
using Microsoft.EntityFrameworkCore;
using System;
using System.Collections.Generic;
using System.Linq;
using BO;
using DA;

namespace BL
{

 /// <summary>
 /// Repository class for Flight entities
 /// </summary>
 public class FlightManager  : IDisposable
 {
  public FlightManager()
  {
   // create instance of context when FlightManager is created
   ctx = new WWWingsContext();
  }

  // keep one EFCore context per instance
  private WWWingsContext ctx;
```

```csharp
/// <summary>
/// Dispose context if FlightManager is disposed
/// </summary>
public void Dispose() { ctx.Dispose(); }

/// <summary>
/// Get one flight
/// </summary>

public Flight GetFlight(int flightID)
{
    return ctx.FlightSet.Find(flightID);
}

/// <summary>
/// Get all flights on a route
/// </summary>
public List<Flight> GetFlightSet(string departure, string destination)
{
 var query = GetAllAvailableFlightsInTheFuture();
 if (!String.IsNullOrEmpty(departure)) query = from f in query
                                                where f.Departure ==
                                                departure
                                                select f;
 if (!String.IsNullOrEmpty(destination)) query = query.Where(f =>
 f.Destination == destination);
 List<Flight> result = query.ToList();
 return result;
}

/// <summary>
/// Base query that callre can extend
/// </summary>
```

```
public IQueryable<Flight> GetAllAvailableFlightsInTheFuture()
{
 var now = DateTime.Now;
 var query = (from x in ctx.FlightSet
              where x.FreeSeats > 0 && x.Date > now
              select x);
 return query;
}

/// <summary>
/// Get the combined list of all departures and all destinations
/// </summary>
/// <returns></returns>
public List<string> GetAirports()
{
 var l1 = ctx.FlightSet.Select(f => f.Departure).Distinct();
 var l2 = ctx.FlightSet.Select(f => f.Destination).Distinct();
 var l3 = l1.Union(l2).Distinct();
 return l3.OrderBy(z => z).ToList();
}

/// <summary>
/// Delegate SaveChanges() to the context class
/// </summary>
/// <returns></returns>
public int Save()
{
 return ctx.SaveChanges();
}

/// <summary>
/// This overload checks if there are objects in the list that do not
belong to the context. These are inserted with Add().
/// </summary>
public int Save(List<Flight> flightSet)
{
 foreach (Flight f in flightSet)
```

```
{
 if (ctx.Entry(f).State == EntityState.Detached)
 {
  ctx.FlightSet.Add(f);
 }
}
return Save();
}

/// <summary>
/// Remove flight (Delegated to context class)
/// </summary>
/// <param name="f"></param>
public void RemoveFlight(Flight f)
{
 ctx.Remove(f);
}

/// <summary>
/// Add flight (Delegated to context class)
/// </summary>
/// <param name="f"></param>
public void Add(Flight f)
{
 ctx.Add(f);
}

/// <summary>
///   Reduces the number of free seats on the  flight, if seats are still
available. Returns true if successful, false otherwise.
/// </summary>
/// <param name="flightID"></param>
/// <param name="numberOfSeats"></param>
/// <returns>true, wenn erfolgreich</returns>
public bool ReducePlatzAnzahl(int flightID, short numberOfSeats)
{
 var f = GetFlight(flightID);
```

```
  if (f != null)
  {
   if (f.FreeSeats >= numberOfSeats)
   {
    f.FreeSeats -= numberOfSeats;
    ctx.SaveChanges();
    return true;
   }
  }
  return false;
 }
 }
}
```

***Listing 8-4.*** Using the Repository Class from Listing 8-3

```
public static void LINQ_RepositoryPattern()
{
 using (var fm = new BL.FlightManager())
 {
  IQueryable<Flight> query = fm.GetAllAvailableFlightsInTheFuture();
  // Extend base query now
  query = query.Where(f => f.Departure == "Berlin");
  // Execute the query now
  var flightSet = query.ToList();
  Console.WriteLine("Number of loaded flights: " + flightSet.Count);
 }
}
```

# LINQ Queries with Paging

Paging means that only one specific range of records should be delivered from a result set. This can be realized in LINQ with the methods Skip() and Take() (or the language elements Skip and Take in Visual Basic .NET).

Listing 8-5 shows a more complex LINQ query. It will search the fights in which

- There is at least one free seat

- There is at least one booking

- There is a passenger named Müller

- The pilot was born before January 1, 1972

- There is a copilot

From the result set, the first 50 data records are then skipped by paging in the database management system, and only the following 10 data records are delivered (that is, data records 51 to 60).

***Listing 8-5.*** Complex LINQ Query

```
[EFCBook("Paging")]
  public static void LINQ_QueryWithPaging()
  {
  CUI.MainHeadline(nameof(LINQ_QueryWithPaging));
  string name = "Müller";
  DateTime date = new DateTime(1972, 1, 1);
  // Create context instance
  using (var ctx = new WWWingsContext())
  {

   // Define query and execute
   var flightSet = (from f in ctx.FlightSet
                    where f.FreeSeats > 0 &&
                        f.BookingSet.Count > 0 &&
                        f.BookingSet.Any(b => b.Passenger.Surname ==
                        name) &&
                        f.Pilot.Birthday < date &&
                        f.Copilot != null
                   select f).Skip(5).Take(10).ToList();

   // Count number of loaded objects
   var c = flightSet.Count;
   Console.WriteLine("Number of found flights: " + c);
```

```
    // Print objects
    foreach (var f in flightSet)
    {
     Console.WriteLine($"Flight Nr {f.FlightNo} from {f.Departure} to
     {f.Destination} has {f.FreeSeats} free seats!");
    }
   } // End using-Block -> Dispose()
  }
```

The following SQL command is the SQL command sent in Listing 8-5, which is much more complex than its LINQ counterpart. This command was sent to Microsoft SQL Server 2017 and was retrieved using the SQL Server Profiler tool included with SQL Server. The SQL Server version is actually important here; the row-limiting clauses with the keywords OFFSET, FETCH FIRST, and FETCH NEXT from the 2008 SQL ANSI standard (http:// /www.iso.org/iso/home/store/catalogue_tc/catalogue_tc_browse. htm?commid=45342) have been supported by Microsoft SQL Server since version 2012. Oracle has been offering this support since version 12*c* (released on 1.7.2013). For DBMSs that do not support this new syntax, Entity Framework Core needs to create an even more complex query with the rownumber() function and subselects to implement Skip().

---

**Note**    An nice improvement in Entity Framework Core is the use of the variable names from the LINQ query (here f and b) in the SQL command. In the classic Entity Framework, names such as extend1, extend2, extend3, and so on, were used instead. If Entity Framework Core in SQL needs an alias multiple times for a table, the ORM appends a number to the variable name (see [b0] in the following SQL code).

---

```
exec sp_executesql N'SELECT [f].[FlightNo], [f].[AircraftTypeID], [f].
[AirlineCode], [f].[CopilotId], [f].[FlightDate], [f].[Departure], [f].
[Destination], [f].[FreeSeats], [f].[LastChange], [f].[Memo], [f].
[NonSmokingFlight], [f].[PilotId], [f].[Price], [f].[Seats], [f].
[Strikebound], [f].[Timestamp], [f].[Utilization]
FROM [Flight] AS [f]
INNER JOIN [Employee] AS [f.Pilot] ON [f].[PilotId] = [f.Pilot].[PersonID]
```

```
WHERE ([f.Pilot].[Discriminator] = N''Pilot'') AND (((((([f].[FreeSeats] > 0)
AND ((
    SELECT COUNT(*)
    FROM [Booking] AS [b]
    WHERE [f].[FlightNo] = [b].[FlightNo]
) > 0)) AND EXISTS (
    SELECT 1
    FROM [Booking] AS [b0]
    INNER JOIN [Passenger] AS [b.Passenger] ON [b0].[PassengerID] =
    [b.Passenger].[PersonID]
    WHERE ([b.Passenger].[Surname] = @__name_0) AND ([f].[FlightNo] =
    [b0].[FlightNo]))) AND ([f.Pilot].[Birthday] < @__date_1)) AND [f].
    [CopilotId] IS NOT NULL)
ORDER BY (SELECT 1)
OFFSET @__p_2 ROWS FETCH NEXT @__p_3 ROWS ONLY',N'@__name_0
nvarchar(4000),@__date_1 datetime2(7),@__p_2 int,@__p_3 int',
@__name_0=N'Müller',@__date_1='1972-01-01 00:00:00',@__p_2=5,@__p_3=10
```

# Projections

In a relational database, the restriction to selected columns is referred to as a *projection* (see `https://en.wikipedia.org/wiki/Set_theory`). It is often a serious performance error to load all the columns of a table if not all columns are really needed.

## Projection to an Entity Type

The LINQ queries shown so far always actually load and materialize all columns of the Flight table. Listing 8-6 shows a projection with `select new Flight()` and only the desired columns. After executing the `ToList()` method, you receive a list of `Flight` objects that have all the properties (since the class is defined as such), but only the specified properties are filled.

***Listing 8-6.*** LINQ Query with Projection

```
public static void Projection_Read()
  {
   using (var ctx = new WWWingsContext())
   {
    CUI.MainHeadline(nameof(Projection_Read));

    var query = from f in ctx.FlightSet
                 where f.FlightNo > 100
                 orderby f.FlightNo
                 select new Flight()
                 {
                  FlightNo = f.FlightNo,
                  Date = f.Date,
                  Departure = f.Departure,
                  Destination = f.Destination,
                  FreeSeats = f.FreeSeats
                 };

    var flightSet = query.ToList();

    foreach (var f in flightSet)
    {
     Console.WriteLine($"Flight Nr {f.FlightNo} from {f.Departure} to
     {f.Destination} has {f.FreeSeats} free seats!");
    }
   }
  }
```

The following SQL output of Listing 8-6 proves that Entity Framework Core is indeed requesting the desired columns in the database management system only:

```
SELECT [f].[FlightNo], [f].[FlightDate] AS [Date], [f].[Departure], [f].
[Destination], [f].[FreeSeats]
FROM [Flight] AS [f]
WHERE [f].[FlightNo] > 100
ORDER BY [f].[FlightNo]
```

> **Note**    The direct support of projections to an entity class is a major advantage
> of Entity Framework Core over the classic Entity Framework. In the classic Entity
> Framework, projections were not possible for entity classes and complex types;
> only anonymous types and nonentity classes could be used with projections.
> Because of the limitation of anonymous types (instances are read-only and
> cannot be used in methods as a return value), it was usually necessary to
> copy in instances of the entity class. For this you could use the NuGet package
> AutoMapper.EF6 (`https://github.com/AutoMapper/AutoMapper.EF6`) with
> the extension method `ProjectTo<T>()`.

## Projections to an Anonymous Type

A projection to an anonymous type is possible. In this case, no class name should be
specified after the new operator. If the names of the properties will not change, then in
the initialization block instead of the assignment `{Departure = f.Departure, ...}`
only a simple mention of the properties is necessary, as follows: `{f.Departure,
f.Destination, ...}`.

> **Note**    The anonymous type is unknown to Entity Framework Core. If you try to query
> for an anonymous state with `ctx.Entry(f).State` or call `ctx.Attach(f)`,
> you get the following runtime error: "The entity type '<>f__AnonymousType8<int,
> DateTime, string, string, Nullable<short>, byte[]>' was not found. Ensure that the
> entity type has been added to the model."

```
f.FreeSeats--
```

Chapter 20 covers how to map anonymous types to other types using object-to-
object mapping. Listing 8-7 shows the projection to an anonymous type.

***Listing 8-7.*** Projection to an Anonymous Type

```
public static void Projection_AnonymousType()
{
  using (var ctx = new WWWingsContext())
```

```
{
 CUI.MainHeadline(nameof(Projection_AnonymousType));

 var q = (from f in ctx.FlightSet
          where f.FlightNo > 100
          orderby f.FlightNo
          select new
          {
           FlightID = f.FlightNo,
           f.Date,
           f.Departure,
           f.Destination,
           f.FreeSeats,
           f.Timestamp
          }).Take(2);

 var flightSet = q.ToList();

 foreach (var f in flightSet)
 {
  Console.WriteLine($"Flight Nr {f.FlightID} from {f.Departure} to
  {f.Destination} has {f.FreeSeats} free seats!");
 }

 Console.WriteLine("Number of flights: " + flightSet.Count);

 foreach (var f in flightSet)
 {
  Console.WriteLine(f.FlightID);
  // not posssible:  Console.WriteLine("Before attach: " + f + " State: " +
  ctx.Entry(f).State + " Timestamp: " + ByteArrayToString(f.Timestamp));
  // not posssible:   ctx.Attach(f);
  // not posssible:   Console.WriteLine("After attach: " + f + " State: " +
  ctx.Entry(f).State + " Timestamp: " + ByteArrayToString(f.Timestamp));
  // not posssible:
 // f.FreeSeats--;
  // not posssible:   Console.WriteLine("After Änderung: " + f + " State: " +
  ctx.Entry(f).State + " Timestamp: " + ByteArrayToString(f.Timestamp));
```

```
    var count = ctx.SaveChanges(); // no changes can be saved
    Console.WriteLine("Number of saved changes: " + count);
    // not posssible:  Console.WriteLine("After saving: " + f + " State: " +
    ctx.Entry(f).State + " Timestamp: " + ByteArrayToString(f.Timestamp));
  }
 }
}
```

# Projections to an Arbitrary Type

The target of the projection can also be any other class that, depending on the structure of the software architecture, would be called a *business object* (BO) or *data transfer object* (DTO). You have to mention the class name after new as with the entity class projection, and full assignments are necessary for the initialization, as follows: {Departure = f.Departure, ...}. Listing 8-8 shows the projection to a DTO.

---

**Note**    As with anonymous types, Entity Framework Core does not know the class in this case. Asking for the state with Attach() and saving changes is therefore not possible.

---

*Listing 8-8.*  Projection to a DTO

```
class FlightDTO
  {
    public int FlightID { get; set; }
    public DateTime Date { get; set; }
    public string Departure { get; set; }
    public string Destination { get; set; }
    public short? FreeSeats { get; set; }
    public byte[] Timestamp { get; set; }
  }
```

```
public static void Projection_DTO()
{
 using (var ctx = new WWWingsContext())
 {
  CUI.MainHeadline(nameof(Projection_DTO));

  var q = (from f in ctx.FlightSet
           where f.FlightNo > 100
           orderby f.FlightNo
           select new FlightDTO()
           {
            FlightID = f.FlightNo,
            Date = f.Date,
            Departure = f.Departure,
            Destination = f.Destination,
            FreeSeats = f.FreeSeats,
            Timestamp = f.Timestamp
           }).Take(2);

  var flightSet = q.ToList();

  foreach (var f in flightSet)
  {
   Console.WriteLine($"Flight Nr {f.FlightID} from {f.Departure} to
   {f.Destination} has {f.FreeSeats} free seats!");
  }

  Console.WriteLine("Number of flights: " + flightSet.Count);

  foreach (var f in flightSet)
  {
   Console.WriteLine(f.FlightID);
   // not posssible:  Console.WriteLine("Before attach: " + f + " State: " +
   ctx.Entry(f).State + " Timestamp: " + ByteArrayToString(f.Timestamp));
   // not posssible:   ctx.Attach(f);
   // not posssible:   Console.WriteLine("After attach: " + f + " State: " +
   ctx.Entry(f).State + " Timestamp: " + ByteArrayToString(f.Timestamp));
```

```
// not posssible:
// f.FreeSeats--;
// not posssible:   Console.WriteLine("After Änderung: " + f + " State: "
+ ctx.Entry(f).State + " Timestamp: " + ByteArrayToString(f.Timestamp));

var anz = ctx.SaveChanges(); // no changes can be saved
Console.WriteLine("Number of saved changes: " + anz);
// not posssible:   Console.WriteLine("After saving: " + f + " State: " +
ctx.Entry(f).State + " Timestamp: " + ByteArrayToString(f.Timestamp));
  }
 }
}
```

# Querying for Single Objects

LINQ offers four operations to select the first or only element of a set, listed here:

- `First()`: The first element of a set. If there are multiple elements in the set, all but the first are discarded. If there is no element, a runtime error occurs.

- `FirstOrDefault()`: The first element of a set or a default value (for reference types `null` or `Nothing`) when the amount is empty. If there are multiple elements in the set, all but the first are discarded.

- `Single()`: The only element of a set. If there is no element or there are multiple elements in the set, a runtime error occurs.

- `SingleOrDefault()`: The only element of a set. If there is no element, the default value (for reference types `null` or `Nothing`) is returned. If there are multiple items in the set, a runtime error occurs.

`First()` and `FirstOrDefault()` limit the output quantity on the database side with the SQL operator `TOP(1)`. `Single()` and `SingleOrDefault()` use `TOP(2)` to determine whether there is more than one element, which results in a runtime error. Listing 8-9 shows the code for the LINQ query.

***Listing 8-9.*** LINQ Query for a Single Object with SingleOrDefault()

```
public static void LINQ_SingleOrDefault()
  {
  using (var ctx = new WWWingsContext())
  {
   var FlightNr = 101;

   var f = (from x in ctx.FlightSet
            where x.FlightNo == FlightNr
            select x).SingleOrDefault();

   if (f != null)
   {
    Console.WriteLine($"Flight Nr {f.FlightNo} from {f.Departure} to
    {f.Destination} has {f.FreeSeats} free seats!");
   }
   else
   {
    Console.WriteLine("Flight not found!");
   }
  } // End using-Block -> Dispose()
  }
```

# Loading Using the Primary Key with Find()

The DbSet<T> class in the classic Entity Framework offers a Find() method as an alternative to loading an object using the primary key with LINQ. Find() passes the value of the primary key. If there is a multipart primary key, you can also pass several values, such as Find ("Holger", "Schwichtenberg", 12345) if the primary key consists of two strings and one number. Find() was not available in Entity Framework Core version 1.0 but has already been integrated into version 1.1. Listing 8-10 shows the LINQ query.

---

**Note**    Find() has the special behavior of first looking for an object in the first-level cache of the Entity Framework Core context and starting a database query only if the object is not there. The methods Single(), SingleOrDefault(), First(), and FirstOrDefault() always ask the database, even if the object exists in the local cache!

---

*Listing 8-10.* LINQ Query for a Single Object with Find

```
public static void LINQ_Find()
{
CUI.MainHeadline(nameof(LINQ_Find));
using (var ctx = new WWWingsContext())
{
 ctx.FlightSet.ToList(); // Caching all flights in context (here as an
 example only to show the caching effect!)

 var FlightNr = 101;
 var f = ctx.FlightSet.Find(FlightNr); // Flight is loaded from cache!

 if (f != null)
 {
  Console.WriteLine($"Flight Nr {f.FlightNo} from {f.Departure} to
  {f.Destination} has {f.FreeSeats} free seats!");
 }
 else
 {
  Console.WriteLine("Flight not found!");
 }
} // End using-Block -> Dispose()

}
```

# Using LINQ in RAM Instead of in the Database (Client Evaluation)

Listing 8-11 shows a LINQ query with grouping using the LINQ grouping operator (group by or GroupBy()). This query provides the desired result (the number of flights per departure), but the query takes a lot of time for large amounts of data. When researching causes, you will find out that in the SQL command sent to the database management system, the grouping is completely missing. Entity Framework Core has loaded all records and grouped them in RAM, which is bad and unexpected.

---

**Attention**   In fact, Entity Framework Core versions 1.*x* and 2.0 do not support the translation of a LINQ grouping into the GROUP  BY syntax of SQL, which is a frightening gap in Entity Framework Core (see the "Working Around the GroupBy Problem" section). An improvement is planned for version 2.1 of Entity Framework Core; see Appendix C.

---

***Listing 8-11.***   Determine the Number of Flights per Departure

```
using (var ctx = new WWWingsContext())
{
 Console.WriteLine(ctx.Database.GetType().FullName);
 ctx.Log();

 var groups = (from p in ctx.FlightSet
            group p by p.Departure into g
            select new { City = g.Key, Count =  g.Count() }).Where
            (x => x.Count > 5).OrderBy(x => x.Count);

 // First roundtrip to the database (done intentionally here!)
 var count = groups.Count();
 Console.WriteLine("Number of groups: " + count);
 if (count == 0) return;
```

```
// Second roundtrip to the database
foreach (var g in groups.ToList())
{
 Console.WriteLine(g.City + ": " + g.Count);
 }
}
```

Listing 8-11 shows the SQL command sent to the database management system
(twice: once for Count() and once for ToList()).

```
SELECT [p0].[FlightNo],
       [p0].[AircraftTypeID],
       [p0].[AirlineCode],
       [p0].[CopilotId],
       [p0].[FlightDate],
       [p0].[Departure],
       [p0].[Destination],
       [p0].[FreeSeats],
       [p0].[LastChange],
       [p0].[Memo],
       [p0].[NonSmokingFlight],
       [p0].[PilotId],
       [p0].[Price],
       [p0].[Seats],
       [p0].[Strikebound],
       [p0].[Timestamp],
       [p0].[Utilization]
FROM   [Flight] AS [p0]
ORDER  BY [p0].[Departure]
```

Unfortunately, there are a number of other cases where Entity Framework Core
performs the operations in RAM rather than in the database.

For the following query, in Entity Framework Core 1.*x*, only filtering via `FlightNo` took place in the database.

```
var q2 = from f to ctx.FlightSet
            where f.FlightNo > 100
            && f.FreeSeats.ToString().Contains("1")
            orderby f.FlightNo
            select f;
```

`ToString().Contains()` could not be translated and executes this condition in RAM. In version 2.0, the entire LINQ command is translated to SQL.

For the following query, `AddDays()` could not be translated in Entity Framework Core 1.*x*, so only the filter on the free seats, but not the date filter, took place in the database management system.

```
var q3 = from f to ctx.FlightSet
            where f.FreeSeats> 0 &&
            f.Date > DateTime.Now.AddDays (10)
            orderby f.FlightNo
            select f;
```

This is also fixed in Entity Framework Core 2.0.

Unfortunately, the following query with the LINQ operator `Union()` also happens in RAM in Entity Framework Core 2.0:

```
var all places = (from f in ctx.FlightSet select f.Departure.Union(from f
in ctx.FlightSet select f.Destination).Count();
```

Although here only one number is needed, Entity Framework Core will execute the following:

```
SELECT [f]. [Departure]
FROM [Flight] AS [f]
SELECT [f0]. [Destination]
FROM [Flight] AS [f0]
```

**Note**    It should be mentioned that some of the previous LINQ queries were not executable at all in the classic Entity Framework. They were compiled but caused runtime errors. It's arguable whether the new solution in Entity Framework Core is a better solution. Although the orders are now possible, there is a big trap lurking. After all, Microsoft has announced in the road map (`https://github.com/aspnet/EntityFrameworkCore/wiki/Roadmap`) that in future versions of Entity Framework Core you can perform more operations in the database management system.

When running in RAM, significant performance problems threaten because too many records are loaded. Developers can run into a hard case if they use such a query and then do not test with a large number of records. This is all the more surprising since Microsoft always talks about Big Data but then with LINQ in Entity Framework Core provides a tool that is just not Big Data capable in some respects.

Such RAM operations are possible only in Entity Framework Core through a new provider architecture, which leaves the provider with the decision to perform certain operations in RAM. Microsoft calls this *client evaluation*.

Software developers can protect against these performance issues by turning off client evaluation. This is possible with the ConfigureWarnings() method, which provides the DbContextOptionsBuilder object that the developer gets in the OnConfiguring() method. The following configuration causes each client evaluation to trigger a runtime error, as shown in Figure 8-2. By default, Entity Framework Core logs client evaluation only (see Chapter 12 Logging).

```
protected override void OnConfiguring(DbContextOptionsBuilder builder)
  {
  builder.UseSqlServer(_DbConnection);

builder.ConfigureWarnings(warnings => warnings.Throw(RelationalEventId.
QueryClientEvaluationWarning));
}
```

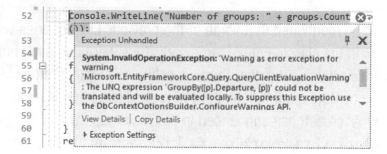

*Figure 8-2.* *Runtime error raised on a client evaluation if disabled*

# Using the Wrong Order of Commands

Sometimes, however, software developers themselves make operations in RAM rather than in the database management system. In Listing 8-12, ToList() is called too early, and the query contains an object of the type List<Flight> instead of the type IQueryable<Flight>. As a result, the filters based on the place of departure and destination and sorting occur in the RAM with LINQ to Object.

---

**Note**    LINQ uses the same syntax for queries in RAM (LINQ to Objects) and LINQ in Entity Framework/Entity Framework Core. Therefore, you cannot see in a program code line, whether it is executed in RAM or in the database management system. This always depends on the data type of the base set (that is, what is after in in the LINQ query).

---

*Listing 8-12.* ToList() Is Set Too Early and Causes the Following Queries to Execute in RAM

```
public static void LINQ_CompositionWrongOrder()
{
 CUI.MainHeadline(nameof(LINQ_Composition));

 var departure = "";
 var destination = "Rome";

 // Create context instance
 using (var ctx = new WWWingsContext())
```

```
{
 // Define query (ToList() ist WRONG here!)
 var query = (from x in ctx.FlightSet
  where x.FreeSeats > 0
  select x).ToList();

 // Conditional addition of further conditions
 if (!String.IsNullOrEmpty(departure)) query = query.Where(x =>
 x.Departure == departure).ToList();
 if (!String.IsNullOrEmpty(destination)) query = query.Where(x =>
 x.Destination == destination).ToList();

 // Sorting
 var querySorted = from x in query
  orderby x.Date, x.Departure
  select x;

 // The query shoud execute here, but it is already executed
 List<Flight> flightSet = querySorted.ToList();

 // Count loaded objects
 long c = flightSet.Count;
 Console.WriteLine("Number of loaded flights: " + c);

 // Print result
 foreach (var f in flightSet)
 {
  Console.WriteLine($"Flight Nr {f.FlightNo} from {f.Departure} to
  {f.Destination} has {f.FreeSeats} free seats!");
 }
} // End using-Block -> Dispose()
}
```

# Using Custom Function in LINQ

The new provider architecture in Entity Framework Core opens up the possibility of incorporating your own functions in LINQ queries. Of course, this part of the query is executed in RAM. For example, the query in Listing 8-13 includes its own GetNumberofDaysUntil() method. Also, in this case, only the filter is executed over the FreeSeats column in the database.

---

**Note**   Local functions that have existed since C# 6.0 cannot be called in LINQ commands.

---

***Listing 8-13.***  Custom Functions in LINQ

```
private static int GetNumberOfDaysUntil(DateTime t)
{
 return (t - DateTime.Now).Days;
}

public static void LINQ_CustomFunction()
{
 CUI.MainHeadline("Query with Custom Function - RAM :-(");
 using (var ctx = new WWWingsContext())
 {

  var q4 = from f in ctx.FlightSet
   where f.FreeSeats > 0 &&
         GetNumberOfDaysUntil(f.Date) > 10
   orderby f.FlightNo
   select f;

  List<Flight> l4 = q4.Take(10).ToList();

  Console.WriteLine("Count: " + l4.Count);
```

```
 foreach (var f in l4)
 {
  Console.WriteLine(f);
 }
 }
}
```

# Working Around the GroupBy Problem

The lack of translation of LINQ groupings to SQL in Entity Framework Core 1.*x* and 2.0 and the grouping of the records in RAM instead are absolutely unacceptable for many real-world scenarios.

In practice, a solution is needed that executes the grouping in the respective database management system. Unfortunately, with LINQ, this cannot be done in Entity Framework Core 1.*x* and 2.0. But the use of SQL also presents a challenge because Entity Framework Core does not yet support the mapping of results from SQL queries to any type, but only to entity classes.

---

**Note**    Microsoft will introduce the GroupBy translation in version 2.1 of Entity Framework Core (see Appendix C), which will make these work-arounds obsolete.

---

## Mapping to Nonentity Types

The code in Listing 8-14 using FromSql() is unfortunately not a solution. It fails to execute FromSql() and results in this runtime error: "Cannot create a DbSet for 'DepartureGroup' because this type is not included in the model for the context." However, the error message suggests another trick that works (See Below "Challenge: Migrations" section).

***Listing 8-14.*** No Solution to the GroupBy Issue

```
public static void GroupBy_SQL_NonEntityType()
  {
   // Get the number of flights per departure
   using (var ctx = new WWWingsContext())
```

```
    {
     // Map SQL to non-entity class
     Console.WriteLine(ctx.Database.GetType().FullName);
     ctx.Log();
     var sql = "SELECT Departure, COUNT(FlightNo) AS FlightCount FROM Flight
     GROUP BY Departure";
     // ERROR!!! Cannot create a DbSet for 'Group' because this type is not
     included in the model for the context."
     var groupSet = ctx.Set<DepartureGroup>().FromSql(sql);
     // Output
     foreach (var g in groupSet)
     {
      Console.WriteLine(g.Departure + ": " + g.FlightCount);
     }
    }
}
```

# Creating an Entity Class for the Database View Result

Because the mapping with FromSql() is not unrealized on nonentity types, you must create a pseudo-entity class for the result of the grouping whose name and type attributes must match the columns of the grouping result. This entity class also requires a primary key that conforms to the convention (an ID of classnameID), or it must be specified using [Key] or the Fluent API's HasKey(). See Listing 8-15.

*Listing 8-15.* Entity Class with Two Properties for the Grouping Result

```
namespace BO
{
public class DepartureGrouping
 {
  [Key] // must have a PK
  public string Departure { get; set; }
  public int FlightCount { get; set; }
 }
...
}
```

# Including the Entity Class in the Context Class

The pseudo-entity class for the grouping result has to be included as an entity class in the context class via DbSet<T>, as shown in Listing 8-16.

***Listing 8-16.*** Including the Entity Class for the Database View in the Context Class

```
public class WWWingsContext: DbContext
{
  #region tables
  public DbSet<Flight> FlightSet {get; set; }
  public DbSet<Pilot> PilotSet {get; set; }
  public DbSet<Passenger> PassengerSet {get; set; }
  public DbSet<Airport> AirportSet {get; set; }
  public DbSet<Booking> BookingSet {get; set; }
  public DbSet<AircraftType> AircraftTypeSet {get; set; }
  #endregion

  #region grouping results (pseudo-entities)
  public DbSet<DepartureGrouping> DepartureGroupingSet {get; set; }
  // for grouping
  #endregion ...
}
```

# Using the Pseudo-Entity Class

The entity class DepartureGrouping can now be used as the return type in FromSQL(), as shown in Listing 8-17.

***Listing 8-17.*** Use of the Pseudo-Entity Class

```
public static void GroupBy_SQL_Trick()
{
 // Get the number of flights per departure
 using (var ctx = new WWWingsContext())
 {
```

```
Console.WriteLine(ctx.Database.GetType().FullName);
ctx.Log();

// Map SQL to entity class
var sql = "SELECT Departure, COUNT(FlightNo) AS FlightCount FROM Flight
GROUP BY Departure";
var groupSet = ctx.Set<BO.DepartureGrouping>().FromSql(sql).Where(x=>x.
FlightCount>5).OrderBy(x=>x.FlightCount);

// Output
foreach (var g in groupSet)
{
 Console.WriteLine(g.Departure + ": " + g.FlightCount);
}

}
}
```

Figure 8-3 shows the output.

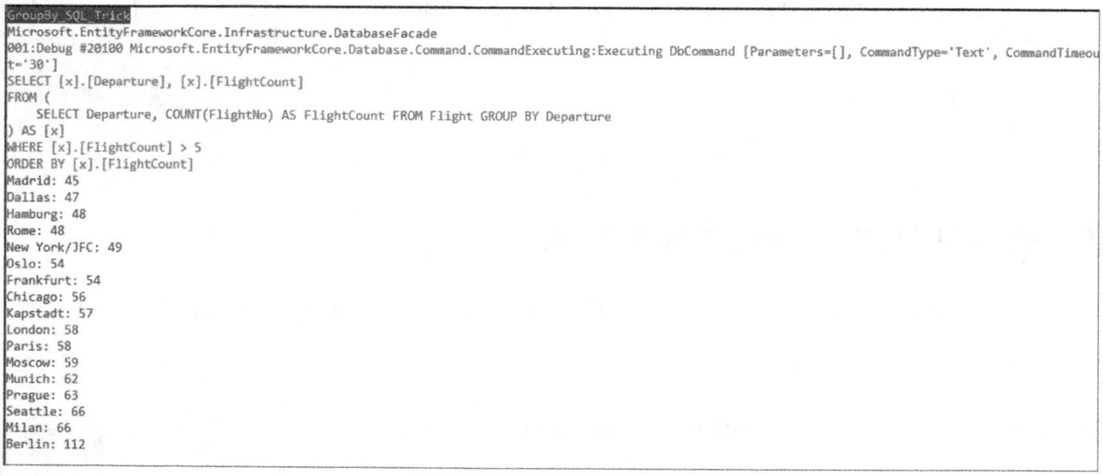

```
GroupBy_SQL_Trick
Microsoft.EntityFrameworkCore.Infrastructure.DatabaseFacade
001:Debug #20100 Microsoft.EntityFrameworkCore.Database.Command.CommandExecuting:Executing DbCommand [Parameters=[], CommandType='Text', CommandTimeou
t='30']
SELECT [x].[Departure], [x].[FlightCount]
FROM (
    SELECT Departure, COUNT(FlightNo) AS FlightCount FROM Flight GROUP BY Departure
) AS [x]
WHERE [x].[FlightCount] > 5
ORDER BY [x].[FlightCount]
Madrid: 45
Dallas: 47
Hamburg: 48
Rome: 48
New York/JFC: 49
Oslo: 54
Frankfurt: 54
Chicago: 56
Kapstadt: 57
London: 58
Paris: 58
Moscow: 59
Munich: 62
Prague: 63
Seattle: 66
Milan: 66
Berlin: 112
```

*Figure 8-3.*  *Output of Listing 8-17*

# Challenge: Migrations

As you've seen, some manual work is necessary to use the grouping result. Unfortunately, there is another challenge in database schema migrations beyond the typing work.

If you create a schema migration class after adding the database view's pseudo-entity class in the context, you will notice that Entity Framework Core undesirably wants to create a table for the pseudo-entity class in the database (see CreateTable() in Listing 8-18). This is correct in that I pretended that DepartureGrouping would be a table. However, it is not desirable to create a table for the grouping result.

*Listing 8-18.* Entity Framework Core Creates a CreateTable() for the Pseudo-Entity in the Schema Migration (Not Desirable)

```
using Microsoft.EntityFrameworkCore.Migrations;
using System;
using System.Collections.Generic;

namespace DA.Migrations
{
    public partial class v3 : Migration
    {
        protected override void Up(MigrationBuilder migrationBuilder)
        {
            migrationBuilder.CreateTable(
                name: "DepartureGrouping",
                columns: table => new
                {
                    Departure = table.Column<string>(nullable: false),
                    FlightCount = table.Column<int>(nullable: false)
                },
                constraints: table =>
                {
                    table.PrimaryKey("PK_DepartureGrouping",
                    x => x.Departure);
                });
        }
```

```
        protected override void Down(MigrationBuilder migrationBuilder)
        {
            migrationBuilder.DropTable(
                name: "DepartureGrouping");
        }
    }
}
```

There are three possible solutions in this situation:

- You can create the table and just not use it.

- You can manually delete CreateTable() in the Up() method and the corresponding DropTable() in Down() from the migration class.

- You can trick Entity Framework Core so that it ignores the entity class DepartureStatistics when creating the migration step at development time but not at runtime.

Listing 8-19 shows how to implement this trick. Entity Framework Core instantiates the context class as part of creating or deleting a schema migration and calls OnModelCreating(). However, at development time this does not happen via the actual starting point of the application (then the application would start) but by hosting the DLL with the context class in the command-line tool ef.exe. In OnModelCreating(), you therefore check whether the current process has the name ef. If so, then the application is not at runtime, and you are in the development environment and want to ignore the database view with Ignore(). At runtime of the application, however, the Ignore() method will not be executed, and thus using the database view through the entity class is possible.

***Listing 8-19.*** Entity Framework Core Should Ignore the Entity Class for the Database View at Development Time Only

```
if (System.Diagnostics.Process.GetCurrentProcess().ProcessName.ToLower()
== "ef")
{
 modelBuilder.Ignore<DepartureGrouping>();
}
```

**Alternative Trick**    If the query of the process name is too uncertain for you because Microsoft could change this name, you can instead use a switch in the context class in the form of a static attribute (e.g., `IsRuntime`). By default, `IsRuntime` is false and ignores the pseudo-entity class. At runtime, however, `IsRuntime` is set to true before the first instantiation of the context class.

## Groupings with Database Views

In Entity Framework Core 1.*x* and 2.0, you solve the `GroupBy` problem a little differently, by using the database views. Here you define a database view that does the grouping and returns the grouping result.

However, because these versions of Entity Framework Core do not support database views either, the same trick still applies to an entity class that represents the database view result. It is important that you then no longer have the option to create the table in the schema migration because there is already a database view with this name, which would lead to a naming conflict.

You can find details about using database views in Chapter 18 (Mapping of Database Views).

## Brief Overview of the LINQ Syntax

This section presents the most important commands of LINQ with meaningful examples as a quick reference for you. All queries are executed on the object model of World Wide Wings version 2, as shown in Figure 8-4. For each of these classes, there is a corresponding table of the same name in the database.

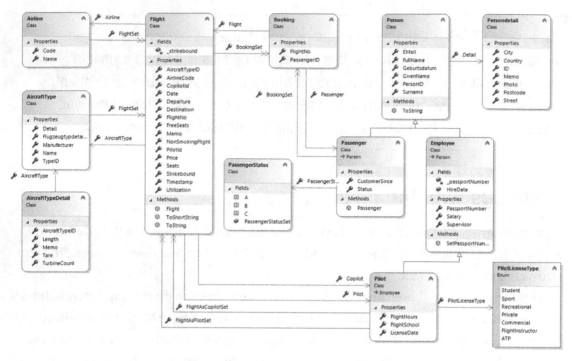

**Figure 8-4.** *Object model for the following LINQ examples*

All commands are based on the previous instantiation of the Entity Framework Core context.

```
WWWingsContext ctx = new WWWingsContext();
```

In addition to the LINQ commands, the alternative lambda notation and the resulting SQL command are displayed. The resulting SQL command is always the same for the LINQ and lambda notation; it is therefore reprinted here only once.

---

**Tip**    For a more detailed sample collection of basic LINQ commands, see
https://code.msdn.microsoft.com/101-LINQ-Samples-3fb9811b.

---

# Simple SELECT Commands (All Records)

Entity Framework Core supports ToArray(), ToList(), ToDictionary(), and ToLookup() to convert a query to a set of objects.

```
CUI.Headline("All records as Array<T>");
Flight[] flightSet0a = (from f in ctx.FlightSet select f).ToArray();
Flight[] flightSet0b = ctx.FlightSet.ToArray();

CUI.Headline("All records as List<T>");
List<Flight> flightSet1a = (from f in ctx.FlightSet select f).ToList();
List<Flight> flightSet1b = ctx.FlightSet.ToList();

CUI.Headline("All records as Dictionary<T, T>");
Dictionary<int, Flight> flightSet2a = (from f in ctx.FlightSet
select f).ToDictionary(f=>f.FlightNo, f=>f);
Dictionary<int, Flight> flightSet2b = ctx.FlightSet.ToDictionary(f =>
f.FlightNo, f => f);

CUI.Headline("All records as ILookup<T, T>");
ILookup<int, Flight> flightSet2c = (from f in ctx.FlightSet
select f).ToLookup(f => f.FlightNo, f => f);
ILookup<int, Flight> flightSet2d = ctx.FlightSet.ToLookup(f =>
f.FlightNo, f => f);
```

In all eight cases, the following SQL statement is sent to the DBMS:

```
SELECT [f].[FlightNo], [f].[AircraftTypeID], [f].[AirlineCode],
[f].[CopilotId], [f].[FlightDate], [f].[Departure], [f].[Destination],
[f].[FreeSeats], [f].[LastChange], [f].[Memo], [f].[NonSmokingFlight],
[f].[PilotId], [f].[Price], [f].[Seats], [f].[Strikebound],
[f].[Timestamp], [f].[Utilization]
FROM [Flight] AS [f]
```

167

# Conditions (where)

```
List<Flight> flightSet3a = (from f in ctx.FlightSet
                            where f.Departure == "Berlin" &&
                            (f.Destination.StartsWith("Rome") ||
                            f.Destination.Contains("Paris"))
                            && f.FreeSeats > 0
                            select f)
                            .ToList();

List<Flight> flightSet3b = ctx.FlightSet.Where(f => f.Departure == "Berlin" &&
                            (f.Destination.StartsWith("Rome") ||
                            f.Destination.Contains("Paris"))
                            && f.FreeSeats > 0)
                            .ToList();
```

The resulting SQL in both cases is as follows:

```
SELECT [f].[FlightNo], [f].[AircraftTypeID], [f].[AirlineCode],
[f].[CopilotId], [f].[FlightDate], [f].[Departure], [f].[Destination],
[f].[FreeSeats], [f].[LastChange], [f].[Memo], [f].[NonSmokingFlight],
[f].[PilotId], [f].[Price], [f].[Seats], [f].[Strikebound],
[f].[Timestamp], [f].[Utilization]
FROM [Flight] AS [f]
WHERE (([f].[Departure] = N'Berlin') AND ((([f].[Destination] LIKE
N'Rome' + N'%' AND (LEFT([f].[Destination], LEN(N'Rome')) = N'Rome')) OR
(CHARINDEX(N'Paris', [f].[Destination]) > 0))) AND ([f].[FreeSeats] > 0)
```

# Contains (in)

```
ist<string> Orte = new List<string>() { "Berlin", "Hamburg", "Köln",
"Berlin" };
   List<Flight> flightSet4a = (from f in ctx.FlightSet
                              where Orte.Contains(f.Departure)
                              select f)
                              .ToList();
   List<Flight> flightSet4b = ctx.FlightSet.Where(f => Orte.Contains
                              (f.Departure)).ToList();
```

The resulting SQL in both cases is as follows:

```
SELECT [f].[FlightNo], [f].[AircraftTypeID], [f].[AirlineCode], [f].[CopilotId],
[f].[FlightDate], [f].[Departure], [f].[Destination], [f].[FreeSeats],
[f].[LastChange], [f].[Memo], [f].[NonSmokingFlight], [f].[PilotId],
[f].[Price], [f].[Seats], [f].[Strikebound], [f].[Timestamp], [f].[Utilization]
FROM [Flight] AS [f]
WHERE [f].[Departure] IN (N'Berlin', N'Hamburg', N'Köln', N'Berlin')
```

# Sorts (orderby)

```
   CUI.Headline("Sorting");
   List<Flight> flightSet5a = (from f in ctx.FlightSet
                              where f.Departure == "Berlin"
                              orderby f.Date, f.Destination, f.FreeSeats
                              descending
                              select f).ToList();
   List<Flight> flightSet5b = ctx.FlightSet.Where(f => f.Departure == "Berlin")
                              .OrderBy(f => f.Date)
                              .ThenBy(f => f.Destination)
                              .ThenByDescending(f => f.FreeSeats)
                              .ToList();
```

The resulting SQL in both cases is as follows:

```
SELECT [f].[FlightNo], [f].[AircraftTypeID], [f].[AirlineCode],
[f].[CopilotId], [f].[FlightDate], [f].[Departure], [f].[Destination],
[f].[FreeSeats], [f].[LastChange], [f].[Memo], [f].[NonSmokingFlight],
[f].[PilotId], [f].[Price], [f].[Seats], [f].[Strikebound],
[f].[Timestamp], [f].[Utilization]
FROM [Flight] AS [f]
WHERE [f].[Departure] = N'Berlin'
ORDER BY [f].[FlightDate], [f].[Destination], [f].[FreeSeats] DESC
```

# Paging (Skip() and Take())

```
CUI.Headline("Paging");
List<Flight> flightSet6a = (from f in ctx.FlightSet
                           where f.Departure == "Berlin"
                           orderby f.Date
                           select f).Skip(100).Take(10).ToList();
List<Flight> flightSet6b = ctx.FlightSet.Where(f => f.Departure == "Berlin")
                           .OrderBy(f => f.Date)
                           .Skip(100).Take(10).ToList();
```

Entity Framework is aware of the database engine in use and will implement paging in the most efficient manner possible. For newer versions, support exists for row-limiting clauses. For older databases, more complex queries and Common Table Express (CTE) style syntax with the `rownumber()` function will be used. This is controlled automatically by Entity Framework Core.

```
exec sp_executesql N'SELECT [f].[FlightNo], [f].[AircraftTypeID], [f].
[AirlineCode], [f].[CopilotId], [f].[FlightDate], [f].[Departure], [f].
[Destination], [f].[FreeSeats], [f].[LastChange], [f].[Memo], [f].
[NonSmokingFlight], [f].[PilotId], [f].[Price], [f].[Seats], [f].
[Strikebound], [f].[Timestamp], [f].[Utilization]
FROM [Flight] AS [f]
```

```
WHERE [f].[Departure] = N''Berlin''
ORDER BY [f].[FlightDate]
OFFSET @__p_0 ROWS FETCH NEXT @__p_1 ROWS ONLY',N'@__p_0 int,@__p_1
int',@__p_0=100,@__p_1=10
```

## Projection

```
List<Flight> flightSet7a = (from f in ctx.FlightSet
                            where f.Departure == "Berlin"
                            orderby f.Date
                            select new Flight()
                            {
                             FlightNo = f.FlightNo,
                             Date = f.Date,
                             Departure = f.Departure,
                             Destination = f.Destination,
                             FreeSeats = f.FreeSeats,
                             Timestamp = f.Timestamp
                            }).ToList();

List<Flight> flightSet7b = ctx.FlightSet
                            .Where(f => f.Departure == "Berlin")
                            .OrderBy(f => f.Date)
                            .Select(f => new Flight()
                            {
                             FlightNo = f.FlightNo,
                             Date = f.Date,
                             Departure = f.Departure,
                             Destination = f.Destination,
                             FreeSeats = f.FreeSeats,
                             Timestamp = f.Timestamp
                            }).ToList();
```

The resulting SQL in both cases is as follows:

```
SELECT [f].[FlightNo], [f].[FlightDate] AS [Date], [f].[Departure],
[f].[Destination], [f].[FreeSeats], [f].[Timestamp]
FROM [Flight] AS [f]
WHERE [f].[Departure] = N'Berlin'
ORDER BY [Date]
```

# Aggregate Functions (Count(), Min(), Max(), Average(), Sum())

```
int agg1a = (from f in ctx.FlightSet select f).Count();
   int? agg2a = (from f in ctx.FlightSet select f).Sum(f => f.FreeSeats);
   int? agg3a = (from f in ctx.FlightSet select f).Min(f => f.FreeSeats);
   int? agg4a = (from f in ctx.FlightSet select f).Max(f => f.FreeSeats);
   double? agg5a = (from f in ctx.FlightSet select f).Average(f => f.FreeSeats);

   int agg1b = ctx.FlightSet.Count();
   int? agg2b = ctx.FlightSet.Sum(f => f.FreeSeats);
   int? agg3b = ctx.FlightSet.Min(f => f.FreeSeats);
   int? agg4b = ctx.FlightSet.Max(f => f.FreeSeats);
   double? agg5b = ctx.FlightSet.Average(f => f.FreeSeats);
```

The resulting SQL is as follows:

```
SELECT COUNT (*)
FROM [Flight] AS [f]

SELECT SUM([f].[FreeSeats])
FROM [Flight] AS [f]

SELECT MIN([f].[FreeSeats])
FROM [Flight] AS [f]
```

```
SELECT MAX([f].[FreeSeats])
FROM [Flight] AS [f]
```

```
SELECT AVG(CAST([f].[FreeSeats] AS float))
FROM [Flight] AS [f]
```

## Groupings (GroupBy)

```
var group1a = (from f in ctx.FlightSet
               group f by f.Departure into g
               select new { City = g.Key, Count = g.Count(),
               Sum = g.Sum(f => f.FreeSeats), Avg = g.Average(f =>
               f.FreeSeats) })
               .ToList();
```

```
var group1b = ctx.FlightSet
               .GroupBy(f => f.Departure)
               .Select(g => new
               {
               City = g.Key,
               Count = g.Count(),
               Sum = g.Sum(f => f.FreeSeats),
               Avg = g.Average(f => f.FreeSeats)
               }).ToList();
```

---

**Note**    LINQ groupings are still running in RAM in version 2.0 of Entity Framework Core. In the upcoming version 2.1 (see Appendix C) these should be translated correctly in SQL. Therefore, groupings should be formulated directly in SQL in versions 1.0 to 2.0 (see Chapter 15).

---

The database management system currently receives the following command:

```
SELECT [f0].[FlightNo], [f0].[AircraftTypeID], [f0].[AirlineCode],
[f0].[CopilotId], [f0].[FlightDate], [f0].[Departure], [f0].[Destination],
[f0].[FreeSeats], [f0].[LastChange], [f0].[Memo], [f0].[NonSmokingFlight],
[f0].[PilotId], [f0].[Price], [f0].[Seats], [f0].[Strikebound],
[f0].[Timestamp], [f0].[Utilization]
FROM [Flight] AS [f0]
ORDER BY [f0].[Departure]
```

## Single Objects (SingleOrDefault(), FirstOrDefault())

```
Flight flight1a = (from f in ctx.FlightSet select f).SingleOrDefault(f
=> f.FlightNo == 101);
```

```
Flight flight1b = ctx.FlightSet.SingleOrDefault(f => f.FlightNo == 101);
```

The resulting SQL in both cases is as follows:

```
SELECT TOP(2) [f].[FlightNo], [f].[AircraftTypeID], [f].[AirlineCode],
[f].[CopilotId], [f].[FlightDate], [f].[Departure], [f].[Destination],
[f].[FreeSeats], [f].[LastChange], [f].[Memo], [f].[NonSmokingFlight],
[f].[PilotId], [f].[Price], [f].[Seats], [f].[Strikebound],
[f].[Timestamp], [f].[Utilization]
FROM [Flight] AS [f]
WHERE [f].[FlightNo] = 101
    Flight flight2a = (from f in ctx.FlightSet
                    where f.FreeSeats > 0
                    orderby f.Date
                    select f).FirstOrDefault();

    Flight flight2b = ctx.FlightSet
                    .Where(f => f.FreeSeats > 0)
                    .OrderBy(f => f.Date)
                    .FirstOrDefault();
```

The resulting SQL in both cases is as follows:

```
SELECT TOP(1) [f].[FlightNo], [f].[AircraftTypeID], [f].[AirlineCode],
[f].[CopilotId], [f].[FlightDate], [f].[Departure], [f].[Destination],
[f].[FreeSeats], [f].[LastChange], [f].[Memo], [f].[NonSmokingFlight],
[f].[PilotId], [f].[Price], [f].[Seats], [f].[Strikebound],
[f].[Timestamp], [f].[Utilization]
FROM [Flight] AS [f]
WHERE [f].[FreeSeats] > 0
ORDER BY [f].[FlightDate]
```

## Related Objects (Include())

```
List<Flight> flightDetailsSet1a = (from f in ctx.FlightSet
                             .Include(f => f.Pilot)
                             .Include(f => f.BookingSet).ThenInclude
                             (b => b.Passenger)
                             where f.Departure == "Berlin"
                             orderby f.Date
                             select f)
                             .ToList();

List<Flight> flightDetailsSet1b = ctx.FlightSet
                             .Include(f => f.Pilot)
                             .Include(f => f.BookingSet).ThenInclude
                             (b => b.Passenger)
                             .Where(f => f.Departure == "Berlin")
                             .OrderBy(f => f.Date)
                             .ToList();
```

---

**Note**  Entity Framework Core executes two SQL statements directly after each other to avoid joins.

---

The resulting SQL in both cases is as follows:

```
SELECT [f].[FlightNo], [f].[AircraftTypeID], [f].[AirlineCode],
[f].[CopilotId], [f].[FlightDate], [f].[Departure], [f].[Destination],
[f].[FreeSeats], [f].[LastChange], [f].[Memo], [f].[NonSmokingFlight],
[f].[PilotId], [f].[Price], [f].[Seats], [f].[Strikebound], [f].[Timestamp],
[f].[Utilization], [f.Pilot].[PersonID], [f.Pilot].[Birthday], [f.Pilot].
[DetailID], [f.Pilot].[Discriminator], [f.Pilot].[EMail], [f.Pilot].
[GivenName], [f.Pilot].[PassportNumber], [f.Pilot].[Salary], [f.Pilot].
[SupervisorPersonID], [f.Pilot].[Surname], [f.Pilot].[FlightHours],
[f.Pilot].[FlightSchool], [f.Pilot].[LicenseDate], [f.Pilot].
[PilotLicenseType]
FROM [Flight] AS [f]
INNER JOIN [Employee] AS [f.Pilot] ON [f].[PilotId] = [f.Pilot].[PersonID]
WHERE ([f.Pilot].[Discriminator] = N'Pilot') AND ([f].[Departure] = N'Berlin')
ORDER BY [f].[FlightDate], [f].[FlightNo]

SELECT [f.BookingSet].[FlightNo], [f.BookingSet].[PassengerID],
[b.Passenger].[PersonID], [b.Passenger].[Birthday], [b.Passenger].
[CustomerSince], [b.Passenger].[DetailID], [b.Passenger].[EMail],
[b.Passenger].[GivenName], [b.Passenger].[Status], [b.Passenger].[Surname]
FROM [Booking] AS [f.BookingSet]
INNER JOIN [Passenger] AS [b.Passenger] ON [f.BookingSet].[PassengerID] =
[b.Passenger].[PersonID]
INNER JOIN (
    SELECT DISTINCT [f0].[FlightNo], [f0].[FlightDate]
    FROM [Flight] AS [f0]
    INNER JOIN [Employee] AS [f.Pilot0] ON [f0].[PilotId] = [f.Pilot0].
    [PersonID]
    WHERE ([f.Pilot0].[Discriminator] = N'Pilot') AND ([f0].[Departure] =
    N'Berlin')
) AS [t] ON [f.BookingSet].[FlightNo] = [t].[FlightNo]
ORDER BY [t].[FlightDate], [t].[FlightNo]
```

# Inner Join

Explicit join operations are not necessary if there are navigation relationships
(see "Related Objects (Include())"). In the following example, to construct a case
without a navigation relationship, all flights that have the same ID as a pilot are searched:

```
var flightDetailsSet2a = (from f in ctx.FlightSet
                          join p in ctx.PilotSet
                          on f.FlightNo equals p.PersonID
                          select new { Nr = f.FlightNo, flight = f,
                          Pilot = p })
                               .ToList();

var flightDetailsSet2b = ctx.FlightSet
                          .Join(ctx.PilotSet, f => f.FlightNo, p =>
                           p.PersonID,
                          (f, p) => new { Nr = f.FlightNo, flight = f,
                          Pilot = p })
                          .ToList();
```

The resulting SQL in both cases is as follows:

```
SELECT [f].[FlightNo] AS [Nr], [f].[AircraftTypeID], [f].[AirlineCode],
[f].[CopilotId], [f].[FlightDate], [f].[Departure], [f].[Destination],
[f].[FreeSeats], [f].[LastChange], [f].[Memo], [f].[NonSmokingFlight],
[f].[PilotId], [f].[Price], [f].[Seats], [f].[Strikebound], [f].[Timestamp],
[f].[Utilization], [p].[PersonID], [p].[Birthday], [p].[DetailID],
[p].[Discriminator], [p].[EMail], [p].[GivenName], [p].[PassportNumber],
[p].[Salary], [p].[SupervisorPersonID], [p].[Surname], [p].[FlightHours],
[p].[FlightSchool], [p].[LicenseDate], [p].[PilotLicenseType]
FROM [Flight] AS [f]
INNER JOIN [Employee] AS [p] ON [f].[FlightNo] = [p].[PersonID]
WHERE [p].[Discriminator] = N'Pilot'
```

# Cross Join (Cartesian Product)

```
var flightDetailsSet3a = (from f in ctx.FlightSet
                          from b in ctx.BookingSet
                          from p in ctx.PassengerSet
                          where f.FlightNo == b.FlightNo && b.PassengerID
                          == p.PersonID && f.Departure == "Rome"
                          select new { flight = f, passengers = p })
                          .ToList();

var flightDetailsSet3b = ctx.FlightSet
    .SelectMany(f => ctx.BookingSet, (f, b) => new  { f = f,  b = b})
    .SelectMany(z => ctx.PassengerSet, (x, p) => new {x = x, p = p})
    .Where(y => ((y.x.f.FlightNo == y.x.b.FlightNo) &&
                       (y.x.b.PassengerID == y.p.PersonID)) &&
                     y.x.f.Departure == "Rome")
    .Select(z => new {flight = z.x.f, passengers = z.p } )
```

The resulting SQL in both cases is as follows:

```
SELECT [f].[FlightNo], [f].[AircraftTypeID], [f].[AirlineCode],
[f].[CopilotId], [f].[FlightDate], [f].[Departure], [f].[Destination],
[f].[FreeSeats], [f].[LastChange], [f].[Memo], [f].[NonSmokingFlight],
[f].[PilotId], [f].[Price], [f].[Seats], [f].[Strikebound], [f].[Timestamp],
[f].[Utilization], [p].[PersonID], [p].[Birthday], [p].[CustomerSince],
[p].[DetailID], [p].[EMail], [p].[GivenName], [p].[Status], [p].[Surname]
FROM [Flight] AS [f]
CROSS JOIN [Booking] AS [b]
CROSS JOIN [Passenger] AS [p]
WHERE ((([f].[FlightNo] = [b].[FlightNo]) AND ([b].[PassengerID] = [p].
[PersonID])) AND ([f].[Departure] = N'Rome')
```

# Join with a Grouping

```
var flightDetailsSet4a = (from b in ctx.BookingSet
                          join f in ctx.FlightSet on b.FlightNo equals
                          f.FlightNo
                          join p in ctx.PassengerSet on b.PassengerID
                          equals p.PersonID
                          where f.Departure == "Berlin"
                          group b by b.Flight into g
                          select new { flight = g.Key, passengers =
                          g.Select(x => x.Passenger) })
                          .ToList();

var flightDetailsSet4b = ctx.BookingSet
                          .Join(ctx.FlightSet, b => b.FlightNo, f =>
                          f.FlightNo, (b, f) => new { b = b, f = f })
                          .Join(ctx.PassengerSet, x => x.b.PassengerID,
                          p => p.PersonID, (x, p) => new { x = x, p = p })
                          .Where(z => (z.x.f.Departure == "Berlin"))
                          .GroupBy(y => y.x.b.Flight, y => y.x.b)
                          .Select(g => new { flight = g.Key, passengers
                          = g.Select(x => x.Passenger) })
                          .ToList();
```

The resulting SQL in both cases is as follows:

```
SELECT [b0].[FlightNo], [b0].[PassengerID], [b.Flight0].[FlightNo],
[b.Flight0].[AircraftTypeID], [b.Flight0].[AirlineCode], [b.Flight0].
[CopilotId], [b.Flight0].[FlightDate], [b.Flight0].[Departure],
[b.Flight0].[Destination], [b.Flight0].[FreeSeats], [b.Flight0].
[LastChange], [b.Flight0].[Memo], [b.Flight0].[NonSmokingFlight],
[b.Flight0].[PilotId], [b.Flight0].[Price], [b.Flight0].[Seats],
[b.Flight0].[Strikebound], [b.Flight0].[Timestamp], [b.Flight0].
[Utilization], [f0].[FlightNo], [f0].[AircraftTypeID], [f0].[AirlineCode],
[f0].[CopilotId], [f0].[FlightDate], [f0].[Departure], [f0].[Destination],
[f0].[FreeSeats], [f0].[LastChange], [f0].[Memo], [f0].[NonSmokingFlight],
```

179

```
[f0].[PilotId], [f0].[Price], [f0].[Seats], [f0].[Strikebound], [f0].
[Timestamp], [f0].[Utilization], [p0].[PersonID], [p0].[Birthday], [p0].
[CustomerSince], [p0].[DetailID], [p0].[EMail], [p0].[GivenName], [p0].
[Status], [p0].[Surname]
FROM [Booking] AS [b0]
INNER JOIN [Flight] AS [b.Flight0] ON [b0].[FlightNo] = [b.Flight0].[FlightNo]
INNER JOIN [Flight] AS [f0] ON [b0].[FlightNo] = [f0].[FlightNo]
INNER JOIN [Passenger] AS [p0] ON [b0].[PassengerID] = [p0].[PersonID]
WHERE [f0].[Departure] = N'Berlin'
ORDER BY [b.Flight0].[FlightNo]
```

## Subqueries (Subselects)

---

**Note**   Both Entity Framework and Entity Framework Core subqueries are sent individually for each result data record of the main database management system query. This can lead to significant performance problems!

---

```
List<Flight> flightDetailsSet5a = (from f in ctx.FlightSet
        where f.FlightNo == 101
        select new Flight()
        {
          FlightNo = f.FlightNo,
          Date = f.Date,
          Departure = f.Departure,
          Destination = f.Destination,
          FreeSeats = f.FreeSeats,
          Timestamp = f.Timestamp,
          Pilot = (from p in ctx.PilotSet where
                  p.PersonID == f.PilotId select p)
                  .FirstOrDefault(),
```

```
        Copilot = (from p in ctx.PilotSet where
                    p.PersonID == f.CopilotId select p)
                .FirstOrDefault(),
        }).ToList();

    List<Flight> flightDetailsSet5b = ctx.FlightSet.Where(f => f.FlightNo
    == 101)
    .Select(f =>new Flight()
        {
            FlightNo = f.FlightNo,
            Date = f.Date,
            Departure = f.Departure,
            Destination = f.Destination,
            FreeSeats = f.FreeSeats,
            Timestamp = f.Timestamp,
            Pilot = ctx.PilotSet
                .Where(p => (p.PersonID == f.PilotId))
                .FirstOrDefault(),
            Copilot = ctx.PilotSet
                .Where(p => (p.PersonID) == f.CopilotId)
                .FirstOrDefault()
        }
    ).ToList();
```

The resulting SQL in both cases is as follows:

```
SELECT [f].[FlightNo], [f].[FlightDate] AS [Date], [f].[Departure],
[f].[Destination], [f].[FreeSeats], [f].[Timestamp], [f].[PilotId],
[f].[CopilotId]
FROM [Flight] AS [f]
WHERE [f].[FlightNo] = 101

exec sp_executesql N'SELECT TOP(1) [p].[PersonID], [p].[Birthday],
[p].[DetailID], [p].[Discriminator], [p].[EMail], [p].[GivenName],
[p].[PassportNumber], [p].[Salary], [p].[SupervisorPersonID], [p].[Surname],
[p].[FlightHours], [p].[FlightSchool], [p].[LicenseDate],
[p].[PilotLicenseType]
```

```
FROM [Employee] AS [p]
WHERE ([p].[Discriminator] = N''Pilot'') AND ([p].[PersonID] = @_outer_
PilotId)',N'@_outer_PilotId int',@_outer_PilotId=23
```

```
exec sp_executesql N'SELECT TOP(1) [p0].[PersonID], [p0].[Birthday],
[p0].[DetailID], [p0].[Discriminator], [p0].[EMail], [p0].[GivenName],
[p0].[PassportNumber], [p0].[Salary], [p0].[SupervisorPersonID],
[p0].[Surname], [p0].[FlightHours], [p0].[FlightSchool],
[p0].[LicenseDate], [p0].[PilotLicenseType]
FROM [Employee] AS [p0]
WHERE ([p0].[Discriminator] = N''Pilot'') AND ([p0].[PersonID] = @_outer_
CopilotId)',N'@_outer_CopilotId int',@_outer_CopilotId=3
```

# CHAPTER 9

# Object Relationships and Loading Strategies

An object model describes the relationships between instances of different classes (for example, between Flight and Pilot) or other instances of the same class (see, for example, the Supervisor property in the Employees class). The question of when and how relational objects are loaded is crucial not only to the software developer but also to the performance of the application.

## Overview of Loading Strategies

The classic Entity Framework supports four connected object loading strategies: lazy loading automatically, explicit loading, eager loading, and preloading versus relationship fixup (see Figure 9-1). In Entity Framework Core 1.0, there was only eager loading and preloading. Entity Framework Core 1.1 introduced explicit loading. Lazy loading does not yet exist in Entity Framework Core 2.0, but it will be introduced in version 2.1 (see Appendix C).

© Holger Schwichtenberg 2018
H. Schwichtenberg, *Modern Data Access with Entity Framework Core*,
https://doi.org/10.1007/978-1-4842-3552-2_9

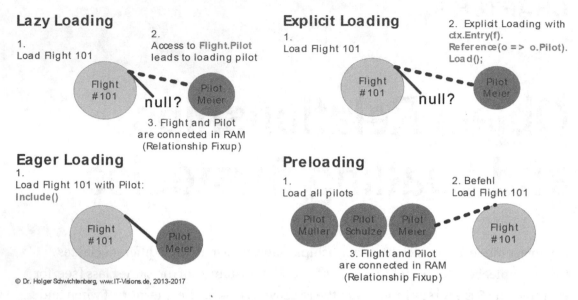

*Figure 9-1.* *Loading strategies in the Entity Framework 1.0 to 6.x. Entity Framework Core currently supports only three strategies; lazy loading is missing.*

# Seeing the Default Behavior

By default, Entity Framework Core limits itself to loading the actual requested objects in a query and does not automatically load linked objects. The following LINQ query loads only flight objects. Objects of type `Pilot`, `Booking`, `Airline`, and `AircraftType` that are associated with the flights are not automatically loaded.

```
List<Flight> list = (from x in ctx.FlightSet
                    where x.Departure == "Berlin" &&
                        x.FreeSeats > 0
                    orderby x.Date, x.Departure
                    select x).ToList();
```

Loading linked records (called *eager loading*) is not a good idea for the default setting because in this case data would be loaded that is not needed later. In addition, the associated records have relationships; for example, bookings are related to passengers. Passengers also have bookings on other flights. If you recursively loaded all these related records, then in the example of the object model in Figure 9-2, you would almost certainly load almost all records into RAM because many passengers are connected to other passengers via shared flights. So Eager Loading would not be good as a default setting.

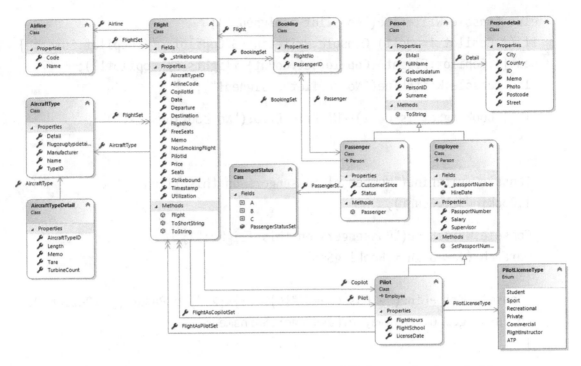

***Figure 9-2.*** *Object model for the management of flights and related objects*

Even if you load a single object using Find(), no linked records are loaded (see
Listing 9-1). Figure 9-3 shows the output.

***Listing 9-1.*** Unsuccessful Attempt to Access Connected Objects in Entity
Framework Core

```
public static void Demo_LazyLoading()
{
 CUI.MainHeadline(nameof(Demo_LazyLoading));

 using (var ctx = new WWWingsContext())
 {
  // Load only the flight
  var f = ctx.FlightSet.SingleOrDefault(x => x.FlightNo == 101);

  Console.WriteLine($"Flight Nr {f.FlightNo} from {f.Departure} to
  {f.Destination} has {f.FreeSeats} free seats!");
  if (f.Pilot != null) Console.WriteLine($"Pilot: {f.Pilot.Surname} has
  {f.Pilot.FlightAsPilotSet.Count} flights as pilot!");
```

```
else Console.WriteLine("No pilot assigned!");
if (f.Copilot != null) Console.WriteLine($"Copilot: {f.Copilot.Surname}
has {f.Copilot.FlightAsCopilotSet.Count} flights as copilot!");
else Console.WriteLine("No copilot assigned!");

if (f.BookingSet is null) CUI.PrintError("No bookings :-(");
else
{
 Console.WriteLine("Number of passengers on this flight: " +
 f.BookingSet.Count);

 Console.WriteLine("Passengers on this flight:");
 foreach (var b in f.BookingSet)
 {
  Console.WriteLine("- Passenger #{0}: {1} {2}", b.Passenger.PersonID,
  b.Passenger.GivenName, b.Passenger.Surname);
 }
}
}
}
}
```

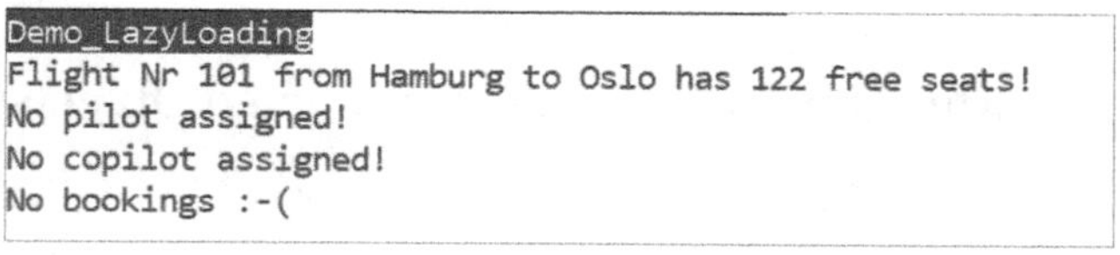

```
Demo_LazyLoading
Flight Nr 101 from Hamburg to Oslo has 122 free seats!
No pilot assigned!
No copilot assigned!
No bookings :-(
```

*Figure 9-3.* *Output of Listing 9-1*

# No Lazy Loading Yet

The previous example would load only the explicitly requested flight in the classic Entity Framework in the first step, but then in the following program code lines the pilot and copilot information (with their other flights) as well as the bookings with the passenger data would be loaded via lazy loading. Entity Framework would send a large number of SELECT commands to the database one after the other. How many commands are used depends on the number of passengers on this flight.

In the case of Entity Framework Core, however, Listing 9-1 does not load the pilot, copilot, or passengers. Microsoft has not yet implemented lazy loading for Entity Framework Core.

---

**Preview**   Basic support for lazy loading is planned for Entity Framework Core version 2.1; see Appendix C.

---

Lazy loading involves a special implementation challenge because the OR mapper must catch any access to any object reference in order to reload the connected objects as needed. This interception is done by using specific classes for individual references and sets of classes.

In the classic Entity Framework, you can realize lazy loading by using certain lazy loading–capable classes and mostly invisible runtime proxy objects. Both will be available in Entity Framework Core 2.1.

When you do not need to preload linked records in Entity Framework Core, it would be great to have lazy loading. A typical example is a master-detail view on the screen. If there are many master records, it would waste time to load the detail records for each master record in advance. Rather, you will request the detail records for the master record that the user has just clicked. In the classic Entity Framework Core you could realize the master-detail display via lazy loading without further program code when clicking the master data record. In Entity Framework Core 1.x and 2.0, unfortunately you have to intercept the click and explicitly load the detail data records.

# Explicit Loading

In Entity Framework Core version 1.1, which was released in 2016, Microsoft retrofitted the explicit reloading feature. You use the methods Reference() (for single objects), Collection() (for sets), and then Load() to specify that related objects should be loaded.

However, these methods are not available on the entity object itself but are part of the class EntityEntry<T>, which is obtained by the method Entry() in the class DbContext (see Listing 9-2). With IsLoaded() you can check whether the object has already been loaded. IsLoaded() returns true even if there was no matching object in the database. It therefore does not indicate whether a navigation relationship has a counter object; it indicates whether in the current context instance a suitable object has ever been loaded for the instance. So if in Listing 9-2 Flight 101 already has an assigned pilot (Mrs. Merkel) but no copilot, this leads to the output in Figure 9-4.

---

**Important**    It is important to understand that every execution of Load() will result in an explicit submission of a SQL command to the database management system.

---

*Listing 9-2.* With Explicit Reloading, Entity Framework Core Sends a Lot of Individual SQL Commands to the Database

```
public static void Demo_ExplizitLoading_v11()
{
 CUI.MainHeadline(nameof(Demo_ExplizitLoading_v11));

 using (var ctx = new WWWingsContext())
 {
  // Load only the flight
  var f = ctx.FlightSet
         .SingleOrDefault(x => x.FlightNo == 101);

  Console.WriteLine($"Flight Nr {f.FlightNo} from {f.Departure} to
  {f.Destination} has {f.FreeSeats} free seats!");

  // Now load the pilot and copilot
  if (!ctx.Entry(f).Reference(x => x.Pilot).IsLoaded)
   ctx.Entry(f).Reference(x => x.Pilot).Load();
  if (!ctx.Entry(f).Reference(x => x.Copilot).IsLoaded)
   ctx.Entry(f).Reference(x => x.Copilot).Load();

  // Check if loaded
  if (ctx.Entry(f).Reference(x => x.Pilot).IsLoaded) Console.
  WriteLine("Pilot is loaded!");
  if (ctx.Entry(f).Reference(x => x.Copilot).IsLoaded) Console.
  WriteLine("Copilot is loaded!");
```

```
if (f.Pilot != null) Console.WriteLine($"Pilot: {f.Pilot.Surname} has
{f.Pilot.FlightAsPilotSet.Count} flights as pilot!");
else Console.WriteLine("No pilot assigned!");
if (f.Copilot != null) Console.WriteLine($"Copilot: {f.Copilot.Surname}
has {f.Copilot.FlightAsCopilotSet.Count} flights as copilot!");
else Console.WriteLine("No copilot assigned!");

// No download the booking list
if (!ctx.Entry(f).Collection(x => x.BookingSet).IsLoaded)
 ctx.Entry(f).Collection(x => x.BookingSet).Load();

Console.WriteLine("Number of passengers on this flight: " +
f.BookingSet.Count);
Console.WriteLine("Passengers on this flight:");
foreach (var b in f.BookingSet)
{
 // Now load the passenger object for this booking
 if (!ctx.Entry(b).Reference(x => x.Passenger).IsLoaded)
  ctx.Entry(b).Reference(x => x.Passenger).Load();
 Console.WriteLine("- Passenger #{0}: {1} {2}", b.Passenger.PersonID,
 b.Passenger.GivenName, b.Passenger.Surname);
 }
 }
}
```

```
Demo_ExplizitLoading
Flight Nr 101 from Berlin to Seattle has 122 free seats!
Pilot is loaded!
Copilot is loaded!
Pilot: Merkel has 1 flights as pilot!
Copilot: Gabriel has 1 flights as copilot!
Number of passengers on this flight: 18
Passengers on this flight:
- Passenger #549: Lukas Koch
- Passenger #622: Lisa Schäfer
- Passenger #650: Leonie Richter
- Passenger #888: Lisa Schulz
- Passenger #973: Hannah Schmidt
- Passenger #1208: Laura Müller
- Passenger #1260: Sarah Richter
- Passenger #1396: Hannah Schneider
- Passenger #1722: Jan Schwarz
- Passenger #1888: Lisa Richter
- Passenger #2068: Anna Wolf
- Passenger #2102: Hannah Wolf
- Passenger #2109: Anna Schulz
- Passenger #2402: Jan Fischer
- Passenger #2457: Lisa Schröder
- Passenger #2694: Leon Bauer
- Passenger #2819: Anna Becker
- Passenger #3297: Anna Schwarz
```

*Figure 9-4.*  *Output of Listing 9-2*

# Eager Loading

Like the classic Entity Framework, Entity Framework Core supports eager loading. However, the syntax has changed a little bit.

In the classic Entity Framework versions 1.0 and 4.0 (there were never versions 2.0 and 3.0), you could specify a string with Include() only with the name of a navigation property; the string was not checked by the compiler. From the third version on (version number 4.1), it was then possible to specify a robust lambda expression for the navigation properties instead of the character string. For multilevel loading paths, you had to nest the lambda expressions and also use the Select() method.

In Entity Framework Core, there are still strings and lambda expressions, but the syntax for the lambda expressions has been slightly modified. Instead of using Select(), the new extension method ThenInclude() can be for nested relations, like ThenOrderBy() does for sorting across multiple columns. Listing 9-3 shows the eager loading of a flight with the following linked data:

- The bookings and passenger information for every booking: Include(b => b.Bookings).ThenInclude (p => p.Passenger)

- The pilot and the pilot's other flights as a pilot: Include(b => b.Pilot).ThenInclude (p => p.FlightAsPilotSet)

- The copilot and the copilot's other flights as copilot: Include (b => b.Co-Pilot).ThenInclude (p => p.FlightAsCopilotSet)

*Listing 9-3.* With Eager Loading You Can Use the Connected Objects in Entity Framework Core

```
public static void Demo_EagerLoading()
{
 CUI.MainHeadline(nameof(Demo_EagerLoading));

 using (var ctx = new WWWingsContext())
 {
  var flightNo = 101;

  // Load the flight and some connected objects via Eager Loading
  var f = ctx.FlightSet
         .Include(b => b.BookingSet).ThenInclude(p => p.Passenger)
         .Include(b => b.Pilot).ThenInclude(p => p.FlightAsPilotSet)
         .Include(b => b.Copilot).ThenInclude(p => p.FlightAsCopilotSet)
         .SingleOrDefault(x => x.FlightNo == flightNo);

 Console.WriteLine($"Flight Nr {f.FlightNo} from {f.Departure} to
 {f.Destination} has {f.FreeSeats} free seats!");
 if (f.Pilot != null) Console.WriteLine($"Pilot: {f.Pilot.Surname} has
 {f.Pilot.FlightAsPilotSet.Count} flights as a pilot!");
 else Console.WriteLine("No pilot assigned!");
```

```
if (f.Copilot != null) Console.WriteLine($"Copilot: {f.Copilot.Surname}
has {f.Copilot.FlightAsCopilotSet.Count} flights as a Copilot!");
else Console.WriteLine("No Copilot assigned!");

Console.WriteLine("Number of passengers on this flight: " +
f.BookingSet.Count);
Console.WriteLine("Passengers on this flight:");
foreach (var b in f.BookingSet)
{
 Console.WriteLine("- Passenger #{0}: {1} {2}", b.Passenger.PersonID,
 b.Passenger.GivenName, b.Passenger.Surname);
}
}
}
```

Figure 9-5 shows the output. Both the Pilot and Copilot information and the list of booked passengers are available.

---

**Attention**   The compiler checks with Include() and ThenInclude() only if the class has an appropriate property or field. It does not check whether this is also a navigation property to another entity class. If it is not a navigation property, the following error will not occur until runtime: "The property xy is not a navigation property of entity type 'ab'. The 'Include(string)' method can only be used with a '.' separated list of navigation property names."

---

```
Demo_EagerLoading
Flight Nr 101 from Berlin to Seattle has 122 free seats!
Pilot: Merkel has 31 flights as a pilot!
Copilot: Gabriel has 19 flights as a Copilot!
Number of passengers on this flight: 18
Passengers on this flight:
- Passenger #549: Lukas Koch
- Passenger #622: Lisa Schäfer
- Passenger #650: Leonie Richter
- Passenger #888: Lisa Schulz
- Passenger #973: Hannah Schmidt
- Passenger #1208: Laura Müller
- Passenger #1260: Sarah Richter
- Passenger #1396: Hannah Schneider
- Passenger #1722: Jan Schwarz
- Passenger #1888: Lisa Richter
- Passenger #2068: Anna Wolf
- Passenger #2102: Hannah Wolf
- Passenger #2109: Anna Schulz
- Passenger #2402: Jan Fischer
- Passenger #2457: Lisa Schröder
- Passenger #2694: Leon Bauer
- Passenger #2819: Anna Becker
- Passenger #3297: Anna Schwarz
```

*Figure 9-5.*  *Output of Listing 9-3*

However, there is another crucial difference to the classic Entity Framework. While Entity Framework versions 1.0 to 6.*x* sent only a single, large SELECT command to the database management system, Entity Framework Core decides to split the query into four steps (see Figure 9-6), as shown here:

1.  First, the flight is loaded with join on the employee table, which also contains the pilot information (a table via hierarchy mapping).

2.  In the second step, Entity Framework Core loads the other flights of the copilot.

3.  In the third step, Entity Framework Core loads the other flights of the pilot.

4.  In the last step, Entity Framework Core loads the passenger details.

***Figure 9-6.*** *SQL Server Profiler shows the four SQL commands that the eager loading example triggers in Entity Framework Core*

This strategy can be faster than executing a large SELECT command that returns a large set of results that duplicate records and that the OR mapper must then disassemble and clean up the duplicates. The strategy of separate SELECT commands from Entity Framework Core can also be slower because each round-trip of the database management system takes time. In the classic Entity Framework, you had the choice of how large you wanted to cut an eager loading instruction and where you wanted to load it. In Entity Framework Core, you lose control over the number of round-trips to the database management system.

# Relationship Fixup

Relationship fixup is a mechanism of Entity Framework Core that already existed in the classic Entity Framework. The relationship fixup does the following between two objects in RAM:

- *Case 1*: When two objects from the database that are related through a foreign key in the database are loaded independently, Entity Framework Core establishes the relationship between the two objects through their defined navigation properties.

- *Case 2*: When an object is created in RAM or modified to be related to another object in RAM by foreign key, Entity Framework Core establishes the relationship between the two through their defined navigation properties.

- *Case 3a*: When an object in RAM is connected to another object in RAM by navigation and there is also a bidirectional relationship with navigation in the other direction, Entity Framework Core also updates the other navigation property.

- *Case 3b*: When an object in RAM is connected to another object in RAM using a foreign key property, Entity Framework Core updates the other navigation properties on the two objects.

---

**Note**    In cases 3a and 3b, the relationship fixup will not be executed until `ctx.ChangeTracker.DetectChanges()` is called. Unlike the classic Entity Framework, Entity Framework Core no longer automatically calls the `DetectChanges()` call on almost all API functions, which was a performance problem. Entity Framework Core runs `DetectChanges()` only on `ctx.SaveChanges()`, `ctx.Entry()`, and `ctx.Entries()` as well as the method `DbSet<T>().Add()`.

---

# Example for Case 1

In Listing 9-4, a flight is first loaded into the flight variable. Then for the flight, the PilotId value of the Pilot object is printed. However, the Pilot object is not yet available at this point because it was not loaded with the flight, and Entity Framework Core does not currently support lazy loading.

Then the Pilot object for the Flight object is individually loaded into the variable pilot using the ID. The Pilot object and the Flight object would normally now be detached from each other. However, the output of flight.Pilot shows now that Entity Framework Core has established the relationship via relationship fixup. Likewise, the backward relationship has been recorded; pilot.FlightAsPilotSet shows the previously loaded flight.

---

**Note**   Further flights of this pilot, which may be included in the database, do not appear here because they were not loaded.

---

*Listing 9-4.*  Relationship Fixup in Case 1

```
public static void RelationshipFixUp_Case1()
{
 CUI.MainHeadline(nameof(RelationshipFixUp_Case1));

 using (var ctx = new WWWingsContext())
 {

  int flightNr = 101;

  // 1. Just load the flight
  var flight = ctx.FlightSet.Find(flightNr);

  // 2. Output of the pilot of the Flight
  Console.WriteLine(flight.PilotId + ": " + (flight.Pilot != null ?
  flight.Pilot.ToString() : "Pilot not loaded!"));

  // 3. Load the pilot separately
  var pilot = ctx.PilotSet.Find(flight.PilotId);
```

```
// 4. Output of the Pilot of the Flight: Pilot now available
Console.WriteLine(flight.PilotId + ": " + (flight.Pilot != null ?
flight.Pilot.ToString() : "Pilot not loaded!"));

// 5. Output the list of flights of this pilot
foreach (var f in pilot.FlightAsPilotSet)
{
 Console.WriteLine(f);
 }
 }
}
```

As with the classic Entity Framework, Entity Framework Core also assigns an instance of a collection type to the navigation property as part of the relationship fixup if the navigation property is null.

This automation exists regardless of whether you used an interface or a class as the type when declaring the navigation property. If a collection class is used, Entity Framework Core instantiates the collection class. In the case of the declaration of the navigation property with an interface type, Entity Framework Core chooses an appropriate collection class. With ICollection<T>, the class HashSet<T> is selected. With IList<T>, the class List<T> is selected.

# Example for Case 2

In Listing 9-5 for case 2, a pilot is loaded. At first, there are no flights from the pilot in RAM. Then a new flight is created, and this flight is assigned the PilotID of the loaded Pilot. When calling ctx.FlightSet.Add(), Entity Framework Core performs a relationship fixup so that the navigation property called FlightAsPilotSet of the Pilot object and the navigation property called flying.Pilot are filled.

*Listing 9-5.* Relationship Fixup in Case 2

```
public static void RelationshipFixUp_Case2()
{
 CUI.MainHeadline(nameof(RelationshipFixUp_Case2));
 void PrintPilot(Pilot pilot)
 {
```

```csharp
 CUI.PrintSuccess(pilot.ToString());
 if (pilot.FlightAsPilotSet != null)
 {
  Console.WriteLine("Flights of this pilot:");
  foreach (var f in pilot.FlightAsPilotSet)
  {
   Console.WriteLine(f);
  }
 }
 else
 {
  CUI.PrintWarning("No flights!");
 }
}

using (var ctx = new WWWingsContext())
{
 // Load a Pilot
 var pilot = ctx.PilotSet.FirstOrDefault();

 // Print pilot and his flights
 PrintPilot(pilot);
 // Create a new flight for this pilot
 var flight = new Flight();
 flight.Departure = "Berlin";
 flight.Destination = "Berlin";
 flight.Date = DateTime.Now.AddDays(10);
 flight.FlightNo = ctx.FlightSet.Max(x => x.FlightNo) + 1;
 flight.PilotId = pilot.PersonID;
 ctx.FlightSet.Add(flight);

 // Print pilot and his flights
 PrintPilot(pilot);
 // Print pilot of the new flight
 Console.WriteLine(flight.Pilot);
}
}
```

# Example for Case 3

Table 9-1 optionally includes cases 3a and 3b. 'In the example, a flight is loaded first'. Then any pilot is loaded and assigned to the loaded flight by PilotID (case 3a) or a navigation property (case 3b). However, Entity Framework Core does not automatically run a relationship fixup operation here. In case 3a, flight.Pilot and pilot.FlightsAsPilotSet are empty. This changes only with the call ctx.ChangeTracker.DetectChanges(). In case 3b, Flight.Pilot is filled manually, and Pilot.FlightAsPilotSet becomes filled after ctx.ChangeTracker.DetectChanges() is called. Listing 9-6 shows the relationship fixup.

***Table 9-1.*** *Comparing the Behavior of Cases 3a and 3b*

|  | Case 3a | Case 3b |
|---|---|---|
| Assignment | flight.Pilot = pilot | flight.PilotId = pilot.PersonID |
| Flight.Pilot | filled | filled after DetectChanges() |
| Flight.PilotID | filled | filled |
| Pilot.FlightAsPilotSet | filled after DetectChanges() | filled after DetectChanges() |

***Listing 9-6.*** Relationship Fixup in Case 3

```
public static void RelationshipFixUp_Case3()
{
 CUI.MainHeadline(nameof(RelationshipFixUp_Case3));

 // Inline helper for output (>= C# 7.0)
 void PrintflightPilot(Flight flight, Pilot pilot)
 {
  CUI.PrintSuccess(flight);
  Console.WriteLine(flight.PilotId + ": " + (flight.Pilot != null ?
  flight.Pilot.ToString() : "Pilot not loaded!"));
  CUI.PrintSuccess(pilot.ToString());
  if (pilot.FlightAsPilotSet != null)
  {
```

```csharp
   Console.WriteLine("Flights of this pilot:");
   foreach (var f in pilot.FlightAsPilotSet)
   {
    Console.WriteLine(f);
   }
  }
  else
  {
   CUI.PrintWarning("No flights!");
  }
 }

 using (var ctx = new WWWingsContext())
 {
  int flightNr = 101;

  CUI.Headline("Load flight");

  var flight = ctx.FlightSet.Find(flightNr);
  Console.WriteLine(flight);
  // Pilot of this flight
  Console.WriteLine(flight.PilotId + ": " + (flight.Pilot != null ?
  flight.Pilot.ToString() : "Pilot not loaded!"));

  CUI.Headline("Load pilot");
  var pilot = ctx.PilotSet.FirstOrDefault();
  Console.WriteLine(pilot);

  CUI.Headline("Assign a new pilot");
  flight.Pilot = pilot;  // Case 3a
  //flight.PilotId = pilot.PersonID; // Case 3b

  // Determine which relationships exist
  PrintflightPilot(flight, pilot);

  // Here you have to trigger the Relationshop fixup yourself
  CUI.Headline("DetectChanges...");
  ctx.ChangeTracker.DetectChanges();
```

```
  // Determine which relationships exist
  PrintflightPilot(flight, pilot);
 }
}
```

# Preloading with Relationship Fixup

Like the classic Entity Framework, Entity Framework Core supports another loading strategy: preloading in conjunction with a relationship fixup operation in RAM. You explicitly send out several LINQ commands for the connected objects, and the OR mapper puts the newly added objects together after their materialization with those objects that are already in RAM. After the following statements, Flight 101 as well the Pilot and Copilot objects of Flight 101 can be found in RAM when accessing flight. Pilot and flight.Copilot:

```
var Flight = ctx.FlightSet.SingleOrDefault(x => x.FlightNo == 101);
ctx.PilotSet.Where(p => p.FlightAsPilotSet.Any(x => x.FlightNo == 101) ||
p.FlightAsCopilotSet.Any(x => x.FlightNo == 101)).ToList();
```

When loading the two pilots Entity Framework Core recognizes that there is already a Flight object in RAM, that needs these two pilots as Pilot or Copilot. It then compiles the Flight object with the two Pilot objects in RAM (via relationship fixup, as covered earlier).

While the Pilot and Copilot objects for Flight 101 were specifically loaded in the previous two lines, you can also use the relationship fixup for caching optimization. Listing 9-7 shows that all pilots and some flights are loaded. For each loaded flight, the Pilot and Copilot objects are available. Of course, as with caching, you will need a bit more RAM here because you will also load Pilot objects that are never needed. In addition, you must be aware that you can have a timeliness problem because the dependent data is on the same level as the main data. But that's always the way caching behaves. However, you can save a round-trip of the database management system and improve its speed.

Listing 9-7 also shows that when loading the information for the two pilots you can avoid the join operator in Entity Framework Core by using the navigation properties and the Any() method. Any() checks whether there is at least one record that meets or does not meet a condition. In the previous case, it is enough that the Pilot object was assigned once as a Pilot or Copilot for the Flight you are looking for. In other cases, you can use the LINQ All() method if you want to address a set of records that all meet or fail a condition.

**Note**    It is noteworthy that neither the previous loading of the two pilots nor the loading of all pilots in the next example assigns the result of the LINQ query to a variable. In fact, this is not necessary because Entity Framework Core (like the classic Entity Framework) contains in its first-level cache a reference in RAM to all objects that have ever been loaded into a specific instance of the context class. The relationship fixup works therefore also without storage in a variable. The assignment to a variable (List<Pilot> allPilot = ctx.PilotSet. ToList()) is of course not harmful but can be useful if you need a list of all pilots in the program flow. It should also be noted that relationship fixup does not work across all instances of the context. A second-level cache required for this purpose is not yet available in Entity Framework Core but is available as an additional component (see Chapters 17 and 20).

*Listing 9-7.* Caching of All Pilots

```
public static void Demo_PreLoadingPilotenCaching()
{
 CUI.MainHeadline(nameof(Demo_PreLoadingPilotenCaching));

 using (var ctx = new WWWingsContext())
 {

  // 1. Load ALL pilots
  ctx.PilotSet.ToList();

  // 2. Load only several flights. The Pilot and Copilot object will then
  be available for every flight!
  var FlightNrListe = new List<int>() { 101, 117, 119, 118 };
  foreach (var FlightNr in FlightNrListe)
  {
   var f = ctx.FlightSet
     .SingleOrDefault(x => x.FlightNo == FlightNr);

   Console.WriteLine($"Flight Nr {f.FlightNo} from {f.Departure} to
   {f.Destination} has {f.FreeSeats} free seats!");
```

```
    if (f.Pilot != null) Console.WriteLine($"Pilot: {f.Pilot.Surname} has
    {f.Pilot.FlightAsPilotSet.Count} flights as pilot!");
    else Console.WriteLine("No pilot assigned!");
    if (f.Copilot != null)
     Console.WriteLine($"Copilot: {f.Copilot.Surname} has {f.Copilot.
     FlightAsCopilotSet.Count} flights as copilot!");
    else Console.WriteLine("No copilot assigned!");
   }
  }
 }
```

Listing 9-8 shows the redesign of the original task of loading all pilot and passenger data into one flight, this time with preloading and relationship fixup instead of eager loading. Here flights, pilots, pilots' other flights, bookings, and passengers are loaded individually. So, the code sends five SELECT commands to the database management system (as opposed to the four SELECT commands that the solution sends with eager loading) but avoids some joins. Figure 9-7 shows the output.

***Listing 9-8.*** Loading Flights, Pilots, Bookings, and Passengers in Separate LINQ Commands

```
/// <summary>
  /// Provides Pilot, booking and passenger information via Preloading /
  RelationshipFixup
  /// </ summary>
  public static void Demo_PreLoading()
  {
   CUI.Headline ( "Demo_PreLoading");
   using (var ctx = new WWWingsContext())
    {
     int Flight no = 101;

     // 1. Just load the Flight
var f = ctx.FlightSet
30.4 SingleOrDefault (x => x.FlightNo == FlightNo);
```

```
    // 2. Load both Pilots
    ctx.PilotSet.Where (p => p.FlightAsPilotSet.Any (x => x.FlightNo ==
    FlightNo) || p.FlightAsCopilotSet.Any (x => x.FlightNo == FlightNo)).
    ToList();

    // 3. Load other Pilots' Flights
    ctx.FlightSet.Where (x => x.PilotId == f.PilotId || x.CopilotId ==
    f.CopilotId).ToList();

    // 4. Loading bookings
    ctx.BuchungSet.Where (x => x.FlightNo == FlightNo).ToList();

    // 5. Load passengers
    ctx.PassengerSet.Where (p => p.BookingsAny (x => x.FlightNo ==
    FlightNo)).ToList();

// not necessary: ctx.ChangeTracker.DetectChanges();

    Console.WriteLine ($ "Flight No {f.FlightNo} from {f.Departure} to
    {f.Destination} has {f.FreeSeats} FreeSeats! ");
if (f.Pilot    != null) Console.WriteLine ($ "Pilot:    {f.Pilot.Name} has
{f.Pilot.FlightAsPilotSet.Count} Flights as a Pilot! ");
else console.WriteLine ("No Pilot assigned!");
if (f.Copilot    != null) Console.WriteLine ($ "Copilot:    {f.Copilot.Name}
has {f.Copilot.FlightAsCopilotSet.Count} Flights as copilot! ");
else console.WriteLine ("No Copilot assigned!");

    Console.WriteLine ("Number of passengers on this Flight:" +
    f.BookingsCount);
    Console.WriteLine ("Passengers on this Flight:");
foreach (var b in f.Bookings
    {
    Console.WriteLine ("- Passenger # {0}: {1} {2}", b.Passenger.
    PersonID,    b.Passenger.First given name,    b.Passenger.Nam
    }
    }
    }
```

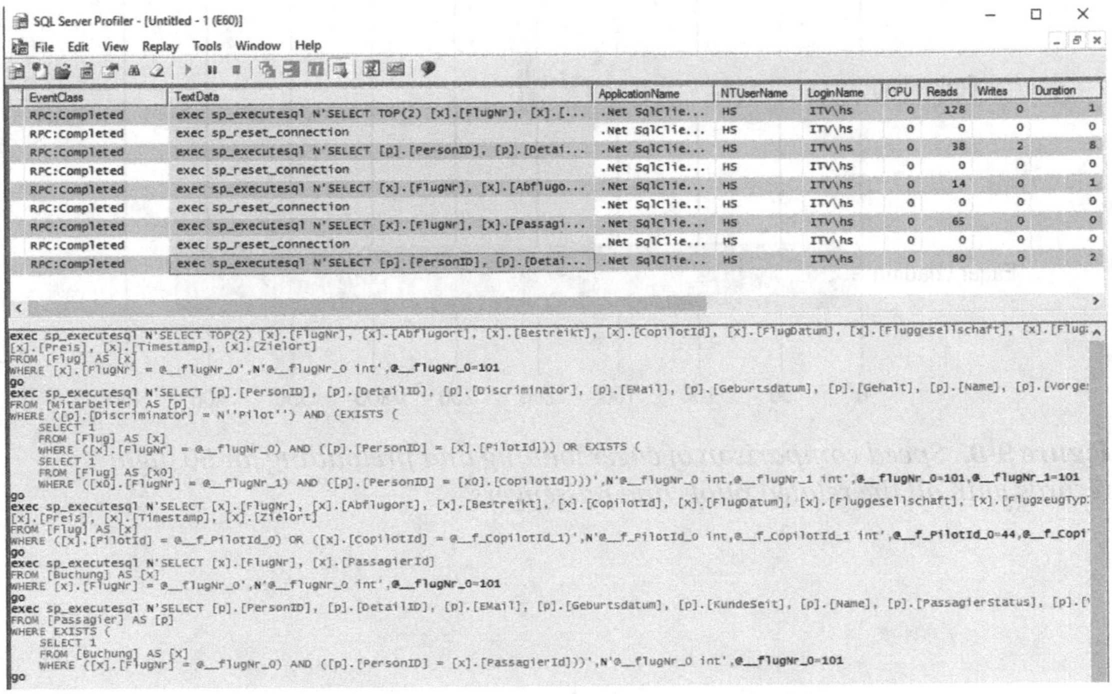

***Figure 9-7.*** *SQL Server Profiler shows the five SQL commands in Listing 9-8*

The relationship fixup trick has a positive effect if one or more of the following conditions are true:

- The result set of the main data is large, and the amount of dependent data is small.

- There are several different dependent data sets that can be preloaded.

- The preloaded objects are rarely variable (parent) data.

- You run multiple queries in a single context instance that have the same dependent data.

In the speed comparison, even in the scenario discussed here of loading a flight with pilots and passengers, it already shows a speed advantage for the preloading. The measurement shown in Figure 9-8 was taken to avoid 51-cycle measurement deviations, with the first pass (a cold start for the Entity Framework Core context and possibly also the database) not taken into account. In addition, all screen editions were expanded.

Of course, you can mix eager loading and preloading as you want. In practice, however, you have to find the optimum ratio for each individual case.

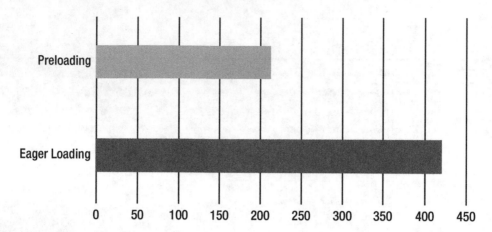

***Figure 9-8.*** *Speed comparison of eager loading and preloading for 50 flight records with all the related pilots and passengers*

# CHAPTER 10

# Inserting, Deleting, and Modifying Records

In many places, the API and approach to inserting, deleting, and modifying records in Entity Framework Core have remained the same compared to the classic Entity Framework. But there are a few changes in the details, especially when merging multiple changes into a batch round-trip to the database management system.

You can write to the entity objects loaded from the database at any time. You do not have to "announce" them before the write operation or "sign them up" afterward. The context class of Entity Framework Core (more precisely, the Change Tracker built into it) tracks all changes to the objects in the standard system (called *change tracking*). However, the change tracking does not take place if the objects were loaded in the no-tracking mode (e.g. with AsNoTracking()), which is specially to be set, or the context instance was destroyed.

## Saving with SaveChanges()

Listing 10-1 shows how to load a Flight object with SingleOrDefault(). In this Flight object, the number of free seats is reduced by two places. In addition, some text is written to the memo attribute of the Flight object.

The SaveChanges() method is used to store changes in the database. It is implemented in the base class DbContext and is inherited from there to the context class that you generate during reverse engineering, or it creates itself during forward engineering.

The SaveChanges() method saves all changes (new records, changed records, and deleted records) since loading or the last SaveChanges() method on all objects loaded in the current context instance. SaveChanges() sends an INSERT, UPDATE, or DELETE command for each saved change to the database.

207

© Holger Schwichtenberg 2018
H. Schwichtenberg, *Modern Data Access with Entity Framework Core*,
https://doi.org/10.1007/978-1-4842-3552-2_10

---

**Note**    Even in Entity Framework Core, it is unfortunately not possible to save only individual changes when several changes exist.

---

Of course, SaveChanges() saves only the changed objects and only the changed attributes for the changed objects. This is proven by the SQL output shown in Figure 10-1. In the SET part of the UPDATE command, only FreeSeats and Memo show up. You can also see that the UPDATE command returns the number of changed records. The caller receives this number from SaveChanges() as the return value.

The UPDATE command contains only the FlightNo value in the WHERE condition. In other words, there is no check here to check whether the object has been changed by another user or background process. The standard for Entity Framework Core is the principle that "the last one wins." However, you can change this default behavior (see Chapter 17). The caller gets a runtime error of type DbConcurrencyException only if the data record to be changed in the database has been deleted. Then, the UPDATE command returns from the database management system that zero records were affected by the change, which Entity Framework Core takes as an indication of a change conflict.

Listing 10-1 prints some information about the Flight object three times (before the change, after the change, and after saving). In addition to FlightNo (the primary key) and the Flight route (the departure and destination), the number of FreeSeats and the current state of the object are printed. The state cannot be determined by the entity object itself, but only by the Entry() method of the context class with ctx.Entry(obj).State.

***Listing 10-1.***  One changed Property of Flights is Saved

```
public static void ChangeFlightOneProperty()
{
 CUI.MainHeadline(nameof(ChangeFlightOneProperty));

 int FlightNr = 101;
 using (WWWingsContext ctx = new WWWingsContext())
 {
  // Load flight
  var f = ctx.FlightSet.Find(FlightNr);

  Console.WriteLine($"Before changes: Flight #{f.FlightNo}:
  {f.Departure}->{f.Destination} has {f.FreeSeats} free seats! State of
  the flight object: " + ctx.Entry(f).State);
```

```
// Change object in RAM
f.FreeSeats -= 2;

Console.WriteLine($"After changes: Flight #{f.FlightNo}:
{f.Departure}->{f.Destination} has {f.FreeSeats} free seats! State
of the flight object: " + ctx.Entry(f).State);

// Persist changes
try
{
 var count = ctx.SaveChanges();
 if (count == 0)
 {
  Console.WriteLine("Problem: No changes saved!");
 }
 else
 {
  Console.WriteLine("Number of saved changes: " + count);
  Console.WriteLine($"After saving: Flight #{f.FlightNo}:
  {f.Departure}->{f.Destination} has {f.FreeSeats} free seats!
  State of the flight object: " + ctx.Entry(f).State);
 }
}
catch (Exception ex)
{
 Console.WriteLine("Error: " + ex.ToString());
}
}
}
```

Figure 10-1 shows how the state of the object changes from the perspective of Entity Framework Core. After loading, it is Unchanged. After the change is set to Modified, Entity Framework Core knows that the object has changed. After saving with SaveChanges(), it is again Unchanged. In other words, the state in RAM again corresponds to the state in the database.

When executing the SaveChanges() method, errors are likely (for example, dbConcurrencyException). Therefore, in Listing 10-1, there is the explicit try-catch to SaveChanges(). Another typical runtime error on SaveChanges() is when the .NET Framework allows unauthorized values to be written from the perspective of the database. This would take place, for example, with the Memo attribute if this column had a length limit in the database. Since the strings in .NET are basically unlimited in length, it would be possible to assign the Memo attribute a string that is too long from the point of view of the database.

---

**Note**    Unlike the classic Entity Framework, Entity Framework Core does not yet do any validation before saving with SaveChanges(). In other words, invalid values are first noticed by the database management system. In this case, you get the following runtime error: "Microsoft.EntityFrameworkCore.DbUpdateException: An error occurred while updating the entries. See the inner exception for details." The inner exception object then provides the actual source of error: "System. DataSqlClient.SqlException: String or binary data would be truncated."

---

```
ChangeFlightOneProperty
Before changes: Flight #101: Berlin->Seattle has 116 free seats! State of the flight object: Unchanged
After changes: Flight #101: Berlin->Seattle has 114 free seats! State of the flight object: Modified
Number of saved changes: 1
After saving: Flight #101: Berlin->Seattle has 114 free seats! Zustand des Flug-Objekts: Unchanged
```

*Figure 10-1.* Output of Listing 10-1

The following is the SQL command sent by Entity Framework Core on SaveChanges() in Listing 10-1:

```
exec sp_executesql N'SET NOCOUNT ON;
UPDATE [Flight] SET [FreeSeats] = @p0
WHERE [FlightNo] = @p1;
SELECT @@ROWCOUNT;
',N'@p1 int,@p0 smallint',@p1=101,@p0=114'
```

# Tracking Changes for Subobjects

Change tracking works in Entity Framework Core (as in its predecessor) for changed subobjects. Listing 10-2 loads the Flight object with its Pilot object. Changes in the Flight object, as well as in the connected Pilot object, are made (the Flight hours of the Pilot are increased). The state of the Pilot object changes analogously to the Flight object from Unchanged to Modified and after the execution of SaveChanges() again to Unchanged.

***Listing 10-2.***  Changes in Subobjects

```
public static void ChangeFlightAndPilot()
{
 CUI.MainHeadline(nameof(ChangeFlightAndPilot));

 int flightNo = 101;
 using (WWWingsContext ctx = new WWWingsContext())
 {

  var f = ctx.FlightSet.Include(x => x.Pilot).SingleOrDefault(x =>
  x.FlightNo == flightNo);

  Console.WriteLine($"After loading: Flight #{f.FlightNo}:
  {f.Departure}->{f.Destination} has {f.FreeSeats} free seats!\nState
  of the flight object: " + ctx.Entry(f).State + " / State of the
  Pilot object: " + ctx.Entry(f.Pilot).State);
  f.FreeSeats -= 2;
  f.Pilot.FlightHours = (f.Pilot.FlightHours ?? 0) + 10;
  f.Memo = $"Changed by User {System.Environment.UserName} on {DateTime.
  Now}.";

  Console.WriteLine($"After changes: Flight #{f.FlightNo}:
  {f.Departure}->{f.Destination} has {f.FreeSeats} free seats!\nState
  of the flight object: " + ctx.Entry(f).State + " / State of the Pilot
  object: " + ctx.Entry(f.Pilot).State);

  try
  {
   var count = ctx.SaveChanges();
```

```
    if (count == 0) Console.WriteLine("Problem: No changes saved!");
    else Console.WriteLine("Number of saved changes: " + count);
    Console.WriteLine($"After saving: Flight #{f.FlightNo}:
    {f.Departure}->{f.Destination} has {f.FreeSeats} free seats!\nState
    of the flight object: " + ctx.Entry(f).State + " / State of the Pilot
    object: " + ctx.Entry(f.Pilot).State);
   }
  catch (Exception ex)
  {
   Console.WriteLine("Error: " + ex.ToString());
  }
 }
}
```

The following is the SQL command sent by Entity Framework Core to
SaveChanges(); it shows that two UPDATE commands are sent to the database
management system. Figure 10-2 shows the output.

```
exec sp_executesql N'SET NOCOUNT ON;
UPDATE [Employee] SET [FlightHours] = @p0
WHERE [PersonID] = @p1;
SELECT @@ROWCOUNT;

UPDATE [Flight] SET [FreeSeats] = @p2, [Memo] = @p3
WHERE [FlightNo] = @p4;
SELECT @@ ROWCOUNT;

',N'@p1 int,@p0 int,@p4 int,@p2 smallint,@p3 nvarchar(4000)',@p1=57,
@p0=40,@p4=101,@p2=104,@p3=N'Changed by User HS on 23/12/2017 00:53:12.'
```

```
ChangeFlightAndPilot
After loading: Flight #101: Berlin->Seattle has 106 free seats!
State of the flight object: Unchanged / State of the Pilot object: Unchanged
After changes: Flight #101: Berlin->Seattle has 104 free seats!
State of the flight object: Modified / State of the Pilot object: Modified
Number of saved changes: 2
After saving: Flight #101: Berlin->Seattle has 104 free seats!
State of the flight object: Unchanged / State of the Pilot object: Unchanged
```

*Figure 10-2.* *Output of the previous code*

# Combining Commands (Batching)

In contrast to the classic Entity Framework, Entity Framework Core does not send every INSERT, UPDATE, or DELETE command in its own round-trip to the database management system; instead, it combines the commands into larger round-trips. This feature is called *batching*.

Entity Framework Core decides on the size of the summary of commands for a round-trip. In a test for the mass insertion of Flight data sets, 300 data sets were used for two round-trips; 1,000 were used for six round-trips; 2,000 were used for 11; and 5,000 were used for 27 round-trips of the database management system.

In addition to the Add() method, there is an AddRange() method on both the context class and the DbSet<EntityClass> class, to which you can pass a list of objects to attach. In the classic Entity Framework, AddRange() was much faster than Add() because it eliminated the need for repeated reviews of the Entity Framework Change Tracker. Entity Framework Core no longer has a performance difference between calling Add() in a loop 1,000 times or calling AddRange() once with a set of 1,000 objects as parameters. A clearly visible performance advantage in Figure 10-3 is created by batching. But if you always call SaveChanges() directly after Add(), no batching is possible (see the third bar in Figure 10-3).

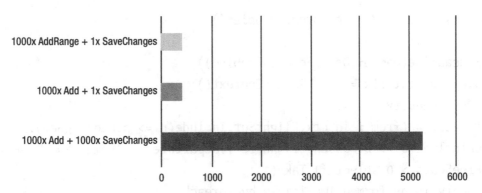

***Figure 10-3.*** *Power measurement during mass insertion of 1,000 records*

# Dealing with foreach Loop Considerations

With Entity Framework, it is not necessary to explicitly materialize a query before an iteration with a conversion operator such as ToList(). A foreach loop over an object with an IQueryable interface is enough to trigger the database query. In this case, however, the database connection remains open while the loop is running, and the records are fetched individually by the iterator of the IQueryable interface. This causes a call to SaveChanges() within a data-fetching foreach loop to result in a runtime error, as shown in Listing 10-3 and Figure 10-4.

There are three solutions, listed here:

- The best solution is to completely materialize the query with ToList() before starting the loop and to place SaveChanges() after the loop. This leads to the transmission of all changes in one or a few round-trips. But there is a transaction over all changes!

- If changes in multiple transactions are requested, then at least ToList() should be executed before the loop.

- Alternatively, you can use SaveChangesAsync() instead of SaveChanges(); see Chapter 13 for more information.

---

**Tip**    Use explicit materialization with ToList().

---

*Listing 10-3.* SaveChanges() Does Not Work Within a foreach Loop Unless You Have Previously Materialized the Records

```
public static void Demo_ForeachProblem()
{
 CUI.Headline(nameof(Demo_ForeachProblem));
 WWWingsContext ctx = new WWWingsContext();
 // Define query
 var query = (from f in ctx.FlightSet.Include(p => p.BookingSet).
 ThenInclude(b => b.Passenger) where f.Departure == "Rome" &&
 f.FreeSeats > 0 select f).Take(1);
 // Query is performed implicitly by foreach
 foreach (var Flight in query)
 {
 // Print results
 CUI.Print("Flight: " + Flight.FlightNo + " from " + Flight.Departure
 + " to " + Flight.Destination + " has " + Flight.FreeSeats + " free
 seats");
```

```
foreach (var p in Flight.BookingSet)
{
 CUI.Print(" Passenger   " + p.Passenger.GivenName + " " + p.Passenger.
 Surname);
}

// Save change to every flight object within the loop
CUI.Print("  Start saving");
Flight.FreeSeats--;
ctx.SaveChangesAsync(); // SaveChanges() will produce ERROR!!!
CUI.Print("  End saving");
}
}
```

***Figure 10-4.*** *Error from running Listing 10-3*

# Adding New Objects

To add a new record (in SQL, using INSERT), you perform the following steps with Entity
Framework Core:

1. Instantiate the object with the new operator (as usual in .NET).
   A Create() factory method, as in the classic Entity Framework,
   does not exist in Entity Framework Core.

2. Fill the object, in particular all mandatory properties, from the
   point of view of the database schema.

3. Append the object to the context either through the Add() method
   in the context class or in the appropriate DbSet<EntityClass> in
   the context class.

4. Call SaveChanges().

Listing 10-4 shows how to create a `Flight` object. Compulsory requirements are the flight number (the primary key is not an auto-increment value here and therefore must be set manually), airline, departure, destination, and date as well as the relationship with a `Pilot` object. The copilot is optional. Even if the airline is a mandatory field, the program code would work without explicitly assigning an enumeration value because Entity Framework Core would use the default value 0 here, which is a valid value for the database.

As an alternative to previously loading the `Pilot` object and then assigning it to the `Flight` object, you could implement the task more efficiently by using the foreign key property `PilotId` in the `Flight` object and directly assigning the primary key of the `Pilot` object there: `f.PilotId = 234`. Here you will see the advantages of an explicit foreign key property, which results in saving a round-trip of the database.

***Listing 10-4.*** Creating a New Flight

```
public static void AddFlight()
{
 CUI.MainHeadline(nameof(AddFlight));

 using (WWWingsContext ctx = new WWWingsContext())
 {
  // Create flight in RAM
  var f = new Flight();
  f.FlightNo = 123456;
  f.Departure = "Essen";
  f.Destination = "Sydney";
  f.AirlineCode = "WWW";
  f.PilotId = ctx.PilotSet.FirstOrDefault().PersonID;
  f.Seats = 100;
  f.FreeSeats = 100;

  Console.WriteLine($"Before adding: Flight #{f.FlightNo}:
  {f.Departure}->{f.Destination} has {f.FreeSeats} free seats! State
  of the Flight object: " + ctx.Entry(f).State);

  // Add flight to context
  ctx.FlightSet.Add(f);
  // or: ctx.Add(f);
```

```
Console.WriteLine($"After adding: Flight #{f.FlightNo}:
{f.Departure}->{f.Destination} has {f.FreeSeats} free seats! State
of the Flight object: " + ctx.Entry(f).State);

try
{
 var count = ctx.SaveChanges();
 if (count == 0) Console.WriteLine("Problem: No changes saved!");
 else Console.WriteLine("Number of saved changes: " + count);
 Console.WriteLine($"After saving: Flight #{f.FlightNo}:
 {f.Departure}->{f.Destination} has {f.FreeSeats} free seats! State
 of the Flight object: " + ctx.Entry(f).State);
}
catch (Exception ex)
{
 Console.WriteLine("Error: " + ex.ToString());
}

}
}
```

The sequence of object states is as follows (see Figure 10-5): Detached (before executing Add(), the Entity Framework core context does not know about the new instance of the Flight and therefore considers it a transient object), Added (after the add()), and then after saving Unchanged. Incidentally, Entity Framework Core does not consider the multiple calls of Add() to be an error, but Add() does not need to be called more than once.

---

**Note**    You can add an object with a primary key value that does not yet exist only in the current context instance. If you want to delete an object and then create a new object under the same primary key value, you have to execute SaveChanges() after Remove() and before Add(); otherwise, Entity Framework Core complains with the following error message: "System. InvalidOperationException: The instance of entity type 'Flight' can not be tracked because another instance with the same key value for {'FlightNo'} is already being tracked."

---

```
AddFlight
Before adding: Flight #123456: Essen->Sydney has 100 free seats! State of the Flight object: Detached
After adding: Flight #123456: Essen->Sydney has 100 free seats! State of the Flight object: Added
Number of saved changes: 1
After saving: Flight #123456: Essen->Sydney has 100 free seats! State of the Flight object: Unchanged
```

***Figure 10-5.*** *Output from Listing 10-4 (creating a Flight object)*

# Creating Related Objects

The Add() method not only considers the object passed as a parameter but also considers objects associated with that object. If there are objects under the object passed as a parameter in the state Detached, they are automatically added to the context and are then in the state Added.

The manipulation of relationship in a 1:N situation is done on the 1 side by manipulating the list with the list-specific methods for adding and deleting, mostly with the methods Add() and Remove(). For bidirectional relationships, the changes can be made either on the 1 side or on the N side. In the specific case of the bidirectional relationship between Flight and Pilot, there are three equivalent ways to establish the relationship, listed here:

- Using the navigation property of the 1 side, so Pilot in Flight:

- flight.Pilot = Pilot;

- Using the foreign key property PersonID in Flight:

- flight.PersonID = 123;

- Using the navigation property of the N side, in other words, FlightAsPilotSet on the Pilot page:

- pilot.FlightAsPilotSet.Add(flight);

Listing 10-5 shows how to create a new Flight with a new Pilot object, a new AircraftType object, and a new AircraftTypeDetail object. It is sufficient to execute Add() for the Flight object. Entity Framework Core then sends SaveChanges() five INSERT commands to the database, one for each of the tables: AircraftType, AircraftTypeDetail, Employees (a common table for instances of the Employee and Pilot classes), Persondetail, and Flight.

***Listing 10-5.*** Creation of a New Pilot's Flight with Persondetail, New AircraftType, and AircraftTypeDetail

```
public static void Demo_CreateRelatedObjects()
{
 CUI.MainHeadline(nameof(Demo_CreateRelatedObjects));
 using (var ctx = new WWWingsContext())
 {
  ctx.Database.ExecuteSqlCommand("Delete from Booking where
  FlightNo = 456789");
  ctx.Database.ExecuteSqlCommand("Delete from Flight where
  FlightNo = 456789");

  var p = new Pilot();
  p.GivenName = "Holger";
  p.Surname = "Schwichtenberg";
  p.HireDate = DateTime.Now;
  p.LicenseDate = DateTime.Now;
  var pd = new Persondetail();
  pd.City = "Essen";
  pd.Country = "DE";
  p.Detail = pd;

  var act = new AircraftType();
  act.TypeID = (byte)(ctx.AircraftTypeSet.Max(x=>x.TypeID)+1);
  act.Manufacturer = "Airbus";
  act.Name = "A380-800";
  ctx.AircraftTypeSet.Add(act);
  ctx.SaveChanges();

  var actd = new AircraftTypeDetail();
  actd.TurbineCount = 4;
  actd.Length = 72.30f;
```

```
    actd.Tare = 275;
    act.Detail = actd;

    var f = new Flight();
    f.FlightNo = 456789;
    f.Pilot = p;
    f.Copilot = null;
    f.Seats = 850;
    f.FreeSeats = 850;
    f.AircraftType = act;

    // One Add() is enough for all related objects!
    ctx.FlightSet.Add(f);
    ctx.SaveChanges();

    CUI.Print("Total number of flights: " + ctx.FlightSet.Count());
    CUI.Print("Total number of pilots: " + ctx.PilotSet.Count());
  }
}
```

Figure 10-6 shows the output of these five INSERT commands.

*Figure 10-6.* *Batch updating makes only three round-trips for five INSERT commands*

# Changing Linked Objects

Entity Framework Core also detects relationship changes between entity objects and automatically saves them in SaveChanges(). As with the initial creation of a Flight object with Pilot, there are three options for a relationship change in the ideal case (bidirectional relationships with foreign key properties):

- Using the navigation property Pilot in Flight: Flight. Pilot = Pilot;

- Using the foreign key property PilotId in Flight: Flight. PersonID = 123;

- Using the navigation property FlightAsPilotSet on the Pilot page: Pilot.FlightAsPilotSet.Add (Flight);

In all three cases, Entity Framework Core sends SaveChanges() to the database. By executing SaveChanges(), Entity Framework Core correctly does not make any changes to the database in the Pilot table but in the Flight table, because the Flight table has the foreign key that establishes the relationship between Pilot and Flight.

```
exec sp_executesql N'SET NOCOUNT ON;
UPDATE [Flight] SET [PilotId] = @p0
WHERE [FlightNo] = @p1;
SELECT @@ ROWCOUNT;
', N' @p1 int, @p0 int ', @p1 = 101, @p0 = 123
```

---

**Tip**   To remove a relationship, you can simply assign zero or nothing.

---

Listing 10-6 shows the assignment of a new Pilot to a Flight. However, this assignment does not take place via Flight101.Pilot = newPilot (the 1 side) but on the Pilot side (the N side) via newPilot.FlightAsPilotSet.Add(flight101). The output of this listing is exciting; see Figure 10-7. You can see that at the beginning, one pilot has 31 flights, and the other has 10 flights. After the assignment, the new Pilot has 11 flights, and the old Pilot still has 31 flights, which is wrong. In addition, flight101.Pilot still refers to the old Pilot, which is also wrong.

However, after running SaveChanges(), the object relations have been corrected. Now the old Pilot has 30 flights only. In addition, flight101.Pilot refers to the new Pilot. This feature of Entity Framework Core is called *relationship fixup.* As part of a relationship fixup operation, Entity Framework Core checks all relations between the objects that are currently in RAM and also changes them on the opposite side if changes have occurred in this side. Entity Framework Core runs the relationship fixup operation when saving with SaveChanges().

You can force the relationship fixup operation at any time with the execution of the method ctx.ChangeTracker.DetectChanges(). If many objects have been loaded into a context instance, DetectChanges() can take many milliseconds. Therefore, in Entity Framework Core, Microsoft does not automatically call DetectChanges() in many places, leaving it up to you to decide when you need a consistent state of the object relationships and to use DetectChanges(). Figure 10-7 shows the output.

***Listing 10-6.*** Making a 1:N Relationship Across the First Page

```
public static void Demo_RelationhipFixup1N()
{
 CUI.MainHeadline(nameof(Demo_RelationhipFixup1N));
 using (var ctx = new WWWingsContext())
 {
  // Load a flight
  var flight101 = ctx.FlightSet.SingleOrDefault(x => x.FlightNo == 101);
  Console.WriteLine($"Flight Nr {flight101.FlightNo} from {flight101.
  Departure} to {flight101.Destination} has {flight101.FreeSeats}
  free seats!");

  // Load the pilot for this flight with the list of his flights
  var oldPilot = ctx.PilotSet.Include(x => x.FlightAsPilotSet).
  SingleOrDefault(x => x.PersonID == flight101.PilotId);
  Console.WriteLine("Pilot: " + oldPilot.PersonID + ": " + oldPilot.
  GivenName + " " + oldPilot.Surname + " has " + oldPilot.
  FlightAsPilotSet.Count + " flights as pilot!");

  // Next pilot in the list load with the list of his flights
  var newPilot = ctx.PilotSet.Include(x => x.FlightAsPilotSet).
  SingleOrDefault(x => x.PersonID == flight101.PilotId + 1);
```

```
Console.WriteLine("Planned Pilot: " + newPilot.PersonID + ": " +
newPilot.GivenName + " " + newPilot.Surname + " has " + newPilot.
FlightAsPilotSet.Count + " flights as pilot!");

// Assign to Flight
CUI.Print("Assignment of the flight to the planned pilot...",
ConsoleColor.Cyan);
newPilot.FlightAsPilotSet.Add(flight101);

// optional:force Relationship Fixup
// ctx.ChangeTracker.DetectChanges();

CUI.Print("Output before saving: ", ConsoleColor.Cyan);
Console.WriteLine("Old pilot: " + oldPilot.PersonID + ": " +
oldPilot.GivenName + " " + oldPilot.Surname + " has " + oldPilot.
FlightAsPilotSet.Count + " flights as a pilot!");
Console.WriteLine("New pilot: " + newPilot.PersonID + ": " +
newPilot.GivenName + " " + newPilot.Surname + " has " + newPilot.
FlightAsPilotSet.Count + " flights as a pilot!");
var pilotAktuell = flight101.Pilot; // Current Pilot in the Flight
object
Console.WriteLine("Pilot for flight " + flight101.FlightNo + " is
currently: " + pilotAktuell.PersonID + ": " + pilotAktuell.GivenName +
" " + pilotAktuell.Surname);

// SaveChanges()()
CUI.Print("Saving... ", ConsoleColor.Cyan);
var count = ctx.SaveChanges();
CUI.MainHeadline("Number of saved changes: " + count);

CUI.Print("Output after saving: ", ConsoleColor.Cyan);
Console.WriteLine("Old Pilot: " + oldPilot.PersonID + ": " +
oldPilot.GivenName + " " + pilotAlt.Surname + " has " + pilotAlt.
FlightAsPilotSet.Count + " flights as pilot!");
Console.WriteLine("New Pilot: " + newPilot.PersonID + ": " +
newPilot.GivenName + " " + newPilot.Surname + " has " + newPilot.
FlightAsPilotSet.Count + " flights as pilot!");
```

```
pilotAktuell = flight101.Pilot; // Current pilot from the perspective
of the Flight object
Console.WriteLine("Pilot for Flight " + flight101.FlightNo + " is now:
" + pilotAktuell.PersonID + ": " + pilotAktuell.GivenName + " " +
pilotAktuell.Surname);
 }
}
```

```
Demo_RelationhipFixup1N
Flight Nr 101 from Berlin to Seattle has 102 free seats!
Pilot: 57: Sahra Merkel has 31 flights as pilot!
Planned Pilot: 58: Cem Gysi has 10 flights as pilot!
Assignment of the flight to the planned pilot...
Control output before saving:
Old pilot: 57: Sahra Merkel has 31 flights as a pilot!
New pilot: 58: Cem Gysi has 11 flights as a pilot!
Pilot for flight 101 is currently: 57: Sahra Merkel
Saving...
Number of saved changes: 1
Control output after save:
Old Pilot: 57: Sahra Merkel has 30 flights as pilot!
New Pilot: 58: Cem Gysi has 11 flights as pilot!
Pilot for Flight 101 is now: 58: Cem Gysi
```

*Figure 10-7.  Output of Listing 10-6 (the relationship fixup works)*

# Dealing with Contradictory Relationships

If, as explained earlier, there are up to three ways to establish a relationship between objects, what happens when multiple options are used in parallel with contradictory data?

Listing 10-7, in connection with the output in Figure 10-8 and Figure 10-9, shows that the priority is as follows:

- The highest priority is the value of the object set on the 1 side, so in the case of the relation Pilot<->Flight (1:N), the value from Pilot. FlightAsPilotSet is first.

- The second highest priority is the value from the single object on the N side. In other words, in the case of Pilot<->Flight (1:N), it's the value from Flight.Pilot.

- Only then is the foreign key property taken into account on the N side. In other words, in the case of Pilot: Flight (1:N), it's the value from Flight.PersonID.

***Listing 10-7.*** Four Test Scenarios for the Question of Which Value Has Priority
When the Relationship Is Contradictory

```
using System;
using System.Linq;
using DA;
using EFC_Console;
using ITVisions;
using Microsoft.EntityFrameworkCore;
namespace EFC_Console
{
 class ContradictoryRelationships
 {
  /// <summary>
  /// Four test scenarios for the question of which value has priority, if
  the relationship is set contradictory
  /// </summary>
  [EFCBook()]
  public static void Demo_ContradictoryRelationships()
  {
   CUI.MainHeadline(nameof(Demo_ContradictoryRelationships));
   Attempt1();
   Attempt2();
   Attempt3();
   Attempt4();
  }

  public static int pilotID = new WWWingsContext().PilotSet.Min(x =>
  x.PersonID);

  public static int GetPilotIdEinesFreienPilots()
  {
   // here we assume that the next one in the list has time for this flight :-)
   pilotID++; return pilotID;
  }
```

```csharp
private static void Attempt1()
{
 using (var ctx = new WWWingsContext())
 {
  CUI.MainHeadline("Attempt 1: first assignment by navigation property,
  then by foreign key property");
  CUI.PrintStep("Load a flight...");
  var flight101 = ctx.FlightSet.Include(f => f.Pilot).SingleOrDefault(x
  => x.FlightNo == 101);
  Console.WriteLine($"Flight Nr {flight101.FlightNo} from {flight101.
  Departure} to {flight101.Destination} has {flight101.FreeSeats} free
  seats!");
  CUI.Print("Pilot object: " + flight101.Pilot.PersonID + " PilotId: " +
  flight101.PilotId);

  CUI.PrintStep("Load another pilot...");
  var newPilot2 = ctx.PilotSet.Find(GetPilotIdEinesFreienPilots());
  // next Pilot
  CUI.PrintStep($"Assign a new pilot #{newPilot2.PersonID} via navigation
  property...");
  flight101.Pilot = newPilot2;
  CUI.Print($"PilotId: {flight101.PilotId} Pilot object: {flight101.
  Pilot?.PersonID}");

  CUI.PrintStep("Reassign a new pilot via foreign key property...");
  var neuePilotID = GetPilotIdEinesFreienPilots();
  CUI.PrintStep($"Assign a new pilot #{neuePilotID} via foreign key
  property...");
  flight101.PilotId = neuePilotID;
  CUI.Print($"PilotId: {flight101.PilotId} Pilot object: {flight101.
  Pilot?.PersonID}");

  CUI.PrintStep("SaveChanges()");
  var anz2 = ctx.SaveChanges();
  CUI.PrintSuccess("Number of saved changes: " + anz2);
```

```
  CUI.PrintStep("Control output after saving: ");
  CUI.Print($"PilotId: {flight101.PilotId} Pilot object: {flight101.
  Pilot?.PersonID}");
 }
}

private static void Attempt2()
{
 using (var ctx = new WWWingsContext())
 {
  CUI.MainHeadline("Attempt 2: First assignment by foreign key property,
  then navigation property");
  CUI.PrintStep("Load a flight...");
  var flight101 = ctx.FlightSet.Include(f => f.Pilot).SingleOrDefault
  (x => x.FlightNo == 101);
  Console.WriteLine($"Flight Nr {flight101.FlightNo} from {flight101.
  Departure} to {flight101.Destination} has {flight101.FreeSeats} free
  seats!");
  CUI.Print("Pilot object: " + flight101.Pilot.PersonID + " PilotId: " +
  flight101.PilotId);

  var neuePilotID2 = GetPilotIdEinesFreienPilots();
  CUI.PrintStep($"Assign a new pilot #{neuePilotID2} via foreign key
  property...");
  flight101.PilotId = neuePilotID2;
  CUI.Print($"PilotId: {flight101.PilotId} Pilot object: {flight101.
  Pilot?.PersonID}");

  CUI.PrintStep("Load another pilot...");
  var newPilot1 = ctx.PilotSet.Find(GetPilotIdEinesFreienPilots());
  // next Pilot
  CUI.PrintStep($"Assign a new pilot #{newPilot1.PersonID} via navigation
  property...");
  flight101.Pilot = newPilot1;
  CUI.Print($"PilotId: {flight101.PilotId} Pilot object: {flight101.
  Pilot?.PersonID}");
```

```csharp
  CUI.PrintStep("SaveChanges()");
  var anz2 = ctx.SaveChanges();
  CUI.PrintSuccess("Number of saved changes: " + anz2);

  CUI.PrintStep("Control output after saving: ");
  CUI.Print($"PilotId: {flight101.PilotId} Pilot object: {flight101.
  Pilot?.PersonID}");
 }
}

private static void Attempt3()
{
 using (var ctx = new WWWingsContext())
 {
  CUI.MainHeadline("Attempt 3: Assignment using FK, then Navigation
  Property at Flight, then Navigation Property at Pilot");
  CUI.PrintStep("Load a flight...");
  var flight101 = ctx.FlightSet.Include(f => f.Pilot).SingleOrDefault
  (x => x.FlightNo == 101);
  Console.WriteLine($"Flight No {flight101.FlightNo} from {flight101.
  Departure} to {flight101.Destination} has {flight101.FreeSeats} free
  seats!");
  CUI.Print("Pilot object: " + flight101.Pilot.PersonID + " PilotId:
  " + flight101.PilotId);

  var neuePilotID3 = GetPilotIdEinesFreienPilots();
  CUI.PrintStep($"Assign a new pilot #{neuePilotID3} via foreign key
  property...");
  flight101.PilotId = neuePilotID3;
  CUI.Print("flight101.PilotId=" + flight101.PilotId);
  CUI.Print($"PilotId: {flight101.PilotId} Pilot object: {flight101.
  Pilot?.PersonID}");

  CUI.PrintStep("Load another pilot...");
  var newPilot3a = ctx.PilotSet.Find(GetPilotIdEinesFreienPilots());
  // next Pilot
```

```
CUI.PrintStep($"Assign a new pilot #{newPilot3a.PersonID} via
navigation property in Flight object...");
flight101.Pilot = newPilot3a;
CUI.Print($"PilotId: {flight101.PilotId} Pilot object: {flight101.
Pilot?.PersonID}");

CUI.PrintStep("Load another Pilot...");
var newPilot3b = ctx.PilotSet.Include(p => p.FlightAsPilotSet).SingleOr
Default(p => p.PersonID == GetPilotIdEinesFreienPilots()); // next Pilot
CUI.PrintStep($"Assign a new pilot #{newPilot3b.PersonID} via
navigation property in Pilot object...");
newPilot3b.FlightAsPilotSet.Add(flight101);
CUI.Print($"PilotId: {flight101.PilotId} Pilot object: {flight101.
Pilot?.PersonID}");

CUI.PrintStep("SaveChanges()");
var anz3 = ctx.SaveChanges();
CUI.PrintSuccess("Number of saved changes: " + anz3);

CUI.PrintStep("Control output after saving: ");
CUI.Print($"PilotId: {flight101.PilotId} Pilot object: {flight101.
Pilot?.PersonID}");
 }
}

private static void Attempt4()
{
 using (var ctx = new WWWingsContext())
 {
  CUI.MainHeadline("Attempt 4: First assignment by FK, then Navigation
  Property at Pilot, then Navigation Property at Flight");
  CUI.PrintStep("Load a flight...");
  var flight101 = ctx.FlightSet.Include(f => f.Pilot).SingleOrDefault
  (x => x.FlightNo == 101);
```

```
Console.WriteLine($"Flight Nr {flight101.FlightNo} from {flight101.
Departure} to {flight101.Destination} has {flight101.FreeSeats} free
seats!");
CUI.Print("Pilot object: " + flight101.Pilot.PersonID + " PilotId:
" + flight101.PilotId);

var neuePilotID4 = GetPilotIdEinesFreienPilots();
CUI.PrintStep($"Assign a new pilot #{neuePilotID4} via foreign key
property...");
flight101.PilotId = neuePilotID4;
CUI.Print("flight101.PilotId=" + flight101.PilotId);
CUI.Print($"PilotId: {flight101.PilotId} Pilot object: {flight101.
Pilot?.PersonID}");

CUI.PrintStep("Load another pilot...");
var newPilot4b = Queryable.SingleOrDefault(ctx.PilotSet.Include(p =>
p.FlightAsPilotSet), p => p.PersonID == GetPilotIdEinesFreienPilots());
// next Pilot
CUI.PrintStep($"Assign a new pilot #{newPilot4b.PersonID} via
navigation property...");
newPilot4b.FlightAsPilotSet.Add(flight101);
CUI.Print($"PilotId: {flight101.PilotId} Pilot object: {flight101.
Pilot?.PersonID}");

CUI.PrintStep("Load another Pilot...");
var newPilot4a = ctx.PilotSet.Find(GetPilotIdEinesFreienPilots());
// next Pilot
CUI.PrintStep($"Assign a new pilot #{newPilot4a.PersonID} via
navigation property in Flight object...");
flight101.Pilot = newPilot4a;
CUI.Print($"PilotId: {flight101.PilotId} Pilot object: {flight101.
Pilot?.PersonID}");

CUI.PrintStep("SaveChanges()");
var anz4 = ctx.SaveChanges();
CUI.PrintSuccess("Number of saved changes: " + anz4);
```

```
    CUI.PrintStep("Control output after saving: ");
    CUI.Print($"PilotId: {flight101.PilotId} Pilot object: {flight101.
    Pilot?.PersonID}");
    }
   }
  }
}
```

```
Demo_ContradictoryRelationships
Attempt 1: first assignment by navigation property, then by foreign key property
Load a flight...
Flight Nr 101 from Berlin to Seattle has 102 free seats!
Pilot object: 60 PilotId: 60
Load annother pilot...
Assign a new pilot #52 via navigation property...
PilotId: 60 Pilot object: 52
Reassign a new pilot via foreign key property...
Assign a new pilot #53 via foreign key property...
PilotId: 53 Pilot object: 52
SaveChanges()
Number of saved changes: 1
Control output after saving:
PilotId: 52 Pilot object: 52
Attempt 2: First assignment by foreign key property, then navigation property
Load a flight...
Flight Nr 101 from Berlin to Seattle has 102 free seats!
Pilot object: 52 PilotId: 52
Assign a new pilot #54 via foreign key property...
PilotId: 54 Pilot object: 52
Load annother pilot...
Assign a new pilot #55 via navigation property...
PilotId: 54 Pilot object: 55
SaveChanges()
Number of saved changes: 1
Control output after saving:
PilotId: 55 Pilot object: 55
```

*Figure 10-8.* *Output of Listing 10-7 (part 1)*

```
Attempt 3: Assigment using FK, then Navigation Property at Flight, then Navigation Property at Pilot
Load a flight...
Flight No 101 from Berlin to Seattle has 102 free seats!
Pilot object: 55 PilotId: 55
Assign a new pilot #56 via foreign key property...
flight101.PilotId=56
PilotId: 56 Pilot object: 55
Load annother pilot...
Assign a new pilot #57 via navigation property bei Flight...
PilotId: 56 Pilot object: 57
Load annother Pilot...  ▌
Assign a new pilot #58 via navigation property bei Pilot...
PilotId: 56 Pilot object: 57
SaveChanges()
Number of saved changes: 1
Control output after saving:
PilotId: 58 Pilot object: 58
Attempt 4: First assignment by FK, then Navigation Property at Pilot, then Navigation Property at Flight
Load a flight...
Flight Nr 101 from Berlin to Seattle has 102 free seats!
Pilot object: 58 PilotId: 58
Assign a new pilot #59 via foreign key property...
flight101.PilotId=59
PilotId: 59 Pilot object: 58
Load annother pilot...
Assign a new pilot #60 via navigation property...
PilotId: 59 Pilot object: 58
Lade noch einen anderen Piloten...
Assign a new pilot #61 via navigation property bei Flight...
PilotId: 59 Pilot object: 61
SaveChanges()
Number of saved changes: 1
Control output after saving:
PilotId: 60 Pilot object: 60
```

*Figure 10-9.* *Output of Listing 10-7 (part 2)*

# Deleting Objects

This section covers different ways of deleting objects and the corresponding rows in the database.

## Deleting Objects with Remove()

To delete an object, you must call the Remove() method, which, like Add(), exists either directly on the context class (inherited from DbContext) or on the DbSet<EntityClass> property in the context class (see Listing 10-8). Calling Remove() causes the loaded Flight object to change from the Unchanged state to the Delete state (see Figure 10-10). However, it has not yet been deleted in the database. Only by calling the SaveChanges() method is a DELETE command sent to the database management system.

***Listing 10-8.***  Deleting a Flight Record

```
public static void RemoveFlight()
{
CUI.MainHeadline(nameof(RemoveFlight));

 using (WWWingsContext ctx = new WWWingsContext())
 {
  var f = ctx.FlightSet.SingleOrDefault(x => x.FlightNo == 123456);
  if (f == null) return;

  Console.WriteLine($"After loading: Flight #{f.FlightNo}:
  {f.Departure}->{f.Destination} has {f.FreeSeats} free seats! State of
  the flight object: " + ctx.Entry(f).State);

  // Remove flight
  ctx.FlightSet.Remove(f);
  // or: ctx.Remove(f);

  Console.WriteLine($"After deleting: Flight #{f.FlightNo}:
  {f.Departure}->{f.Destination} has {f.FreeSeats} free seats! State of
  the flight object: " + ctx.Entry(f).State);

  try
  {
   var count = ctx.SaveChanges();
   if (count == 0) Console.WriteLine("Problem: No changes saved!");
   else Console.WriteLine("Number of saved changes: " + count);
   Console.WriteLine($"After saving: Flight #{f.FlightNo}:
   {f.Departure}->{f.Destination} has {f.FreeSeats} free seats! State
   of the flight object: " + ctx.Entry(f).State);
  }
```

```
  catch (Exception ex)
  {
   Console.WriteLine("Error: " + ex.ToString());
  }
 }
}
```

```
RemoveFlight
After loading: Flight #123456: Essen->Sydney has 100 free seats! State of the flight object: Unchanged
After deleting: Flight #123456: Essen->Sydney has 100 free seats! State of the flight object: Deleted
Number of saved changes: 1
After saving: Flight #123456: Essen->Sydney has 100 free seats! State of the flight object: Detached
```

***Figure 10-10.*** *Output of Listing 10-8*

Here are the SQL commands issued by Entity Framework Core in Listing 10-8:

```
SELECT TOP(2) [x].[FlightNo], [x].[AircraftTypeID], [x].[AirlineCode], [x].
[CopilotId], [x].[FlightDate], [x].[Departure], [x].[Destination], [x].
[FreeSeats], [x].[LastChange], [x].[Memo], [x].[NonSmokingFlight], [x].
[PilotId], [x].[Price], [x].[Seats], [x].[Strikebound], [x].[Utilization]
FROM [Flight] AS [x]
WHERE [x].[FlightNo] = 123456

exec sp_executesql N'SET NOCOUNT ON;
DELETE FROM [Flight]
WHERE [FlightNo] = @p0;
SELECT @@ROWCOUNT;
',N'@p0 int',@p0=123456
```

# Deleting Objects with a Dummy Object

In the previous code, it is inefficient to load the Flight object completely; you send only a delete command. Listing 10-9 shows a solution that avoids this round-trip to the database management system by creating a Flight object in RAM where only the primary key is set to the object to be deleted. You then attach this dummy object to the context with Attach(). This causes the object to change state from Detached to Unchanged. Finally, you execute Remove() and SaveChanges(). The trick works because Entity Framework needs to know only the primary key for deletion.

Note the following about this trick:

- The method `Attach()`, not `Add()`, is called here; otherwise, Entity Framework Core would consider the dummy object as a new object.

- The trick works only if no conflict checks have been configured in Entity Framework Core. If, however, the model is set to compare the values of other columns when saving, these must be filled with the current values in the dummy object. Otherwise, the object cannot be deleted, and a `DbConcurrenyException` occurs.

***Listing 10-9.*** Deleting a Flight Record More Efficiently with a Dummy Object

```
public static void RemoveFlightWithKey()
{
 Console.WriteLine(nameof(RemoveFlightWithKey));

 using (WWWingsContext ctx = new WWWingsContext())
 {
  // Create a dummy object
  var f = new Flight();
  f.FlightNo = 123456;

  Console.WriteLine($"After creation: Flight #{f.FlightNo}:
{f.Departure}->{f.Destination} has {f.FreeSeats} free seats! State of
the flight object: " + ctx.Entry(f).State);

  // Append dummy object to context
  ctx.Attach(f);

  Console.WriteLine($"After attach: Flight #{f.FlightNo}:
{f.Departure}->{f.Destination} has {f.FreeSeats} free seats! State of
the flight object: " + ctx.Entry(f).State);

  // Delete flight
  ctx.FlightSet.Remove(f);
  // or: ctx.Remove(f);

  Console.WriteLine($"After remove: Flight #{f.FlightNo}:
{f.Departure}->{f.Destination} has {f.FreeSeats} free seats! State of
the flight object: " + ctx.Entry(f).State);
```

```
try
{
 var count = ctx.SaveChanges();
 if (count == 0) Console.WriteLine("Problem: No changes saved!");
 else Console.WriteLine("Number of saved changes: " + count);
 Console.WriteLine($"After saving: Flight #{f.FlightNo}:
 {f.Departure}->{f.Destination} has {f.FreeSeats} free seats! State
 of the flight object: " + ctx.Entry(f).State);
}
catch (Exception ex)
{
 Console.WriteLine("Error: " + ex.ToString());
}
}
}
```

## Bulk Deleting

The Remove() method is not suitable for a mass deletion as defined in Delete from Flight where FlightNo> 10000 because Entity Framework Core will generate a DELETE command per object in each case. Entity Framework Core does not recognize that many DELETE commands can be combined into one command. In this case, you should always rely on classic techniques (SQL or stored procedures) since using Remove() would be many times slower here. Another option is the extension EFPlus (see Chapter 20).

# Performing Database Transactions

Note these important points about database transactions:

- When you run SaveChanges(), Entity Framework Core always automatically makes a transaction, meaning that all changes made in the context are persisted or none of them.

- If you need a transaction across multiple calls to the SaveChanges() method, you must do so with ctx.Database.BeginTransaction(), Commit(), and Rollback().

- System.Transactions.Transactions.TransactionScope is not supported yet in Entity Framework Core. It will be supported in version 2.1 of Entity Framework Core; see Appendix C.

---

**Tip**    The best transaction is a transaction that you avoid. Transactions always negatively impact the performance, scalability, and robustness of an application.

---

# Example 1

The following example shows a transaction with two changes to a Flight object that are each persisted independently with SaveChanges():

```
public static void ExplicitTransactionTwoSaveChanges()
{

Console.WriteLine(nameof(ExplicitTransactionTwoSaveChanges));
using (var ctx = new WWWingsContext())
{
// Start transaction. Default is System.Data.IsolationLevel.
ReadCommitted
using (var t = ctx.Database.BeginTransaction(System.Data.
IsolationLevel.ReadCommitted))
{
// Print isolation level
RelationalTransaction rt = t as RelationalTransaction;
DbTransaction dbt = rt.GetDbTransaction();
Console.WriteLine("Transaction with Level: " + dbt.IsolationLevel);

// Read data
int FlightNr = ctx.FlightSet.OrderBy(x => x.FlightNo).
FirstOrDefault().FlightNo;
var f = ctx.FlightSet.Where(x => x.FlightNo == FlightNr).
SingleOrDefault();

Console.WriteLine("Before: " + f.ToString());
```

```
// Change data and save
f.FreeSeats--;
var count1 = ctx.SaveChanges();
Console.WriteLine("Number of saved changes: " + count1);

//  Change data again and save
f.Memo = "last changed at " + DateTime.Now.ToString();
var count2 = ctx.SaveChanges();
Console.WriteLine("Number of saved changes: " + count2);

Console.WriteLine("Commit or Rollback? 1 = Commit, other = Rollback");
var input = Console.ReadKey().Key;
if (input == ConsoleKey.D1)
{ t.Commit(); Console.WriteLine("Commit done!"); }
else
{ t.Rollback(); Console.WriteLine("Rollback done!"); }

Console.WriteLine("After in RAM: " + f.ToString());
ctx.Entry(f).Reload();
Console.WriteLine("After in DB: " + f.ToString());
 }
 }
}
```

# Example 2

The following example shows a transaction with a change to the table Booking (insert a new booking) and to the table Flight (Reduce Number of Free Picks). Here the transaction takes place via two different context instances of a context class. It would also be possible to have a transaction via two different context classes if they refer to the same database.

Note the following:

- The database connection is created and opened separately.

- The transaction is opened on this connection.

- The context instances do not open their own connections but use the open connection. For this purpose, the database connection object is passed into the constructor of the context class, where it is kept in mind. In OnConfiguring(), this database connection object must be used with UseSqlServer() or similar instead of passing the connection string as a parameter!

- After instantiating the transaction object has to be passed to ctx.Database.UseTransaction().

---

**Note**    Failure to open the connection in advance and passing it to the context instances involved will result in the following runtime error: "The specified transaction is not associated with the current connection. Only transactions associated with the current connection may be used."

---

Figure 10-11 shows the output.

```
public static void ExplicitTransactionTwoContextInstances()
{
CUI.MainHeadline(nameof(ExplicitTransactionTwoContextInstances));

  // Open shared connection
  using (var connection = new SqlConnection(Program.CONNSTRING))
  {
   connection.Open();
   // Start transaction. Default is System.Data.IsolationLevel.
   ReadCommitted
   using (var t = connection.BeginTransaction(System.Data.IsolationLevel.
   ReadCommitted))
   {
    // Print isolation level
    Console.WriteLine("Transaction with Level: " + t.IsolationLevel);
    int flightNo;

    using (var ctx = new WWWingsContext(connection))
    {
     ctx.Database.UseTransaction(t);
     var all = ctx.FlightSet.ToList();
```

```
    var flight = ctx.FlightSet.Find(111);
    flightNo = flight.FlightNo;
    ctx.Database.ExecuteSqlCommand("Delete from booking where flightno= "
    + flightNo);
    var pasID = ctx.PassengerSet.FirstOrDefault().PersonID;

    // Create and persist booking
    var b = new BO.Booking();
    b.FlightNo = flightNo;
    b.PassengerID = pasID;
    ctx.BookingSet.Add(b);
    var count1 = ctx.SaveChanges();
    Console.WriteLine("Numer of bookings saved: " + count1);
}

using (var ctx = new WWWingsContext(connection))
{
  ctx.Database.UseTransaction(t);

  // Change free seats and save
  var f = ctx.FlightSet.Find(flightNo);
  Console.WriteLine("BEFORE: " + f.ToString());
  f.FreeSeats--;
  f.Memo = "last changed at " + DateTime.Now.ToString();
  Console.WriteLine("AFTER: " + f.ToString());
  var count2 = ctx.SaveChanges();
  Console.WriteLine("Number of saved changes: " + count2);

  Console.WriteLine("Commit or Rollback? 1 = Commit, other =
  Rollback");
  var input = Console.ReadKey().Key;
  Console.WriteLine();
  if (input == ConsoleKey.D1)
  {t.Commit(); Console.WriteLine("Commit done!");}
  else
  {t.Rollback(); Console.WriteLine("Rollback done!");}
```

```
    Console.WriteLine("After in RAM: " + f.ToString());
    ctx.Entry(f).Reload();
    Console.WriteLine("After in DB: " + f.ToString());
   }
  }
 }
}
```

```
ExplicitTransactionTwoContextInstances
Transaction with Level: ReadCommitted
Numer of bookings saved: 1
BEFORE: Flight #111: from Moscow to Oslo on 19.06.18 20:03: 178 free Seats.
AFTER: Flight #111: from Moscow to Oslo on 19.06.18 20:03: 177 free Seats.
Number of saved changes: 1
Commit or Rollback? 1 = Commit, other = Rollback
2
Rollback done!
After in RAM: Flight #111: from Moscow to Oslo on 19.06.18 20:03: 177 free Seats.
After in DB: Flight #111: from Moscow to Oslo on 19.06.18 20:03: 178 free Seats.
```

***Figure 10-11.***  *Output of the previous example*

# Using the Change Tracker

The Change Tracker built into Entity Framework Core, which monitors changes to all objects connected to the Entity Framework Core context, can be queried at any time by program code.

## Getting the State of an Object

Because Entity Framework Core works with plain old CLR objects (POCOs), which have entity objects and not a base class and which implement an interface, entity objects have no knowledge of their context class or state.

To query the object state, you do not ask the entity object itself but the ChangeTracker object of the context class. The ChangeTracker object has an Entry() method that returns an associated EntryObject<EntityType> for a given Entity object. This object owns the following:

- The ChangeTracker object has a State property of type EntityState, which is an enumeration type with the values Added, Deleted, Detached, Modified, and Unchanged.

- In Properties you find a list of all the properties of the entity object in the form of PropertyEntry objects. The PropertyEntry objects each have an IsModified property that indicates whether the property is changed, as well as the old (OriginalValue) and new (CurrentValue) values.

- Using EntryObject<EntityType>, you can also directly get a specific PropertyEntry object by specifying a lambda expression using the Property method.

- GetDatabaseValues() is used to get the current state of the object from the database.

The subroutine in Listing 10-10 loads a Flight (the first one) and modifies this Flight object. At the beginning of the routine, not only is a variable created for the Flight object itself but also an entryObj variable is created for EntryObject<Flight> and propObj is created for a PropertyEntry object.

After loading Flight, entryObj and propObj are first filled with objects of the ChangeTracker object. The entity object is in the Unchanged state, and the FreeSeats property returns IsModified False. Then the object is changed in the property FreeSeats. The entity object is now in the Modified state, and IsModified for FreeSeats returns True.

---

**Note**   It is important that the information from the ChangeTracker object of the context is retrieved here; the instances of EntryObject<Flight> and PropertyEntry do not automatically update with the entity object change but reflect the current state at the time of the retrieval.

---

Therefore, you must also request these objects for the third time after the SaveChanges() method from the ChangeTracker object. After SaveChanges(), the state of the entity object is again Unchanged, and the property FreeSeats returns IsModified False.

The routine also loops over the `Properties` property of `EntryObject<Flight>`
to return all the modified properties of the entity object with old and new values
and the current value of the database. This value can be determined using the
`GetDatabaseValues()` method in `EntryObject<Flight>`. Then `GetDatabaseValues()`
raises a query of the database and populates a `PropertyValues` list with all the current
values in the database. These values in the database may be different from the values
that Entity Framework Core knows and that are visible in the `OriginalValue` properties,
another process (or another Entity Framework Core context in the same process)
meanwhile persisted a change to the record. In this case a data conflict occured.
Figure 10-12 shows the output.

***Listing 10-10.*** Querying the Change Tracker for a Changed Object

```
/// </summary>
public static void ChangeTracking_OneObject()
{
 CUI.MainHeadline(nameof(ChangeTracking_OneObject));

 Flight flight;
 EntityEntry<BO.Flight> entryObj;
 PropertyEntry propObj;

 using (var ctx = new WWWingsContext())
 {

  CUI.Headline("Loading Object...");
  flight = (from y in ctx.FlightSet select y).FirstOrDefault();

  // Access Change Tracker
  entryObj = ctx.Entry(flight);
  propObj = entryObj.Property(f => f.FreeSeats);
  Console.WriteLine(" Object state: " + entryObj.State);
  Console.WriteLine(" Is FreeSeats modified?: " + propObj.IsModified);

  CUI.Headline("Changing Object...");
  flight.FreeSeats--;
```

```csharp
// Access Change Tracker again
entryObj = ctx.Entry(flight);
propObj = entryObj.Property(f => f.FreeSeats);
Console.WriteLine(" Object state: " + entryObj.State);
Console.WriteLine(" Is FreeSeats modified?: " + propObj.IsModified);

// Print old and new values
if (entryObj.State == EntityState.Modified)
{
 foreach (PropertyEntry p in entryObj.Properties)
 {
  if (p.IsModified)
    Console.WriteLine(" " + p.Metadata.Name + ": " + p.OriginalValue +
    "->" + p.CurrentValue +
                      " / State in database: " + entryObj.
                      GetDatabaseValues()[p.Metadata.Name]);
 }
 }

CUI.Headline("Save...");
int count = ctx.SaveChanges();
Console.WriteLine(" Number of changes: " + count);

// Update of the Objects of the Change Tracker
entryObj = ctx.Entry(flight);
propObj = entryObj.Property(f => f.FreeSeats);
Console.WriteLine(" Object state: " + entryObj.State);
Console.WriteLine(" Is FreeSeats modified?: " + propObj.IsModified);
 }
}
```

```
ChangeTracking_OneObject
Loading Object...
 Object state: Unchanged
 Is FreeSeats modified?: False
Chanhing Object...
 Object state: Modified
 Is FreeSeats modified?: True
 FreeSeats: 240->239 / State in database: 240
Save...
 Number of changes: 1
 Object state: Unchanged
 Is FreeSeats modified?: False
```

*Figure 10-12.* *Output*

# Listing All Changed Objects

The ChangeTracker object can provide not only information about a single object but also a list of all entity objects monitored by it using its Entries() method. You can then filter entity objects for the desired state.

The routine in Listing 10-11 modifies three flights and then creates a flight with the value 123456 if it does not already exist. If the Flight object already exists, it will be deleted. Thereafter, the routine separately asks the ChangeTracker object for the new, changed, and deleted objects (Listing 10-12). All three sets are provided by Entries(). The set is filtered from LINQ to Objects using the Where() operator. In all three cases, the PrintChangedProperties() helper routine is called. But only in the case of the changed objects does it deliver some output. If the object has been added or deleted, the individual properties are considered unchanged.

Figure 10-13 and Figure 10-14 show the output.

*Listing 10-11.* Querying the Change Tracker for Several Changed Objects

```
public static void ChangeTracking_MultipleObjects()
{
 CUI.MainHeadline(nameof(ChangeTracking_MultipleObjects));

 using (var ctx = new WWWingsContext())
 {
```

```csharp
var flightQuery = (from y in ctx.FlightSet select y).OrderBy(f4 =>
f4.FlightNo).Take(3);
foreach (var flight in flightQuery.ToList())
{
 flight.FreeSeats -= 2;
 flight.Memo = "Changed on " + DateTime.Now;
}

var newFlight = ctx.FlightSet.Find(123456);
if (newFlight != null)
{
 ctx.Remove(newFlight);
}
else
{
 newFlight = new Flight();
 newFlight.FlightNo = 123456;
 newFlight.Departure = "Essen";
 newFlight.Destination = "Sydney";
 newFlight.AirlineCode = "WWW";
 newFlight.PilotId = ctx.PilotSet.FirstOrDefault().PersonID;
 newFlight.Seats = 100;
 newFlight.FreeSeats = 100;
 ctx.FlightSet.Add(newFlight);
}
CUI.Headline("New objects");
IEnumerable<EntityEntry> neueObjecte = ctx.ChangeTracker.Entries().
Where(x => x.State == EntityState.Added);
if (neueObjecte.Count() == 0) Console.WriteLine("none");
foreach (EntityEntry entry in neueObjecte)
{
 CUI.Print("Object " + entry.Entity.ToString() + " State: " + entry.
 State, ConsoleColor.Cyan);
 ITVisions.EFCore.EFC_Util.PrintChangedProperties(entry);
}
```

```csharp
CUI.Headline("Changed objects");
IEnumerable<EntityEntry> geaenderteObjecte =
 ctx.ChangeTracker.Entries().Where(x => x.State == EntityState.
 Modified);
if (geaenderteObjecte.Count() == 0) Console.WriteLine("none");
foreach (EntityEntry entry in geaenderteObjecte)
{
 CUI.Print("Object " + entry.Entity.ToString() + " State: " + entry.
 State, ConsoleColor.Cyan);
 ITVisions.EFCore.EFC_Util.PrintChangedProperties(entry);
}

CUI.Headline("Deleted objects");
IEnumerable<EntityEntry> geloeschteObjecte = ctx.ChangeTracker.
Entries().Where(x => x.State == EntityState.Deleted);
if (geloeschteObjecte.Count() == 0) Console.WriteLine("none");
foreach (EntityEntry entry in geloeschteObjecte)
{
 CUI.Print("Object " + entry.Entity.ToString() + " State: " + entry.
 State, ConsoleColor.Cyan);
}
Console.WriteLine("Changes: " + ctx.SaveChanges());
 }
}
```

***Listing 10-12.*** Auxiliary Routine for Querying the Change Tracker

```csharp
/// <summary>
/// Lists the changed properties of an object, including the current
database state
/// </summary>
/// <param name="entry"></param>
public static void PrintChangedProperties(EntityEntry entry)
{
 PropertyValues dbValue = entry.GetDatabaseValues();
 foreach (PropertyEntry prop in entry.Properties.Where(x =>
 x.IsModified))
```

```
    {
      var s = "- " + prop.Metadata.Name + ": " +
        prop.OriginalValue + "->" +
        prop.CurrentValue +
        " State in the database: " + dbValue[prop.Metadata.Name];
      Console.WriteLine(s);
    }
  }
}
```

```
ChangeTracking_MultipleObjects
New object
Object Flight #123456: from Essen to Sydney on 23.12.17 02:38: 100 free Seats. State: Added
Changed objects
Object Flight #100: from Berlin to Paris on 23.08.18 03:31: 221 free Seats. State: Modified
- FreeSeats: 223->221 State in the database: 223
- Memo: Changed on 23/12/2017 02:38:25->Changed on 23/12/2017 02:38:29 State in the database: Changed on 23/12/2017 02:38:25
Object Flight #101: from Berlin to Seattle on 24.04.18 13:31: 84 free Seats. State: Modified
- FreeSeats: 86->84 State in the database: 86
- Memo: Changed on 23/12/2017 02:38:25->Changed on 23/12/2017 02:38:29 State in the database: Changed on 23/12/2017 02:38:25
Object Flight #102: from New York/JFC to Milan on 25.12.17 23:38: -15 free Seats. State: Modified
- FreeSeats: -13->-15 State in the database: -13
- Memo: Changed on 23/12/2017 02:38:25->Changed on 23/12/2017 02:38:29 State in the database: Changed on 23/12/2017 02:38:25
Deleted objects
none
Changes: 4
```

***Figure 10-13.*** *First run of the code: flight 123456 is added*

```
ChangeTracking_MultipleObjects
New object
none
Changed objects
Object Flight #100: from Berlin to Paris on 23.08.18 03:31: 219 free Seats. State: Modified
- FreeSeats: 221->219 State in the database: 221
- Memo: Changed on 23/12/2017 02:38:29->Changed on 23/12/2017 02:38:44 State in the database: Changed on 23/12/2017 02:38:29
Object Flight #101: from Berlin to Seattle on 24.04.18 13:31: 82 free Seats. State: Modified
- FreeSeats: 84->82 State in the database: 84
- Memo: Changed on 23/12/2017 02:38:29->Changed on 23/12/2017 02:38:44 State in the database: Changed on 23/12/2017 02:38:29
Object Flight #102: from New York/JFC to Milan on 25.12.17 23:38: -17 free Seats. State: Modified
- FreeSeats: -15->-17 State in the database: -15
- Memo: Changed on 23/12/2017 02:38:29->Changed on 23/12/2017 02:38:44 State in the database: Changed on 23/12/2017 02:38:29
Deleted objects
Object Flight #123456: from Essen to Sydney on 23.12.17 02:38: 100 free Seats. State: Deleted
Changes: 4
```

***Figure 10-14.*** *Second pass of the code: flight 123456 is deleted again*

# CHAPTER 11

# Preventing Conflicts (Concurrency)

In many production scenarios, it's possible for multiple people or automatic background tasks to access the same records at the same time. This can lead to conflicts in which contradictory data changes take place. This chapter shows how to detect and resolve such conflicts in Entity Framework Core.

## A Look at the History of Concurrency

Like its predecessor Entity Framework and the underlying basic technology ADO.NET, Entity Framework Core does not support the blocking of data records for read access by other processes. This was a deliberate decision by Microsoft in .NET 1.0 (2002) because locks cause a lot of performance issues. In the alpha version of .NET 2.0 (2005), there was a prototype of such a lock function in the then-new class `SqlResultSet`, but this class was never released in an RTM version of .NET.

Therefore, in .NET and the frameworks based on it, such as Entity Framework and Entity Framework Core, there is only so-called optimistic locking. *Optimistic locking* is a euphemism because actually nothing is blocked in the database management system and in RAM. It is only ensured that change conflicts will be noticed later. The first process that wants to write a change wins. All other processes cannot write and will get an error message. To accomplish this, the `UPDATE` and `DELETE` commands in the `WHERE` condition include single or multiple values from the source record.

© Holger Schwichtenberg 2018
H. Schwichtenberg, *Modern Data Access with Entity Framework Core*,
https://doi.org/10.1007/978-1-4842-3552-2_11

A DataSet in conjunction with a DataAdapter and a CommandBuilder object not only queries the primary key column or columns in the WHERE clause of an UPDATE or DELETE command but also queries all columns with their old values from the point of view of the current process values that the process received while reading the record (see Listing 11-1). In the meantime, if another process has changed any individual columns, the UPDATE or DELETE command does not cause a runtime error in the database management system; instead, it results in zero records being affected. This allows the DataAdapter to detect that there was a change conflict.

***Listing 11-1.*** Update Command, As Created by a SqlCommandBuilder for the Flight Table with the Primary Key FlightNo

```
UPDATE [dbo]. [Flight]
SET [FlightNo] = @p1, [Departure] = @p2, [Strikebound] = @p3,
[CopilotId] = @p4, [FlightDate] = @p5, [Flightgesellschaft] = @p6,
[AircraftTypeID] = @p7, [FreeSeats] = @p8, [LastChange] = @p9,
[Memo] = @p10, [NonSmokingFlight] = @p11, [PilotId] = @p12, [Seats] = @p13,
[Price] = @p14, [Timestamp] = @p15, [destination] = @p16
WHERE ((([FlightNo] = @p17) AND ((@p18 = 1 AND [Departure] IS NULL) OR
([Departure] = @p19)) AND ((@p20 = 1 AND [Expires] IS NULL) OR
( [Strikebound] = @p21)) AND ((@p22 = 1 AND [CopilotId] IS NULL) OR
([CopilotId] = @p23)) AND ([FlightDate] = @p24) AND ([Airline] = @p25)
AND ((@p26 = 1 AND [aircraftID_ID] IS NULL) OR ([aircraft_type_ID] =
@p27)) AND ((@p28 = 1 AND [FreeSeats] IS NULL) OR ([FreeSeats] = @p29))
AND ( [Lastchange] = @p30) AND ((@p31 = 1 AND [NonSmokingFlight] IS NULL)
OR ([NonSmokingFlight] = @p32)) AND ([PilotId] = @p33) AND ([Seats] = @p34)
AND ((@p35 = 1 AND [price] IS NULL) OR ([price] = @p36)) AND ((@p37 = 1 AND
[destination] IS NULL) OR ([destination] = @p38)))
```

# No Conflict Detection by Default

Entity Framework Core, like Entity Framework, does not lock at all by default, not even with optimistic locking. The standard is simply "the last who writes wins." Listing 11-2 shows how to change a Flight object in RAM and persist the change with

SaveChanges() via Entity Framework Core. This program code sends the following SQL command to the database management system:

```
UPDATE [Flight] SET [FreeSeats] = @p0
WHERE [FlightNo] = @p1;
SELECT @@ ROWCOUNT;
```

As you can see, only the primary key FlightNo appears in the WHERE condition; the old value of the column FreeSeats or other columns does not appear. Thus, the value is persisted, even if other processes have changed the value in the meantime. As a result, an airline may experience the overbooking of flights. For example, if there are only two free seats left and two processes load this information (almost) simultaneously, each of the two processes can subtract two places from this remaining contingent. The state in the column FreeSeats is then zero in the database. In fact, four passengers were placed in the two seats. That will be a tight fit on the plane!

Although SaveChanges() opens a transaction, it applies only to the one store operation and therefore does not protect against data change conflicts. However, ignoring conflicts is often not viable or acceptable to users. Luckily, you can reconfigure Entity Framework Core, much as you did with Code First in its predecessor Entity Framework.

The only change conflict that Entity Framework Core would notice would be to delete the record with another process because, in this case, the UPDATE command would return the fact that zero records were changed, and then Entity Framework Core would raise a DbUpdateConcurrencyException error.

***Listing 11-2.*** Changing a Flight Object

```
public static void Change.FlightOneProperty()
{
 CUI.MainHeadline(nameof(ChangeFlightOneProperty));

 int FlightNr = 101;
 using (WWWingsContext ctx = new WWWingsContext())
 {

  // Load flight
  var f = ctx.FlightSet.Find(FlightNr);
```

```
Console.WriteLine($"Before changes: Flight #{f.FlightNo}:
{f.Departure}->{f.Destination} has {f.FreeSeats} free seats! State of
the flight object: " + ctx.Entry(f).State);

// Change object in RAM
f.FreeSeats -= 2;

Console.WriteLine($"After changes: Flight #{f.FlightNo}: {f.Departure}->
{f.Destination} has {f.FreeSeats} free seats! State of the flight
object: " + ctx.Entry(f).State);

// Persist changes
try
{
 var count = ctx.SaveChanges();
 if (count == 0)
 {
  Console.WriteLine("Problem: No changes saved!");
 }
 else
 {
  Console.WriteLine("Number of saved changes: " + count);
  Console.WriteLine($"After saving: Flight #{f.FlightNo}:
  {f.Departure}->{f.Destination} has {f.FreeSeats} free seats! Zustand
  des Flight-Objekts: " + ctx.Entry(f).State);
 }
}
catch (Exception ex)
{
 Console.WriteLine("Error: " + ex.ToString());
}
}
}
```

# Detecting Conflicts with Optimistic Locking

Entity Framework Core sends the following SQL command to the database management system:

```
UPDATE [Flight] SET [FreeSeats] = @p0
WHERE [FlightNo] = @p1 AND [FreeSeats] = @p2;
SELECT @@ROWCOUNT;
```

Here, in addition to the query of the primary key, FlightNo takes a query of the old value (the original value when reading) of the column FreeSeats. To implement this conflict detection, it is not the program code that needs to be changed but the Entity Framework Core model.

There are two ways to configure the model.

- Via the data annotation [ConcurrencyCheck]

- Via IsConcurrencyToken() in the Fluent API

Listing 11-3 shows a section from the entity class Flight. Here, FreeSeats was annotated with [ConcurrencyCheck], whereby Entity Framework Core automatically queries the old value in the WHERE condition for all UPDATE and DELETE commands. This is achieved by calling IsConcurrencyToken() on the corresponding PropertyBuilder object in OnModelCreating() in the Entity Framework Core context class (see Listing 11-4).

***Listing 11-3.*** Use of Data Annotation [ConcurrencyCheck]

```
public class Flight
{

  [Key]
  public int FlightNo {get; set; }

  [ConcurrencyCheck]
  public short? FreeSeats {get; set;}

[ConcurrencyCheck]
public decimal? Price {get; set; }
  public short? Seats { get; set; }

...
}
```

***Listing 11-4.***  Using IsConcurrencyToken( ) in the Fluent API

```
public class WWWingsContext: DbContext
{
  public DbSet<Flight> FlightSet { get; set; }
...
  protected override void OnModelCreating (ModelBuilder builder)
  {
    Builder %Entity<Flight>().Property (f => f.FreeSeats).IsConcurrencyToken();
...
}
}
```

Now it may be useful to run the conflict check over several columns. For example, the conflict check can also be performed via the Price column of the Flight object. In terms of content, this would mean that if the price of the Flight has changed, you cannot change the number of seats because this booking would then have been displayed to the user at the old price. You can then annotate the Price property with [ConcurrencyCheck] or add it to the Fluent API.

```
builder.Entity<Flight>().Property(x => x.FreeSeats).ConcurrencyToken();
```

The following SQL command with three parts in the WHERE condition arises from the listing:

```
SET NOCOUNT ON;
UPDATE [Flight] SET [FreeSeats] = @p0
WHERE [FlightNo] = @p1 AND [FreeSeats] = @p2 AND [Price] = @p3;
SELECT @@ ROWCOUNT;
```

# Detecting Conflicts for All Properties

It can be tedious to do this configuration via a data annotation or the Fluent API for all entity classes and all persistent properties. Fortunately, Entity Framework Core lets you do a mass configuration. Listing 11-5 shows how to use Model in OnModelCreating() from the ModelBuilder object. GetEntityTypes() gets a list of all entity classes and therefore all properties in each entity class via GetProperties() so that IsConcurrencyToken = true is set there.

***Listing 11-5.*** Mass Configuration of the ConcurrencyToken for All Properties in All Entity Classes

```
public class WWWingsContext: DbContext
{
  public DbSet<Flight> FlightSet { get; set; }
...
  protected override void OnModelCreating (ModelBuilder builder)
  {
   foreach (IMutableEntityType entity in modelBuilder.Model.
   GetEntityTypes())
   {
    // get all properties
    foreach (var prop in entity.GetProperties())
    {
      prop.IsConcurrencyToken = true;
    }
   }
...
  }
}
```

The listing then creates a SQL command that includes all the columns in the WHERE condition, shown here:

```
SET NOCOUNT ON;
UPDATE [Flight] SET [FreeSeats] = @p0
WHERE [FlightNo] = @p1 AND [Departure] = @p2 AND [Destination] = @p3
AND [CopilotId] = @p4 AND [FlightDate] = @p5 AND [Airline] = @p6 AND
[AircraftTypeID] IS NULL AND [ FreeSeats] = @p7 AND [LastChange] = @p8
AND [Memo] = @p9 AND [NonSmokingFlight] IS NULL AND [PilotId] = @p10 AND
[Seats] = @p11 AND [Price] = @p12 AND [Strikebound] = @p13;
SELECT @@ ROWCOUNT;
```

# Settling Conflicts by Convention

If you want to exclude individual columns, that's also possible. In this case, it makes sense to define a separate annotation, called [ConcurrencyNoCheckAttribute] (see Listing 11-6), which then annotates all persistent properties of the entity class for which Entity Framework Core should not perform a conflict check. Listing 11-7 shows the extension of the example that considers the annotation [ConcurrencyNoCheck]. What is important here is the zero-propagation operator ?. after PropertyInfo; it's important because you can define so-called shadow properties in Entity Framework Core, which exist only in the Entity Framework Core model and not in the entity class. These shadow properties will not have a PropertyInfo object, so in the case of a shadow property without the null-propagating operator, the popular Null Reference runtime error would occur. With ConcurrencyNoCheckAttribute, you can elegantly exclude individual properties from conflict checking as needed.

***Listing 11-6.*** Annotation for Entity Class Properties for Which Entity Framework Core Should Not Run a Concurrency Check

```
using system;
namespace EFCExtensions
{
/// <summary>
/// Annotation for EFCore entity classes and properties for which EFCore
should not run a concurrency check
/// </ summary>
[AttributeUsage (AttributeTargets.Property | AttributeTargets.Class,
AllowMultiple = false)]
public class ConcurrencyNoCheckAttribute: Attributes
{
}
}
```

***Listing 11-7.*** Mass Configuration of the ConcurrencyToken for All Properties in All Entity Classes, Except the Properties Annotated with [ConcurrencyNoCheck]

```
public class WWWingsContext: DbContext
{
  public DbSet<Flight> FlightSet {get; set; }
...
  protected override void OnModelCreating (ModelBuilder builder)
  {
   // Get all entity classes
   foreach (IMutableEntityType entity in modelBuilder.Model.GetEntityTypes())
   {
    // get all properties
    foreach (var prop in entity.GetProperties())
    {
     // Look for annotation [ConcurrencyNoCheck]
     var annotation = prop.PropertyInfo?.GetCustomAttribute<ConcurrencyNo
     CheckAttribute>();
     if (annotation == null)
     {
      prop.IsConcurrencyToken = true;
     }
     else
     {
      Console.WriteLine("No Concurrency Check for" + prop.Name);
     }
     if (prop.Name == "Timestamp")
     {
      prop.ValueGenerated = ValueGenerated.OnAddOrUpdate;
      prop.IsConcurrencyToken = true;
     }
```

```
    foreach (var a in prop.GetAnnotations())
    {
     Console.WriteLine(prop.Name + ":" + a.Name + "=" + a.Value);
    }
   }
  }
 }
}
...
}
}
```

# Setting Up Conflict Checks Individually

Sometimes, in practice, there is a desire to activate or deactivate the conflict check on a case-by-case basis for individual changes for individual properties. Unfortunately, this cannot be fulfilled because the data annotations are compiled, and OnModelCreating() is also called only once per process. Unfortunately, you cannot change the Entity Framework Core model after the end of OnModelCreating(). Although the DbContext class, like the ModelBuilder class, provides a property model, in ModelBuilder the property model has the IMutalModel type (which is a variable, as the name implies). DbContext gets only the IModel type, and IsConcurrencyToken is read-only like many other properties. So, if you want to change optimistic locking columns on a case-by-case basis, you need to send UPDATE and DELETE commands to the database management system yourself (via Entity Framework Core or another means).

# Adding Timestamps

Instead of the original value comparison at the level of individual data columns, it is possible to introduce an additional timestamp column. You can find one such column in Microsoft SQL Server with the type rowversion (https://docs.microsoft.com/en-us/sql/t-sql/data-types/rowversion-transact-sql), called timestamp (see Figure 11-1 and Figure 11-2. It is automatically increased by the database management system for each individual data record change. Consequently, in the case of an UPDATE or DELETE command, it is only necessary to check whether the value is still at the previous value that existed during loading. If so, the entire record is unchanged. If not, another process has changed at least part of the record. However, with a `timestamp` column, you cannot differentiate between columns where changes are relevant and those where changes are not relevant. The database management system adjusts the timestamp each time a column is changed; exceptions are not possible.

---

**Note**    While currently SQL Server Management Studio (SSMS) still displays the old name `timestamp`, SQL Server Data Tools in Visual Studio 2016 shows the current name of `rowversion`.

---

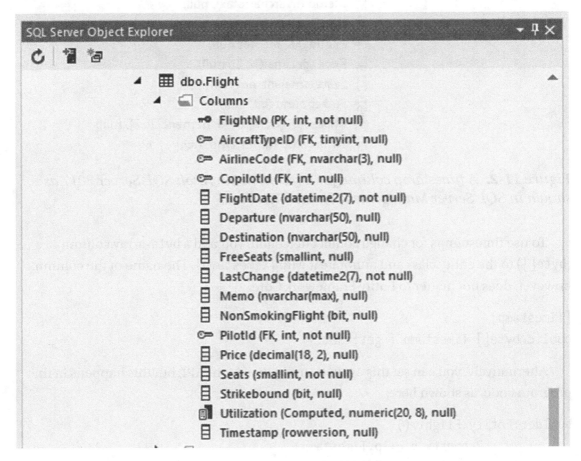

***Figure 11-1.*** *A timestamp column for a record in Microsoft SQL Server 2017 as shown in Visual Studio 2017*

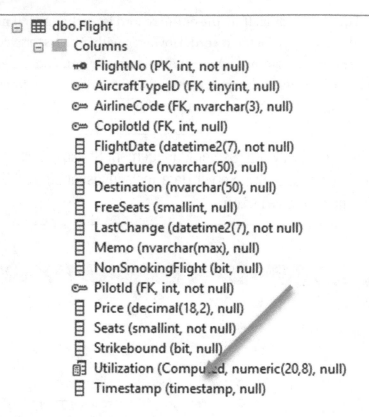

**Figure 11-2.** *A timestamp column for a record in Microsoft SQL Server 2017 as shown in SQL Server Management Studio 17.1*

To use timestamps for change conflict detection, you add a byte-array column (byte[]) to the entity class and annotate it with [timestamp]. The name of the column, however, does not matter to Entity Framework Core.

```
[Timestamp]
public byte[] Timestamp { get; set; }
```

Alternatively, you can set this again by using the Fluent API, but this happens in the program code, as shown here:

```
builder.Entity<Flight>()
        .Property(p => p.Timestamp)
        .ValueGeneratedOnAddOrUpdate()
        .IsConcurrencyToken();
```

Since version 1.1 of Entity Framework Core, you can also use IsRowVersion() as an alternative, as shown here:

```
modelBuilder.Entity<Flight>().Property(x => x.Timestamp).IsRowVersion();
```

---

**Note**   There can be only one `timestamp/rowversion` column per table. Unfortunately, the error message "A table can only have one timestamp column." occurs only when you call `Update-Database`, not with `Add-Migration`.

---

You do not have to implement anything else for the timestamp support. If there is such a property in the object model and the corresponding column in the database table, Entity Framework Core always refers to the previous timestamp value for all DELETE and UPDATE commands in the WHERE condition.

```
SET NOCOUNT ON;
UPDATE [Flight] SET [FreeSeats] = @p0
WHERE [FlightNo] = @p1 AND [Timestamp] IS NULL;
SELECT [Timestamp]
FROM [Flight]
WHERE @@ ROWCOUNT = 1 AND [FlightNo] = @p1;
```

As you can see, Entity Framework Core also uses SELECT [Timestamp] to reload the timestamp that is changed by the database management system after an UPDATE to update the object in RAM accordingly. If that did not happen, then a second update of an object would not be possible because then the timestamp in RAM would be out of date and Entity Framework Core would always report a change conflict, even if there was none (because the first change was the one that changed the timestamp in the database table).

The timestamp configuration can also be automated by convention. The mass configuration shown in Listing 11-8 automatically makes all properties bearing the name `timestamp` timestamps for conflict detection.

***Listing 11-8.*** Automatically Turning Any Properties Called Timestamp into Timestamps for Conflict Detection

```
public class WWWingsContext: DbContext
{
  public DbSet<Flight> FlightSet {get; set; }
  ...
  protected override void OnModelCreating (ModelBuilder builder)
  {
   // Get all entity classes
   foreach (IMutableEntityType entity in modelBuilder.Model.
   GetEntityTypes())
   {
    // Get all properties
    foreach (var prop in entity.GetProperties())
    {
     if (prop.Name == "Timestamp")
     {
      prop.ValueGenerated = ValueGenerated.OnAddOrUpdate;
      prop.IsConcurrencyToken = true;
     }

    }
   }
  ...
  }
}
```

# Resolving Conflicts

This section shows how to validate the conflict detection. Figure 11-3 shows some typical output of the program when it is started twice. First the process with the ID 10596 starts, and then the process 18120 starts. Both read the flight with the number 101, in which there are currently 143 seats left. The process 10596 then reduces the number of places by 5 and persists 138 in the FreeSeats column. Now process 18120 makes a booking for two people, so in RAM the value changes to 141 FreeSeats. However,

process 18120 cannot persist because Entity Framework Core throws an error of type
DbUpdateConcurrencyException because of conflict detection for the FreeSeats column
or based on a timestamp column. The user in process 18120 is given the choice of either
accepting or overwriting the other user's changes or offsetting the two changes, which
may make sense in some cases.

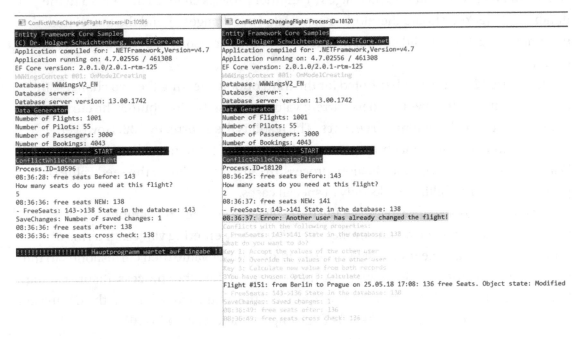

*Figure 11-3.* *Conflict detection and conflict resolution*

Listing 11-9 shows the implementation. SaveChanges() catches the
DbUpdateConcurrencyException. In the error handler, the PrintChangedProperties()
helper function is used to specify which properties of the flight were changed in this
process and what the current database state is. You get the current state of the database
through the method GetDatabaseValues(), which sends a corresponding SQL query
to the database management system. After that, the user has to decide. If the user
chooses to apply the changes to the other process, the Reload() method is called in the
Entity Framework Core API, which discards the changed object in RAM and reloads
it from the database. If the user chooses to override the changes to the other process,
the program code is a little more complicated and indirect. The command chain loads
the current state from the database and sets it in the Entity Framework Core Change

Tracker to the original values of the object: `ctx.Entry(Flight).OriginalValues.`
`SetValues(ctx.Entry(Flight).GetDatabaseValues())`. After that, `SaveChanges()` is
called again, which now works because the original values used in the `WHERE` condition
or the timestamp used there correspond to the current state of the data record in the
database. Theoretically, however, a conflict can occur again if in the short time between
`GetDatabaseValues()` and `SaveChanges()` another process changes the record in the
database table. You should therefore encapsulate `SaveChanges()` and the associated
error handling, but this did not happen here for didactic reasons to better illustrate the
example.

In Figure 11-3, the user has opted for the third option, clearing the two changes. In
addition to its original value and its current value, process 18120 requires the current
database value of the column `FreeSeats`. The result is then correctly 136. However,
the calculation assumes that both processes have started from the same original value.
Without interprocess communication, process 18120 may not know the seed value of
process 10596. The billing works only in special cases.

Of course, it is also possible to let the user solve the conflict in a (graphical) user
interface by not deciding on one or the other side but by entering values. In terms of
implementation, the clearing just like the input of another value then corresponds to the
second case, in other words, overwriting the changes of the other process. Before calling
`SaveChanges()`, simply set the values in the object that you want to have in the database
afterward (see the case `ConsoleKey.D3` in Listing 11-9 and Listing 11-10).

***Listing 11-9.*** Conflict Detection and Conflict Resolution with Entity
Framework Core

```
public static void ConflictWhileChangingFlight()
  {
   CUI.MainHeadline(nameof(ConflictWhileChangingFlight));
   Console.WriteLine("Process.ID=" + Process.GetCurrentProcess().Id);
   Console.Title = nameof(ConflictWhileChangingFlight) + ": Process-ID=" +
   Process.GetCurrentProcess().Id;

   // Flight, where the conflict should arise
   int flightNo = 151;
```

```
using (WWWingsContext ctx = new WWWingsContext())
{
 // --- load flight
 Flight flight = ctx.FlightSet.Find(flightNo);
 Console.WriteLine(DateTime.Now.ToLongTimeString() + ": free seats
 Before: " + flight.FreeSeats);

 short seats = 0;
 string input = "";
 do
 {
  Console.WriteLine("How many seats do you need at this flight?");
  input = Console.ReadLine(); // wait (time to start another process)
 } while (!Int16.TryParse(input, out seats));

 // --- change the free seats
 flight.FreeSeats -= seats;
 Console.WriteLine(DateTime.Now.ToLongTimeString() + ": free seats NEW:
 " + flight.FreeSeats);

 try
 {
  // --- try to save
  EFC_Util.PrintChangedProperties(ctx.Entry(flight));
  var count = ctx.SaveChanges();
  Console.WriteLine("SaveChanges: Number of saved changes: " + count);
 }
 catch (DbUpdateConcurrencyException ex)
 {
  Console.ForegroundColor = ConsoleColor.Red;
  CUI.PrintError(DateTime.Now.ToLongTimeString() + ": Error: Another
  user has already changed the flight!");

  CUI.Print("Conflicts with the following properties:");
  EFC_Util.PrintChangedProperties(ex.Entries.Single());
```

```csharp
// --- Ask the user
Console.WriteLine("What do you want to do?");
Console.WriteLine("Key 1: Accept the values of the other user");
Console.WriteLine("Key 2: Override the values of the other user");
Console.WriteLine("Key 3: Calculate new value from both records");

ConsoleKeyInfo key = Console.ReadKey();
switch(key.Key)
{
 case ConsoleKey.D1: // Accept the values of the other user
  {
  Console.WriteLine("You have chosen: Option 1: Accept");
  ctx.Entry(flight).Reload();
  break;
 }
 case ConsoleKey.D2: // Override the values of the other user
  {
  Console.WriteLine("You have chosen: Option 2: Override");
  ctx.Entry(flight).OriginalValues.SetValues(ctx.Entry(flight).
  GetDatabaseValues());
  // wie RefreshMode.ClientWins bei ObjectContext
  EFC_Util.PrintChangeInfo(ctx);
  int count = ctx.SaveChanges();
  Console.WriteLine("SaveChanges: Saved changes: " + count);
  break;
 }
 case ConsoleKey.D3: // Calculate new value from both records
  {

  Console.WriteLine("You have chosen: Option 3: Calculate");
  var FreeSeatsOrginal = ctx.Entry(flight).OriginalValues.
  GetValue<short?>("FreeSeats");
  var FreeSeatsNun = flight.FreeSeats.Value;
  var FreeSeatsInDB = ctx.Entry(flight).GetDatabaseValues().
  GetValue<short?>("FreeSeats");
```

```
        flight.FreeSeats = (short) (FreeSeatsOrginal -
                           (FreeSeatsOrginal - FreeSeatsNun) -
                           (FreeSeatsOrginal - FreeSeatsInDB));
      EFC_Util.PrintChangeInfo(ctx);
      ctx.Entry(flight).OriginalValues.SetValues(ctx.Entry(flight).
      GetDatabaseValues());
      int count = ctx.SaveChanges();
      Console.WriteLine("SaveChanges: Saved changes: " + count);
      break;
    }
  }
}
Console.WriteLine(DateTime.Now.ToLongTimeString() + ": free seats
after: " + flight.FreeSeats);

// --- Cross check the final state in the database
using (WWWingsContext ctx2 = new WWWingsContext())
{
 var f = ctx.FlightSet.Where(x => x.FlightNo == flightNo).
 SingleOrDefault();
 Console.WriteLine(DateTime.Now.ToLongTimeString() + ": free seats
 cross check: " + f.FreeSeats);

} // End using-Block -> Dispose()
}
}
```

***Listing 11-10.*** Subroutines for Listing 11-9

```
/// <summary>
  /// Print all changed objects and the changed properties
  /// </summary>
  /// <param name="ctx"></param>
  public static void PrintChangeInfo(DbContext ctx)
  {
   foreach (EntityEntry entry in ctx.ChangeTracker.Entries())
   {
```

```
  if (entry.State == EntityState.Modified)
  {
   CUI.Print(entry.Entity.ToString() + " Object state: " + entry.State,
   ConsoleColor.Yellow);
   IReadOnlyList<IProperty> listProp = entry.OriginalValues.Properties;
   PrintChangedProperties(entry);
  }
 }
}

/// <summary>
/// Print the changed properties of an object, including the current
database state
/// </summary>
/// <param name="entry"></param>
public static void PrintChangedProperties(EntityEntry entry)
{
 PropertyValues dbValue = entry.GetDatabaseValues();
 foreach (PropertyEntry prop in entry.Properties.Where(x => x.IsModified))
 {
  var s = "- " + prop.Metadata.Name + ": " +
   prop.OriginalValue + "->" +
   prop.CurrentValue +
   " State in the database: " + dbValue[prop.Metadata.Name];
  Console.WriteLine(s);
 }
}
```

# Pessimistic Locking on Entity Framework Core

Although Microsoft has deliberately not implemented a class in .NET and .NET Core that can be used to block records for read access by others, I am constantly meeting clients who nevertheless desperately want to avoid conflicts in the first place. With a LINQ command, a read lock is not feasible, even with the activation of a transaction. You need a transaction and additionally a database management system–specific SQL command. In Microsoft SQL Server, this is the query hint SELECT ... WITH (UPDLOCK) associated with a transaction. This query hint ensures that a read record is locked until the transaction is completed. It works only within a transaction, so you will find a ctx. Database.BeginTransaction() method in Listing 11-11 and later the call to commit(). The listing also shows the use of the FromSql() method provided by Entity Framework Core, which allows you to send your own SQL command to the database management system and materialize the result into entity objects.

**Listing 11-11.**  A Lock Is Already Set Up When Reading the Data Record

```
public static void UpdateWithReadLock()
 {
  CUI.MainHeadline(nameof(UpdateWithReadLock));
  Console.WriteLine("--- Change flight");
  int flightNo = 101;
  using (WWWingsContext ctx = new WWWingsContext())
  {
   try
   {
    ctx.Database.SetCommandTimeout(10); // 10 seconds
    // Start transaction
    IDbContextTransaction t = ctx.Database.BeginTransaction(IsolationLevel.
    ReadUncommitted); // default is System.Data.IsolationLevel.ReadCommitted
    Console.WriteLine("Transaction with Level: " + t.GetDbTransaction().
    IsolationLevel);

    // Load flight with read lock using  WITH (UPDLOCK)
    Console.WriteLine("Load flight using SQL...");
    Flight f = ctx.FlightSet.FromSql("SELECT * FROM dbo.Flight WITH
    (UPDLOCK) WHERE flightNo = {0}", flightNo).SingleOrDefault();

    Console.WriteLine($"Before changes: Flight #{f.FlightNo}:
    {f.Departure}->{f.Destination} has {f.FreeSeats} free seats! State of
    the flight object: " + ctx.Entry(f).State);

    Console.WriteLine("Waiting for ENTER key...");
    Console.ReadLine();
```

271

```csharp
// Change object in RAM
Console.WriteLine("Change flight...");
f.FreeSeats -= 2;

Console.WriteLine($"After changes: Flight #{f.FlightNo}: {f.Departure}->
{f.Destination} has {f.FreeSeats} free seats! State of the flight
object: " + ctx.Entry(f).State);

// Send changes to DBMS
Console.WriteLine("Save changes...");
var c = ctx.SaveChanges();
t.Commit();
if (c == 0)
{
 Console.WriteLine("Problem: No changes saved!");
}
else
{
 Console.WriteLine("Number of saved changes: " + c);
 Console.WriteLine($"After saving: Flight #{f.FlightNo}: {f.Departure}->
 {f.Destination} has {f.FreeSeats} free seats! State of the flight
  object: " + ctx.Entry(f).State);
 }
}
catch (Exception ex)
{
 CUI.PrintError("Error: " + ex.ToString());
 }
 }
}
```

Figure 11-4 provides evidence that in fact a second process cannot read FlightNo 101 as long as the first process has not yet completed its transaction. In this sample code, processing waits for user input in the middle of the transaction. User input in a transaction is of course a "worst practice" and should never occur in productive code. Here in the example code, however, it is a useful tool to simulate a transaction's runtime over several seconds until the other process times out.

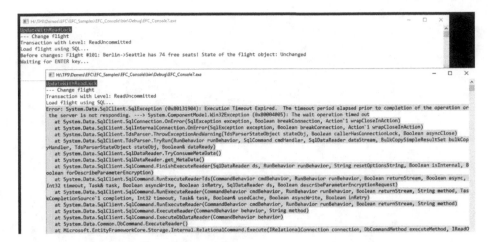

***Figure 11-4.*** *If the program code runs twice in parallel, in the second run the process will time out because the first process has a read lock on the record*

---

**Note**    I must reiterate that such a data record lock in the database management system is not a good practice. Locks, especially read locks, slow down applications and also quickly lead to deadlocks, where processes wait for each other so that processing can no longer take place. Such practices harm the performance, scalability, and stability of software. Why did I even show it in this chapter? Because I know some developers who still want it.

---

Incidentally, a better alternative to record locks in the database management system is to use locks at the application level, where the application manages the locks, possibly also using a self-defined lock table in RAM. This has the advantage that the application can accurately present to the user, who is currently working on a record. For example, user Müller can say, "This record is still available for 4 minutes and 29 seconds for exclusive editing by you." And Mrs. Meier says, "This data set is currently being processed by Mr. Müller. He still has 4 minutes and 29 seconds to save the record. During this time, you cannot make any changes to this record." This offers many more possibilities. For example, you can have the button say "I am a star, and now I want to kick out Mr. Miller immediately from the record." However, there is no predefined mechanism for application-level locking in Entity Framework Core. Here your own creativity is in demand!

# CHAPTER 12

# Logging

In the classic Entity Framework, there are two easy ways to get the SQL commands that the OR mapper sends to the database.

- You can call `ToString()` on a query object (`IQueryable<T>`).
- You can use the `Log` attribute (since Entity Framework version 6.0), as in `ctx.Database.Log = Console.WriteLine;`.

Unfortunately, neither of these options is available in Entity Framework Core. The following command

```
var query = ctx.FlightSet.Where(x => x.FlightNo > 300).OrderBy(x =>
x.- Date).Skip(10).Take(5);
Console.WriteLine (query.ToString());
```

only delivers the following output: `Microsoft.EntityFrameworkCore.Query.Internal.` `EntityQueryable`1 [BO.Flight].`

The `ctx.Database` object has no `log` attribute in Entity Framework Core.

## Using the Extension Method Log()

Logging in Entity Framework Core is possible, but it's much more complicated than in its predecessor. Therefore, I created an extension to the `Database` object in the form of the `Log()` extension method for the `DbContext` class. It is a method, not an attribute, because in .NET unfortunately there are only extension methods and no extension attributes.

---

**Note** The `Log()` method is used in a few listings in this book.

---

275

© Holger Schwichtenberg 2018
H. Schwichtenberg, *Modern Data Access with Entity Framework Core*,
https://doi.org/10.1007/978-1-4842-3552-2_12

You can use the method Log() like this:

```
using (var ctx1 = new WWWingsContext())
  {
   var query1 = ctx1.FlightSet.Where(x => x.FlightNo > 100).OrderBy(x =>
   x.Date).Skip(10).Take(5);
   ctx1.Log(Console.WriteLine);
   var flightSet1 = query1.ToList();
   flightSet1.ElementAt(0).FreeSeats--;
   ctx1.SaveChanges();
  }
```

Similar to the Log attribute in Entity Framework, Log() is a method with no return value; it expects a string as the only parameter. Unlike the classic Entity Framework, you can omit the parameter in Log(). Then the output will be printed automatically to Console.WriteLine() in the color cyan (see Figure 12-1).

```
using (var ctx2 = new WWWingsContext())
  {
   var query2 = ctx2.FlightSet.Where(x => x.FlightNo < 3000).OrderBy(x =>
   x.Date).Skip(10).Take(5);
   ctx2.Log();
   var flightSet2 = query2.ToList();
   flightSet2.ElementAt(0).FreeSeats--;
   ctx2.SaveChanges();
  }
```

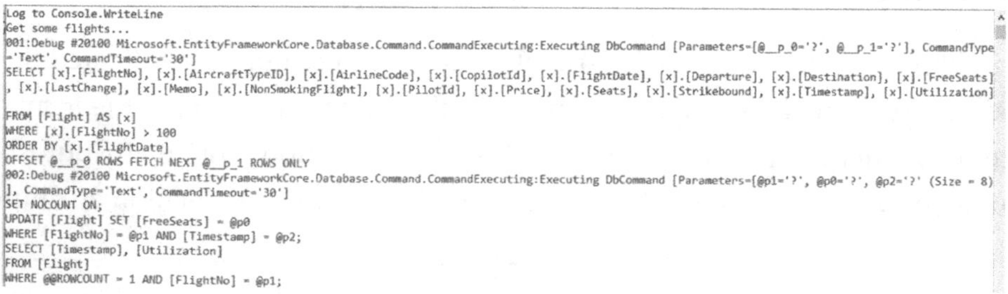

```
Log to Console.WriteLine
Get some flights...
001:Debug #20100 Microsoft.EntityFrameworkCore.Database.Command.CommandExecuting:Executing DbCommand [Parameters=[@__p_0='?', @__p_1='?'], CommandType
='Text', CommandTimeout='30']
SELECT [x].[FlightNo], [x].[AircraftTypeID], [x].[AirlineCode], [x].[CopilotId], [x].[FlightDate], [x].[Departure], [x].[Destination], [x].[FreeSeats]
, [x].[LastChange], [x].[Memo], [x].[NonSmokingFlight], [x].[PilotId], [x].[Price], [x].[Seats], [x].[Strikebound], [x].[Timestamp], [x].[Utilization]

FROM [Flight] AS [x]
WHERE [x].[FlightNo] > 100
ORDER BY [x].[FlightDate]
OFFSET @__p_0 ROWS FETCH NEXT @__p_1 ROWS ONLY
002:Debug #20100 Microsoft.EntityFrameworkCore.Database.Command.CommandExecuting:Executing DbCommand [Parameters=[@p1='?', @p0='?', @p2='?' (Size = 8)
], CommandType='Text', CommandTimeout='30']
SET NOCOUNT ON;
UPDATE [Flight] SET [FreeSeats] = @p0
WHERE [FlightNo] = @p1 AND [Timestamp] = @p2;
SELECT [Timestamp], [Utilization]
FROM [Flight]
WHERE @@ROWCOUNT = 1 AND [FlightNo] = @p1;
```

***Figure 12-1.** Default logging for the Log() method*

If you want to log to a file, you can do that by writing a method with a string parameter and no return value and by passing it to Log().

```
using (var ctx3 = new WWWingsContext())
  {
    Console.WriteLine("Get some flights...");
    var query3 = ctx3.FlightSet.Where(x => x.FlightNo > 100).OrderBy(x =>
    x.Date).Skip(10).Take(5);
    ctx3.Log(LogToFile);
    var flightSet3 = query3.ToList();
    flightSet3.ElementAt(0).FreeSeats--;
    ctx3.SaveChanges();
  }
}

  public static void LogToFile(string s)
  {
   Console.WriteLine(s);
   var sw = new StreamWriter(@"c:\temp\log.txt");
   sw.WriteLine(DateTime.Now + ": " + s);
   sw.Close();
  }
```

By default, the Log() method logs only those commands sent to the DBMS. Two other parameters of the Log() method affect the amount of logging, shown here:

- Parameter 2 is a list of logging categories (strings).

- Parameter 3 is a list of event numbers (numbers).

The next command will print all the log output from Entity Framework Core (which produces a lot of screen output for each command):

```
ctx1.Log(Console.WriteLine, new List<string>(), new List<int>());
```

The next command will print only certain logging categories and event numbers:

```
ctx1.Log(Console.WriteLine, new List<string>() { "Microsoft.
EntityFrameworkCore.Database.Command" }, new List<int>() { 20100, 20101});
```

Event 20100 is Executing, and 20101 is Executed.

**Note**    Because Entity Framework Core assigns the internally used Logger Factory classes not to one context instance but to all context instances, an established logging method on a particular instance also applies to other instances of the same context class.

# Implementing the Log() Extension Method

Listing 12-1 shows the implementation of the extension method Log().

- The Log() extension method adds an instance of a logger provider to the ILoggerFactory service.

- A logger provider is a class that implements ILoggerProvider. In this class, Entity Framework Core calls CreateLogger() once for each logging category.

- CreateLogger() must then supply a logger instance for each logging category.

- A logger is a class that implements ILogger.

- Listing 12-1 has a FlexLogger class that sends a string to the method specified by Log(). If no method is specified, ConsoleWriteLineColor() is called.

- A second logger class is NullLogger, which discards the log output for all log categories that have nothing to do with the SQL output.

*Listing 12-1.*  Entity Framework Core Extensions for Easy Logging

```
// Logging for EF Core
// (C) Dr. Holger Schwichtenberg, www.IT-Visions.de 2016-2017

using Microsoft.EntityFrameworkCore;
using Microsoft.EntityFrameworkCore.Infrastructure;
using Microsoft.Extensions.DependencyInjection;
using Microsoft.Extensions.Logging;
using System;
using System.Collections.Generic;
using System.Reflection;
```

```csharp
namespace ITVisions.EFCore
{
 /// <summary>
 /// Enhancement for the DbContext class for easy logging of the SQL
 commands sent by EF Core to a method that expects a string (C) Dr. Holger
 Schwichtenberg, www.IT-Visions.de
 /// </summary>
 public static class DbContextExtensionLogging
 {
  public static Dictionary<string, ILoggerProvider> loggerFactories = new
  Dictionary<string, ILoggerProvider>();

  public static bool DoLogging = true;
  public static bool DoVerbose = true;
  private static Version VERSION = new Version(4, 0, 0);

  private static List<string> DefaultCategories = new List<string>
  {
   "Microsoft.EntityFrameworkCore.Storage.
   IRelationalCommandBuilderFactory", // für EFCore 1.x
   "Microsoft.EntityFrameworkCore.Database.Sql", // für EFCore 2.0Preview1
   "Microsoft.EntityFrameworkCore.Database.Command", // für EFCore >=
   2.0Preview2
  };

  private static List<int> DefaultEventIDs = new List<int>
  {
   20100 // 20100 = "Executing"
 };

  public static void ClearLog(this DbContext ctx)
  {
   var serviceProvider = ctx.GetInfrastructure<IServiceProvider>();
   // Add Logger-Factory
   var loggerFactory = serviceProvider.GetService<ILoggerFactory>();
   (loggerFactory as LoggerFactory).Dispose();
  }
```

```
/// <summary>
/// Extension Method for Logging to a method expecting a string
/// </summary>
/// <example>Log() or Log(Console.WriteLine) for console logging
</example>
public static void Log(this DbContext ctx, Action<string> logMethod =
null, List<string> categories = null, List<int> eventsIDs = null, bool
verbose = false)
{
 DbContextExtensionLogging.DoVerbose = verbose;
 if (eventsIDs == null) eventsIDs = DefaultEventIDs;
 if (categories == null) categories = DefaultCategories;
 var methodName = logMethod?.Method?.Name?.Trim();
 if (string.IsNullOrEmpty(methodName)) methodName = "Default (Console.
 WriteLine)";

 if (DbContextExtensionLogging.DoVerbose)
 {
  Console.WriteLine("FLEXLOGGER EFCore " + VERSION.ToString() + " (C)
  Dr. Holger Schwichtenberg 2016-2017 " + methodName);
  Console.WriteLine("FLEXLOGGER Start Logging to " + methodName);
  Console.WriteLine("FLEXLOGGER Event-IDs: " + String.Join(";",
  eventsIDs));
  Console.WriteLine("FLEXLOGGER Categories: " + String.Join(";",
  categories));
 }
 // Make sure we only get one LoggerFactory for each LogMethod!
 var id = ctx.GetType().FullName + "_" + methodName.Replace(" ", "");
 if (!loggerFactories.ContainsKey(id))
 {
  if (verbose) Console.WriteLine("New Logger Provider!");
  var lp = new FlexLoggerProvider(logMethod, categories, eventsIDs);
  loggerFactories.Add(id, lp);
  // Get ServiceProvider
  var serviceProvider = ctx.GetInfrastructure();
```

```
  // Get Logger-Factory
  var loggerFactory = serviceProvider.GetService<ILoggerFactory>();
  // Add Provider to Factory
  loggerFactory.AddProvider(lp);
 }

}
}

/// <summary>
/// LoggerProvider for FlexLogger (C) Dr. Holger Schwichtenberg www.IT-
Visions.de
/// </summary>
public class FlexLoggerProvider : ILoggerProvider
{

public Action<string> _logMethod;
 public List<int> _eventIDs = null;
 public List<string> _categories = null;

 public FlexLoggerProvider(Action<string> logMethod = null, List<string>
 categories = null, List<int> eventIDs = null)
 {
  _logMethod = logMethod;
  _eventIDs = eventIDs;
  _categories = categories;
  if (_eventIDs == null) _eventIDs = new List<int>();
  if (_categories == null) _categories = new List<string>();
 }

 /// <summary>
 /// Constructor is called for each category. Here you have to specify
 which logger should apply to each category
 /// </summary>
 /// <param name="categoryName"></param>
 /// <returns></returns>
```

```
public ILogger CreateLogger(string categoryName)
{
 if (_categories == null || _categories.Count == 0 || _categories.
 Contains(categoryName))
 {
  if (DbContextExtensionLogging.DoVerbose) Console.WriteLine("FLEXLOGGER
  CreateLogger: " + categoryName + ": Yes");
  return new FlexLogger(this._logMethod, this._eventIDs);
 }
 if (DbContextExtensionLogging.DoVerbose) Console.WriteLine("FLEXLOGGER
 CreateLogger: " + categoryName + ": No");
 return new NullLogger(); // return NULL nicht erlaubt :-(
}

public void Dispose()
{ }

/// <summary>
/// Log output to console or custom method
/// </summary>
private class FlexLogger : ILogger
{
 private static int count = 0;

 readonly Action<string> logMethod;
 readonly List<int> _EventIDs = null;
 public FlexLogger(Action<string> logMethod, List<int> eventIDs)
 {
  count++;
  this._EventIDs = eventIDs;
  if (logMethod is null) this.logMethod = ConsoleWriteLineColor;
  else this.logMethod = logMethod;
 }

 private static void ConsoleWriteLineColor(object s)
 {
  var farbeVorher = Console.ForegroundColor;
  Console.ForegroundColor = ConsoleColor.Cyan;
```

```csharp
  Console.WriteLine(s);
  Console.ForegroundColor = farbeVorher;
 }

 public bool IsEnabled(LogLevel logLevel) => true;

 private static long Count = 0;

 public void Log<TState>(LogLevel logLevel, EventId eventId, TState
 state, Exception exception, Func<TState, Exception, string> formatter)
 {
  if (!DbContextExtensionLogging.DoLogging) return;

  if (Assembly.GetAssembly(typeof(Microsoft.EntityFrameworkCore.
  DbContext)).GetName().Version.Major == 1 || (this._EventIDs != null
  && (this._EventIDs.Contains(eventId.Id) || this._EventIDs.Count == 0)))
  {
   Count++;
   string text = $"{Count:000}:{logLevel} #{eventId.Id} {eventId.
   Name}:{formatter(state, exception)}";
   // Call log method now
   logMethod(text);
  }
 }

 public IDisposable BeginScope<TState>(TState state)
 {
  return null;
 }
}

/// <summary>
/// No Logging
/// </summary>
private class NullLogger : ILogger
{
 public bool IsEnabled(LogLevel logLevel) => false;
```

```
public void Log<TState>(LogLevel logLevel, EventId eventId, TState
state, Exception exception, Func<TState, Exception, string> formatter)
{ }

public IDisposable BeginScope<TState>(TState state) => null;
    }
  }
}
```

# Logging Categories

Unfortunately, Microsoft changed the names of the logging categories between Entity Framework Core versions 1.*x* and 2.0.

The following are the logging categories in Entity Framework Core 1.*x*:

- `Microsoft.EntityFrameworkCore.Storage.Internal.`
  `SQLServerConnection`

- `Microsoft.EntityFrameworkCore.Storage.IExecutionStrategy`

- `Microsoft.EntityFrameworkCore.Internal.`
  `RelationalModelValidator`

- `Microsoft.EntityFrameworkCore.Query.Internal.`
  `SqlServerQueryCompilationContextFactory`

- `Microsoft.EntityFrameworkCore.Query.Translators expression.`
  `Internal.SqlServerCompositeMethodCallTranslator`

- `Microsoft.EntityFrameworkCore.Storage.`
  `IRelationalCommandBuilderFactory`

- `Microsoft.EntityFrameworkCore.Query.Internal.QueryCompiler`

- `Microsoft.EntityFrameworkCore.DbContext`

The following are the logging categories in Entity Framework Core 2.0:

- `Microsoft.EntityFrameworkCore.Infrastructure`

- `Microsoft.EntityFrameworkCore.Update`

- `Microsoft.EntityFrameworkCore.Database.Transaction`

- `Microsoft.EntityFrameworkCore.Database.Connection`

- `Microsoft.EntityFrameworkCore.Model.Validation`

- `Microsoft.EntityFrameworkCore.Query`

- `Microsoft.EntityFrameworkCore.Database.Command`

The implementation of the `Log()` extension method takes this change into account; it also considers the category `Microsoft.EntityFrameworkCore.Query`, which has two events: `Executing` (event ID 20100) and `Executed` (event ID 20101). `Log()` outputs only the event ID 20100 in the standard system. However, categories and event IDs are controllable through the parameters of `Log()`.

# CHAPTER 13

# Asynchronous Programming

.NET has supported simplified asynchronous, task-based programming with `async` and `await` since .NET Framework 4.5. The classic ADO.NET Entity Framework has supported related asynchronous operations since version 6.0. In Entity Framework Core, you have been able to use asynchronous operations since version 1.0.

## Asynchronous Extension Methods

Entity Framework Core allows both the reading and writing of data with the asynchronous design pattern. For this Microsoft extended the result set delivering LINQ conversion and aggregate operators in the `Microsoft.EntityFrameworkCore.EntityFrameworkQueryableExtensions` class with asynchronous variants. Now you can find here for example, the `EntityFrameworkQueryableExtensions` class includes the extension methods `ToListAsync()` and `ToArrayAsync()`; the methods `SingleAsync()`, `FirstAsync()`, `SingleOrDefaultAsync()`, and `FirstOrDefaultAsync()`; and aggregate functions such as `CountAsync()`, `AllAsync()`, `AnyAsync()`, `AverageAsync()`, `MinAsync()`, `MaxAsync()`, and `SumAsyn()`. The `SaveChangesAsync()` method is used for saving. Putting a `using Microsoft.EntityFrameworkCore` line in your code is a prerequisite to using these extension methods.

© Holger Schwichtenberg 2018
H. Schwichtenberg, *Modern Data Access with Entity Framework Core*,
https://doi.org/10.1007/978-1-4842-3552-2_13

# ToListAsync()

Listing 13-1 shows how to use ToListAsync(). The only differences from a synchronous call with the same result set are in these two lines:

- The subprocedure DataReadingAsync() is declared as async.

- Instead of query.ToList(), now await query.ToListAsync() is called.

Figure 13-1 shows that database queries and object materialization by Entity Framework Core are actually asynchronous with these marginal changes. This shows that the main program is already waiting for an input before the first Flight is issued. The output Start Database Query is still in thread 10. Then it returns to the main program, while the database query and the remainder of the procedure Read DataReadingAsync() continue in thread 13.

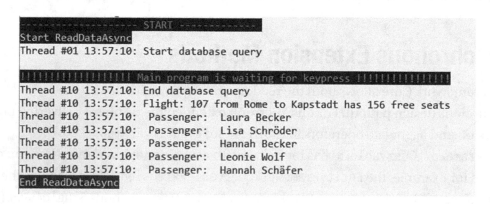

**Figure 13-1.**  *Output of Listing 13-1*

**Listing 13-1.**  Using ToListAsync()

```
public static async void ReadDataAsync()
{
CUI.MainHeadline("Start " + nameof(ReadDataAsync)); using (var ctx =
new WWWingsContext())
 {
  // Define query
```

```
  var query = (from f in ctx.FlightSet.Include(p => p.BookingSet).
  ThenInclude(b => b.Passenger) where f.Departure == "Rome" &&
  f.FreeSeats > 0 select f).Take(1);
  // Execute Query asynchronously
  CUI.PrintWithThreadID("Start database query");
  var flightSet = await query.ToListAsync();
  CUI.PrintWithThreadID("End database query");
  // Print results
  foreach (Flight flight in flightSet)
  {
   CUI.PrintWithThreadID("Flight: " + flight.FlightNo + " from " +
   flight.Departure + " to " + flight.Destination + " has " +
   flight.FreeSeats + " free seats");

   foreach (var p in flight.BookingSet.Take(5))
   {
    CUI.PrintWithThreadID(" Passenger:  " + p.Passenger.GivenName + " " +
    p.Passenger.Surname);
   }
  }
  CUI.MainHeadline("End " + nameof(ReadDataAsync));
 }
}
```

# SaveChangesAsync()

There is also an asynchronous operation for saving. Listing 13-2 shows how the threads switch to ToList() and SaveChangesAsync(), and Figure 13-2 shows the output.

***Listing 13-2.***  Using SaveChangesAsync()

```
public static async void ChangeDataAsync()
{
 CUI.MainHeadline("Start " + nameof(ChangeDataAsync));
 using (var ctx = new WWWingsContext())
 {
  // Define query
  var query = (from f in ctx.FlightSet.Include(p => p.BookingSet).
  ThenInclude(b => b.Passenger) where f.Departure == "Rome" &&
  f.FreeSeats > 0 select f).Take(1);
  // Query aynchron ausführen
  CUI.PrintWithThreadID("Start database query");
  var flightSet = await query.ToListAsync();
  CUI.PrintWithThreadID("End database query");
  // Print results
  foreach (Flight flight in flightSet)
  {
   CUI.PrintWithThreadID("Flight: " + flight.FlightNo + " from " +
   flight.Departure + " to " + flight.Destination + " has " + flight.
   FreeSeats + " free seats");

   foreach (var b in flight.BookingSet.Take(5))
   {
    CUI.PrintWithThreadID(" Passenger:   " + b.Passenger.GivenName + " " +
    b.Passenger.Surname);
    CUI.PrintWithThreadID("   Start saving");
    b.Passenger.Status = "A";
    var count = await ctx.SaveChangesAsync();
    CUI.PrintWithThreadID($"   {count} Changes saved!");
   }
  }
  CUI.Headline("End " + nameof(ChangeDataAsync));
 }
}
```

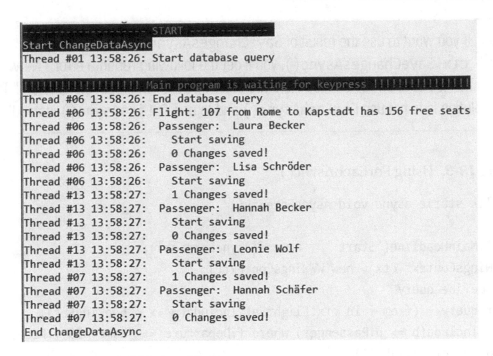

*Figure 13-2.* *Output of Listing 13-2*

# ForEachAsync()

With Entity Framework, it is not necessary to explicitly materialize a query before an iteration with a conversion operator such as ToList(). A foreach loop over an object with an IQueryable interface is enough to trigger the database query. In this case, however, the database connection remains open while the loop is running, and the records are fetched individually by the iterator of the IQueryable interface.

You can also execute this construct asynchronously using the method ForEachAsync(), which retrieves a result set step-by-step and executes the method body of a method (an anonymous method in the form of a lambda expression in Listing 13-3) over all elements of the set.

As mentioned in Chapter 10, you have to use SaveChangesAsync() instead of SaveChanges() to avoid a transaction problem between the still-running read operation and the write operation.

---

**Note**   If you want to use the result of SaveChangesAsync(): var count = await ctx.SaveChangesAsync(), you'll get the following runtime error: "New transaction is not allowed because there are other threads running in the session." The best solution is to not use ForEachAsync() and to use ToListAsync() instead!

---

***Listing 13-3.***  Using ForEachAsync( )

```
public static async void AsyncForeach()
{
 CUI.MainHeadline("Start " + nameof(AsyncForeach));
 WWWingsContext ctx = new WWWingsContext();
 // Define query
 var query = (from f in ctx.FlightSet.Include(p => p.BookingSet).
 ThenInclude(b => b.Passenger) where f.Departure == "Rome" && f.FreeSeats
 > 0 select f).Take(1);
 // Executing and iterate query with ForEachAsync
 CUI.PrintWithThreadID("Print objects");
 await query.ForEachAsync(async flight =>
{
 // Print results
 CUI.PrintWithThreadID("Flight: " + flight.FlightNo + " from " + flight.
 Departure + " to " + flight.Destination + " has " + flight.FreeSeats + "
 free Seats");

 foreach (var p in flight.BookingSet)
 {
  CUI.PrintWithThreadID(" Passenger: " + p.Passenger.GivenName + " " +
  p.Passenger.Surname);
 }

 // Save changes to each flight object within the loop
 CUI.PrintWithThreadID("  Start saving");
 flight.FreeSeats--;
 await ctx.SaveChangesAsync();
```

```
//not possible: var count  = await ctx.SaveChangesAsync(); --> "New
transaction is not allowed because there are other threads running in
the session."

CUI.PrintWithThreadID("   Changes saved!");
});

CUI.Headline("End " + nameof(AsyncForeach));
}
```

# CHAPTER 14

# Dynamic LINQ Queries

One of the benefits of Language Integrated Query (LINQ) is that compilers can validate the instructions at design time. However, there are some real-world cases in which the executing command is not completely established at design time, for example, because the user selects from a number of drop-down filters on a screen. This chapter shows ways to make LINQ commands dynamic partially or fully in conjunction with Entity Framework Core.

## Creating LINQ Queries Step-by-Step

LINQ queries are not executed until you execute a conversion operator (called *deferred execution*). Previously, you can extend a LINQ query with more conditions, but also with sorting and paging. Listing 14-1 shows how occasionally a condition for a departure or a destination, occasionally a filtering for flights with free seats and occasionally a sorting is to be added. The cases are controlled by the variables at the beginning, which represents a user input here.

---

**Tip** As long as you do not apply a conversion operator to a query, such as `ToList()`, `To Dictionary()`, `ToArray()`, or `First()`/`Single()`, or access on the enumerator or a single element (e.g., with `ElementAt()`), the query lingers in RAM and can still be modified.

---

© Holger Schwichtenberg 2018
H. Schwichtenberg, *Modern Data Access with Entity Framework Core*,
https://doi.org/10.1007/978-1-4842-3552-2_14

***Listing 14-1.*** Composing LINQ Commands

```
public static void LINQComposition()
{
 CUI.MainHeadline(nameof(LINQComposition));
 using (WWWingsContext ctx = new WWWingsContext())
 {
  ctx.Log();
  string departure = "Paris";
  string destination = "";
  bool onlyWithFreeSeats = true;
  bool sortieren = true;

  // Base query
  IQueryable<Flight> flightQuery = (from f in ctx.FlightSet select f);

  // Adding optional condition
  if (!String.IsNullOrEmpty(departure)) flightQuery = from f in
  flightQuery where f.Departure == departure select f;
  if (!String.IsNullOrEmpty(destination)) flightQuery = from f in
  flightQuery where f.Destination == destination select f;
  // Adding optional condition using a method
  if (onlyWithFreeSeats) flightQuery = FreeSeatsMustBeGreaterZero(flight
  Query);
  // Optional sorting
  if (sortieren) flightQuery = flightQuery.OrderBy(f => f.Date);

  // Send to the database now!
  List<Flight> flightSet = flightQuery.ToList();

  // Print the result set
  Console.WriteLine("Flights found:");
  foreach (var f in flightSet)
```

```
    {
     Console.WriteLine($"Flight Nr {f.FlightNo} from {f.Departure} to
     {f.Destination}: {f.FreeSeats} free seats! Pilot: {f.PilotId} ");
    }
   }
  }

  static public IQueryable<Flight> FreeSeatsMustBeGreaterZero(IQueryable
  <Flight> query)
  {
   return query.Where(f => f.FreeSeats > 0);
  }
```

The following SQL command is generated from this program code, in this case with a filter about Departure and FreeSeats and the sorting by flight date:

```
SELECT [f].[FlightNo], [f].[AircraftTypeID], [f].[AirlineCode], [f].
[CopilotId], [f].[FlightDate], [f].[Departure], [f].[Destination], [f].
[FreeSeats], [f].[LastChange], [f].[Memo], [f].[NonSmokingFlight], [f].
[PilotId], [f].[Price], [f].[Seats], [f].[Strikebound], [f].[Timestamp],
[f].[Utilization]
FROM [Flight] AS [f]
WHERE ([f].[Departure] = @__departure_0) AND ([f].[FreeSeats] > 0)
ORDER BY [f].[FlightDate]
```

# Expression Trees

Expression trees (.NET namespace System.Linq.Expressions) are the basis for all LINQ queries. Entity Framework Core converts every LINQ query into an expression tree. You can also work with expression trees directly, but it is usually very time-consuming.

The C# program code in Listing 14-2 first defines a LINQ query with a condition. Then, depending on the values of two local variables, the query is extended by two additional conditions. This program code also takes advantage of the fact that LINQ commands are not executed until the result set is actually needed (again, called *deferred execution*).

***Listing 14-2.*** Extending LINQ Commands with Expression Trees

```
public static void ExpressionTreeTwoConditions()
{
 CUI.MainHeadline(nameof(ExpressionTreeTwoConditions));

 string destination = "Rome";
 short? minNumberOfFreeSeats = 10;

 using (WWWingsContext ctx = new WWWingsContext())
 {
  // Base query
  IQueryable<BO.Flight> query = from flight in ctx.FlightSet where
  flight.FlightNo < 300 select flight;

  // Optional conditions
  if (!String.IsNullOrEmpty(destination) && minNumberOfFreeSeats > 0)
  {
   // Define query variable
   ParameterExpression f = Expression.Parameter(typeof(BO.Flight), "f");

   // Add first condition
   Expression left = Expression.Property(f, "Destination");
   Expression right = Expression.Constant(destination);
   Expression condition1 = Expression.Equal(left, right);

   // Add second condition
   left = Expression.Property(f, "FreeSeats");
   right = Expression.Constant((short?)minNumberOfFreeSeats,
   typeof(short?));
   Expression condition2 = Expression.GreaterThan(left, right);

   // Connect conditions with AND operator
   Expression predicateBody = Expression.And(condition1, condition2);

   // Build expression tree
   MethodCallExpression whereCallExpression = Expression.Call(
       typeof(Queryable),
```

```
    "Where",
    new Type[] { query.ElementType },
    query.Expression,
    Expression.Lambda<Func<BO.Flight, bool>>(predicateBody, new
    ParameterExpression[] { f }));

  // Create query from expression tree
  query = query.Provider.CreateQuery<BO.Flight>(whereCallExpression);
}

ctx.Log();
// Print the result set
Console.WriteLine("Flights found:");
foreach (BO.Flight f in query.ToList())
{
  Console.WriteLine($"Flight No {f.FlightNo} from {f.Departure} to
  {f.Destination}: {f.FreeSeats} free seats! Pilot: {f.PilotId} ");
  }
 }
}
```

The listing generates the following SQL command:

```
SELECT [flight].[FlightNo], [flight].[AircraftTypeID], [flight].
[AirlineCode], [flight].[CopilotId], [flight].[FlightDate], [flight].
[Departure], [flight].[Destination], [flight].[FreeSeats], [flight].
[LastChange], [flight].[Memo], [flight].[NonSmokingFlight], [flight].
[PilotId], [flight].[Price], [flight].[Seats], [flight].[Strikebound],
[flight].[Timestamp], [flight].[Utilization]
FROM [Flight] AS [flight]
WHERE ([flight].[FlightNo] < 300) AND ((CASE
    WHEN [flight].[Destination] = N'Rome'
    THEN CAST(1 AS BIT) ELSE CAST(0 AS BIT)
END & CASE
    WHEN [flight].[FreeSeats] > 10
    THEN CAST(1 AS BIT) ELSE CAST(0 AS BIT)
END) = 1)
```

Of course, this also works with any number of conditions that are not fixed at runtime, as Listing 14-3 shows, in which the conditions are passed as SortedDictionary.

***Listing 14-3.*** Extending LINQ Commands with Expression Trees

```
public static void ExpressionTreeNumerousConditions()
{
 CUI.MainHeadline(nameof(ExpressionTreeNumerousConditions));

 // Input data
 var filters = new SortedDictionary<string, object>() { { "Departure",
 "Berlin" }, { "Destination", "Rome" }, { "PilotID", 57 } };

 using (WWWingsContext ctx = new WWWingsContext())
 {
  ctx.Log();
  // Base query
  var baseQuery = from flight in ctx.FlightSet where flight.FlightNo <
  1000 select flight;

  ParameterExpression param = Expression.Parameter(typeof(BO.Flight), "f");

  Expression completeCondition = null;
  foreach (var filter in filters)
  {
   // Define condition
   Expression left = Expression.Property(param, filter.Key);
   Expression right = Expression.Constant(filter.Value);
   Expression condition = Expression.Equal(left, right);
   // Add to existing conditions using AND operator
   if (completeCondition == null) completeCondition = condition;
   else completeCondition = Expression.And(completeCondition, condition);
  }

  // Create query from expression tree
  MethodCallExpression whereCallExpression = Expression.Call(
      typeof(Queryable),
      "Where",
```

```
        new Type[] { baseQuery.ElementType },
        baseQuery.Expression,
        Expression.Lambda<Func<BO.Flight, bool>>(completeCondition, new
        ParameterExpression[] { param }));

    // Create query from expression tree
    var Q_Endgueltig = baseQuery.Provider.CreateQuery<BO.
    Flight>(whereCallExpression);

    // Print the result set
    Console.WriteLine("Flights found:");
    foreach (var f in Q_Endgueltig)
    {
     Console.WriteLine($"Flight Nr {f.FlightNo} from {f.Departure} to
     {f.Destination}: {f.FreeSeats} free seats! Pilot: {f.PilotId} ");
    }
   }
  }
```

Listing 14-3 results in the following SQL command:

```
SELECT [flight].[FlightNo], [flight].[AircraftTypeID], [flight].
[AirlineCode], [flight].[CopilotId], [flight].[FlightDate], [flight].
[Departure], [flight].[Destination], [flight].[FreeSeats], [flight].
[LastChange], [flight].[Memo], [flight].[NonSmokingFlight], [flight].
[PilotId], [flight].[Price], [flight].[Seats], [flight].[Strikebound],
[flight].[Timestamp], [flight].[Utilization]
FROM [Flight] AS [flight]
WHERE ([flight].[FlightNo] < 1000) AND ((((CASE
    WHEN [flight].[Departure] = N'Berlin'
    THEN CAST(1 AS BIT) ELSE CAST(0 AS BIT)
END & CASE
    WHEN [flight].[Destination] = N'Rome'
    THEN CAST(1 AS BIT) ELSE CAST(0 AS BIT)
END) & CASE
    WHEN [flight].[PilotId] = 57
    THEN CAST(1 AS BIT) ELSE CAST(0 AS BIT)
END) = 1)
```

# Using Dynamic LINQ

If creating LINQ queries step-by-step is not sufficient, you do not necessarily have to use expression trees. An alternative is to use the library Dynamic LINQ. Dynamic LINQ is not part of the .NET Framework and is not an official add-on. Dynamic LINQ is just an example that Microsoft used in a sample collection (see `http://weblogs.asp.net/scottgu/archive/2008/01/07/dynamic-linq-part-1-using-the-linq-dynamic-query-library.aspx`). However, this example was "ennobled" by a blog post by Scott Guthrie (`http://weblogs.asp.net/scottgu/archive/2008/01/07/dynamic-linq-part-1-using-the-linq-dynamic-query-library.aspx`) and has been widely used ever since. The original example had `System.Linq.Dynamic` in the root namespace `System`, which was unusual and suggested that Microsoft might integrate this into the .NET Framework in the future. This has not happened yet and does not seem to be on the agenda anymore. Dynamic LINQ consists of several classes with around 2,000 lines of code. The most important class is `DynamicQueryable`. This class provides many extension methods for the `IQueryable` interface, such as `Where()`, `OrderBy()`, `GroupBy()`, and `Select()`, all of which accept strings.

Listing 14-4 shows a solution with Dynamic LINQ that is more elegant than the expression tree solution.

---

**Note**   Unfortunately, there are no dynamic joins in Dynamic LINQ., but you can find a solution on the Internet (`http://stackoverflow.com/questions/389094/how-to-create-a-dynamic-linq-join-extension-method`).

---

*Listing 14-4.* Use of Dynamic LINQ

```
public static void DynamicLINQNumerousCondition()
{
 CUI.MainHeadline(nameof(DynamicLINQNumerousCondition));
 // input data
 var filters = new SortedDictionary<string, object>() { { "Departure",
 "Berlin" }, { "Destination", "Rome" }, { "PilotID", 57 } };
 string sorting = "FreeSeats desc";
```

```csharp
using (WWWingsContext ctx = new WWWingsContext())
{
 ctx.Log();
 // base query
 IQueryable<BO.Flight> query = from flight in ctx.FlightSet where
 flight.FlightNo < 1000 select flight;

 // Add conditions
 foreach (var filter in filters)
 {
  Console.WriteLine(filter.Value.GetType().Name);
  switch (filter.Value.GetType().Name)
  {
   case "String":
    query = query.Where(filter.Key + " = \"" + filter.Value + "\""); break;

   default:
     query = query.Where(filter.Key + " = " + filter.Value); break;
  }
 }

 // optional sorting
 if (!String.IsNullOrEmpty(sorting)) query = query.OrderBy(sorting);

 // Print the result set
 Console.WriteLine("Flights found:");
 foreach (var f in query)
 {
  Console.WriteLine($"Flight Nr {f.FlightNo} from {f.Departure} to
  {f.Destination}: {f.FreeSeats} free seats!");
 }
}
}
```

The following SQL command results from this program code:

```
SELECT [flight].[FlightNo], [flight].[AircraftTypeID], [flight].
[AirlineCode], [flight].[CopilotId], [flight].[FlightDate], [flight].
[Departure], [flight].[Destination], [flight].[FreeSeats], [flight].
[LastChange], [flight].[Memo], [flight].[NonSmokingFlight], [flight].
[PilotId], [flight].[Price], [flight].[Seats], [flight].[Strikebound],
[flight].[Timestamp], [flight].[Utilization]
FROM [Flight] AS [flight]
WHERE (((([flight].[FlightNo] < 1000) AND ([flight].[Departure] = N'Berlin'))
AND ([flight].[Destination] = N'Rome')) AND ([flight].[PilotId] = 57)
ORDER BY [flight].[FreeSeats] DESC
```

# Reading and Modifying Data with SQL, Stored Procedures, and Table-Valued Functions

If LINQ and the Entity Framework Core API are not sufficient in terms of functionality or performance, you can send any SQL commands directly to the database, including invoking stored procedures and using table-value functions (TVFs).

LINQ and the Entity Framework Core API (Add(), Remove(), SaveChanges(), and so on) are abstractions of SQL. Entity Framework Core (or the respective database provider) converts LINQ and API calls to SQL. The abstraction that Entity Framework Core offers here is, in many cases, well-suited to sending efficient, robust database management system–neutral commands to the database. But LINQ and the API cannot do everything that SQL can do, and not everything that Entity Framework Core sends to the database is always powerful enough.

Even in the classic Entity Framework, you were able to send SQL commands to the database instead of using LINQ. In Entity Framework Core, you have some of these same abilities, but they take a different form. In some cases there are more options, but in other situations there are fewer options than in the classic Entity Framework.

© Holger Schwichtenberg 2018
H. Schwichtenberg, *Modern Data Access with Entity Framework Core*,
https://doi.org/10.1007/978-1-4842-3552-2_15

# Writing Queries with FromSql()

For SQL queries that return an entity type known to the Entity Framework Core
context, the FromSql() and FromSql<EntityType>() methods are available in the
DbSet<EntityType> class. The result is an IQueryable<EntityType> (see Listing 15-1).

---

**Note**   You should not compose the SQL commands as a string because this
carries the risk of a SQL injection attack.

---

*Listing 15-1.* SQL Query in Entity Framework Core, Risking a SQL Injection Attack

```
public static void Demo_SQLDirect1()
{
 CUI.MainHeadline(nameof(Demo_SQLDirect1));
 string departure = "Berlin";
 using (var ctx = new WWWingsContext())
 {
  ctx.Log();
  IQueryable<Flight> flightSet = ctx.FlightSet.FromSql("Select * from
  Flight where Departure='" + departure + "'");
  Console.WriteLine(flightSet.Count());
  foreach (var flight in flightSet)
  {
   Console.WriteLine(flight);
  }
  Console.WriteLine(flightSet.Count());
 }
}
```

It is better to use the .NET placeholders {0}, {1}, {2}, and so on (see Listing 15-2).
These placeholders are handled by Entity Framework Core as parameterized SQL
commands so that a SQL injection attack is not possible.

***Listing 15-2.*** SQL Query in Entity Framework Core Without the Risk of SQL Injection Attack

```
public static void Demo_SQLDirect2()
{
 CUI.MainHeadline(nameof(Demo_SQLDirect2));
 string departure = "Berlin";
 using (var ctx = new WWWingsContext())
 {
  ctx.Log();
  IQueryable<Flight> flightSet = ctx.FlightSet.FromSql("Select * from
  Flight where Departure={0}", departure);
  Console.WriteLine(flightSet.Count());
  foreach (var flight in flightSet)
  {
   Console.WriteLine(flight);
  }
  Console.WriteLine(flightSet.Count());
 }
}
```

Since Entity Framework Core 2.0, it is even possible to use string interpolation, which has existed since C# 6.0 (see Listing 15-3).

***Listing 15-3.*** Third Variant of SQL Query in Entity Framework

```
public static void Demo_SQLDirect3()
{
 CUI.MainHeadline(nameof(Demo_SQLDirect3));
 string departure = "Berlin";
 string destination = "Rome";
 using (var ctx = new WWWingsContext())
 {
  ctx.Log();
  IQueryable<Flight> flightSet = ctx.FlightSet.FromSql($@"Select * from
  Flight where Departure={departure} and Destination={destination}");
  Console.WriteLine(flightSet.Count());
```

```
  foreach (var flight in flightSet)
  {
   Console.WriteLine(flight);
  }
  Console.WriteLine(flightSet.Count());
 }
}
```

In Listing 15-2 and Listing 15-3 you do not have to put the placeholders in single quotes. Strictly speaking, you should not use single quotes here because Entity Framework Core converts the query into a parameterized query (the parameters become an instance of the class dbParameter). In both listings, the database receives a SQL command with a parameter, as shown here:

```
Select * from Flight where departure = @p0
```

or with two parameters, as shown here:

```
Select * from Flight where departure= @p0 and destination = @p1
```

You are therefore protected against SQL injection attacks; you should never compose the SQL command as a string to avoid such vulnerabilities!

As with the classic Entity Framework, the SQL command must supply all columns to completely populate the entity object. A partial filling (projection) is not supported yet. Any attempt at partial filling fails because Entity Framework Core complains at runtime with the following error message: "The required column xy was not present in the results of a FromSql." This happens until all the properties of the entity object have counterparts in the result set.

# Using LINQ and SQL Together

The implementation in the classic Entity Framework was called DbSet<EntityClass>. SqlQuery() and did not return an IQueryable<EntityClass> as the return object; instead it returned an instance of DbRawSqlQuery<EntityType>. Returning IQueryable<EntityClass> has the advantage that you can now mix SQL and LINQ in one query.

Here's an example:

```
IQueryable<Flight> Flightlist = ctx.FlightSet.FromSql("Select * from Flight
where departure = {0}", location);
Console.WriteLine (Flight list.Count());
foreach(var Flight in Flight list) {...}
```

Here Entity Framework Core will convert flightSet.Count() to the following:

```
SELECT COUNT(*)
FROM (
    Select * from Flight where departure = 'Berlin'
) AS [f]
```

The self-written SQL command is thus embedded as a subquery in the query that is generated by the LINQ operation.

In this case, there are three round-trips to the database, two for counting and one for collecting the records. As with LINQ, you should make sure to use a conversion operator such as ToList() when using FromSql(). In Listing 15-4, the result set will now be fetched directly, and instead of counting three-times, you have only one query of the length of the materialized object set in RAM, which is much faster!

***Listing 15-4.*** ToList() Ensures That There Is Only One Round-Trip to the Database

```
public static void Demo_SQLDirect4()
{
 CUI.MainHeadline(nameof(Demo_SQLDirect4));
 string departure = "Berlin";
 using (var ctx = new WWWingsContext())
 {
  ctx.Log();
  List<Flight> flightSet = ctx.FlightSet.FromSql($@"Select * from Flight
  where Departure={departure}").ToList();
  Console.WriteLine(flightSet.Count());
  foreach (var flight in flightSet)
  {
   Console.WriteLine(flight);
  }
```

```
Console.WriteLine(flightSet.Count());
 }
}
```

Even more impressively, Listing 15-5 shows the possibility to combine SQL and LINQ together in Entity Framework Core, where Include() even loads linked records. This otherwise Listing is impossible with FromSql().

***Listing 15-5.*** Compiling SQL and LINQ in Entity Framework Core

```
public static void Demo_SQLDirectAndLINQComposition()
{
 CUI.MainHeadline(nameof(Demo_SQLDirectAndLINQComposition));
 string departure = "Berlin";
 using (var ctx = new WWWingsContext())
 {
  ctx.Log();
  var flightSet = ctx.FlightSet.FromSql("Select * from Flight where
  Departure={0}", departure).Include(f => f.Pilot).Where(x => x.FreeSeats >
  10).OrderBy(x => x.FreeSeats).ToList();
  Console.WriteLine(flightSet.Count());
  foreach (var flight in flightSet)
  {
   Console.WriteLine(flight);
  }
  Console.WriteLine(flightSet.Count());
 }
}
```

With the program code from Listing 15-5, the database receives the following SQL command:

```
SELECT [x].[FlightNo], [x].[AircraftTypeID], [x].[AirlineCode], [x].
[CopilotId], [x].[FlightDate], [x].[Departure], [x].[Destination], [x].
[FreeSeats], [x].[LastChange], [x].[Memo], [x].[NonSmokingFlight], [x].
[PilotId], [x].[Price], [x].[Seats], [x].[Strikebound], [x].[Timestamp],
[x].[Utilization], [x.Pilot].[PersonID], [x.Pilot].[Birthday], [x.Pilot].
```

```
[DetailID], [x.Pilot].[Discriminator], [x.Pilot].[EMail], [x.Pilot].
[GivenName], [x.Pilot].[PassportNumber], [x.Pilot].[Salary], [x.Pilot].
[SupervisorPersonID], [x.Pilot].[Surname], [x.Pilot].[FlightHours],
[x.Pilot].[FlightSchool], [x.Pilot].[LicenseDate], [x.Pilot].
[PilotLicenseType]
FROM (
    Select * from Flight where Departure=@p0
) AS [x]
INNER JOIN [Employee] AS [x.Pilot] ON [x].[PilotId] = [x.Pilot].[PersonID]
WHERE ([x.Pilot].[Discriminator] = N'Pilot') AND ([x].[FreeSeats] > 10)
ORDER BY [x].[FreeSeats]
```

# Using Stored Procedures and Table-Valued Functions

With FromSql() you can also call stored procedures, which provide a result set, and you can call table-valued functions elegantly (Listing 15-6). It should be noted, however, that the composability works only with table-valued functions.

***Listing 15-6.*** Using a Stored Procedure That Delivers Flight Records

```
/// <summary>
/// Use of a stored procedure that delivers Flight records
/// </summary>
public static void Demo_SP()
{
 CUI.MainHeadline(nameof(Demo_SP));

 using (var ctx = new WWWingsContext())
 {
  ctx.Log();
  var flightSet = ctx.FlightSet.FromSql("EXEC GetFlightsFromSP {0}",
  "Berlin").Where(x => x.FreeSeats > 0).ToList();
  Console.WriteLine(flightSet.Count());
  foreach (var flight in flightSet)
```

```
    {
     Console.WriteLine(flight);
    }
    Console.WriteLine(flightSet.Count());
  }
}
```

This listing runs the following statement in the database:

```
EXEC GetFlightsFromSP @p0
```

The supplementary conditions (here, FreeSeats) are executed in RAM.

In Listing 15-7, however, the additional condition is executed in the database management system because a table-valued function is called here.

***Listing 15-7.*** Using a Table-Valued Function That Delivers Flight Records

```
public static void Demo_TVF()
{
 CUI.MainHeadline(nameof(Demo_TVF));
 using (var ctx = new WWWingsContext())
 {
  ctx.Log();
  var flightSet = ctx.FlightSet.FromSql("Select * from
  GetFlightsFromTVF({0})", "Berlin").Where(x => x.FreeSeats > 10).ToList();
  Console.WriteLine(flightSet.Count());
  foreach (var flight in flightSet)
  {
   Console.WriteLine(flight);
  }
  Console.WriteLine(flightSet.Count());
 }
}
```

Listing 15-7 runs the following in the database:

```
SELECT [x].[FlightNo], [x].[AircraftTypeID], [x].[AirlineCode], [x].
[CopilotId], [x].[FlightDate], [x].[Departure], [x].[Destination], [x].
[FreeSeats], [x].[LastChange], [x].[Memo], [x].[NonSmokingFlight], [x].
```

```
[PilotId], [x].[Price], [x].[Seats], [x].[Strikebound], [x].[Timestamp],
[x].[Utilization]
FROM (
    Select * from GetFlightsFromTVF(@p0)
) AS [x]
WHERE [x].[FreeSeats] > 10
```

However, in Entity Framework Core, there is neither a program code generator for wrapper methods for stored procedures and table-valued functions (as is available in Database First in the classic Entity Framework) nor a SQL generator for stored procedures for INSERT, UPDATE, and DELETE (as available with Code First in the classic Entity Framework).

---

**Tip**   The third-party tool Entity Developer (see Chapter 20) provides a program code generator for wrapper methods for stored procedures and table-valued functions.

---

# Using Nonentity Classes as Result Sets

Unfortunately, there is also (at least so far) a big limitation in Entity Framework Core compared to its predecessor. In the classic Entity Framework, SqlQuery() was not only offered on instances of the class DbSet<EntityType> but also in the Database object in the context class. You could also specify other types that were not entity types, in other words, other self-defined classes or even elementary data types, in which the classic Entity Framework would materialize the query results.

Unfortunately, Entity Framework Core cannot do this yet (https://github.com/aspnet/EntityFramework/issues/1862). Any attempt to specify an object type other than FromSql() as the entity type does not compile.

```
var flightSet = ctx.FromSql<FligthDTO>("Select FlightNo, Departure,
Destination, Date from Flight");
```

An experiment with the following method Set <T>():

```
flightSet = ctx.Set<FligthDTO>().FromSql("Select FlightNo, Departure,
Destination, Date from Flight");
```

compiles, but Entity Framework Core then says at runtime that this is not supported: "Cannot create a dbSet for 'FlightDTO' because this type is not included in the model for the context."

---

**Preview**   Microsoft will introduce mapping to arbitrary types in Entity Framework Core version 2.1; see Appendix C.

---

Listing 15-8 shows the retrofitting of an ExecuteSqlQuery() method in the Database object, which returns only one DbDataReader object and does not allow materialization. This extension method will then be used in Listing 15-9.

***Listing 15-8.***  Database Extension Method.ExecuteSqlQuery()

```
public static class RDFacadeExtensions
{
    public static RelationalDataReader ExecuteSqlQuery(this DatabaseFacade
    databaseFacade, string sql, params object[] parameters)
  {
   var concurrencyDetector = databaseFacade.GetService<IConcurrencyDetector>();

   using (concurrencyDetector.EnterCriticalSection())
   {
    var rawSqlCommand = databaseFacade
        .GetService<IRawSqlCommandBuilder>()
        .Build(sql, parameters);

    return rawSqlCommand
        .RelationalCommand
        .ExecuteReader(
            databaseFacade.GetService<IRelationalConnection>(),
            parameterValues: rawSqlCommand.ParameterValues);
  }
  }
}
```

***Listing 15-9.*** Using Database.ExecuteSqlQuery()

```
public static void Demo_Datareader()
{
 CUI.MainHeadline(nameof(Demo_Datareader));
 string Ort = "Berlin";
 using (var ctx = new WWWingsContext())
 {
  RelationalDataReader rdr = ctx.Database.ExecuteSqlQuery("Select * from
  Flight where Departure={0}", Ort);
  DbDataReader dr = rdr.DbDataReader;
  while (dr.Read())
  {
   Console.WriteLine("{0}\t{1}\t{2}\t{3} \n", dr[0], dr[1], dr[2], dr[3]);
  }
  dr.Dispose();
 }
}
```

# Using SQL DML Commands Without Result Sets

SQL Data Manipulation Language (DML) commands that do not return a result set such as INSERT , UPDATE, and DELETE can be executed in Entity Framework Core exactly like in the classic Entity Framework with ExecuteSqlCommand() in the Database object. You will get back the number of affected records (see Listing 15-10).

***Listing 15-10.*** Using Database.ExecuteSqlCommand()

```
public static void Demo_SqlCommand()
{
 CUI.MainHeadline(nameof(Demo_SqlCommand));
 using (var ctx = new WWWingsContext())
 {
  var count = ctx.Database.ExecuteSqlCommand("Delete from Flight where
  flightNo > {0}", 10000);
  Console.WriteLine("Number of deleted records: " + count);
 }
}
```

# CHAPTER 16

# Tips and Tricks for Mapping

This chapter describes additional ways to influence the mapping of entity classes to the database schema. Many of these possibilities are not shown in the example of World Wide Wings because they do not fit into it well.

## Shadow Properties

Entity Framework Core can create *shadow properties* (aka *shadow state properties*) for those columns of the database table for which there is no corresponding property or field in the entity class.

## Automatic Shadow Properties

A shadow property is created automatically if there is no matching foreign key property in the entity class for a navigation relationship. Because the database schema needs a foreign key column for most relations, this column automatically becomes a shadow property. If the counterpart to a navigation relationship has several primary key columns, then several foreign key columns with one shadow property each are created accordingly.

© Holger Schwichtenberg 2018
H. Schwichtenberg, *Modern Data Access with Entity Framework Core*,
https://doi.org/10.1007/978-1-4842-3552-2_16

The foreign key column and the shadow property are formed from the name of the navigation property and the primary key name of the main class. In doing so, Entity Framework Core avoids word duplications (see Table 16-1).

***Table 16-1.*** *Automatic Naming for Shadow Properties*

| Primary Key of the Main Class | Name of the Navigation Property | Name of the Foreign Key Column/Shadow Property |
|---|---|---|
| MainClassID | MainClass | MainClassID |
| ID | MainClass | MainClassID |
| ID | Anything | AnythingID |
| MainClassID | Anything | AnythingMainClassID |
| MainClassID1 and MainClassID2 (composite key) | Anything | AnythingMainClassID1 and AnythingMainClassID2 |

# Defining a Shadow Property

You can also define shadow properties manually using the `Property()` method. Listing 16-1 shows how to add a `DateTime` shadow property named `LastChange`. This name, however, is stored in a variable and thus can be changed.

---

**Note**   Changing the content of the variable `ShadowPropertyName` must be done before the first instantiation of the context class because `OnModelCreating()` is called only once, the first time the context class is used.

---

***Listing 16-1.*** Defining a Shadow Property in OnModelCreating() of the Context Class

```
public class WWWingsContext: DbContext
{
  static public string ShadowPropertyName = "LastChange";
  ...
  protected override void OnModelCreating (ModelBuilder builder)
```

```
  {
  ...
  builder.Entity<Flight>().Property<DateTime>(ShadowPropertyName);
  }
}
```

## Getting the Output of All Shadow Properties of an Entity Class

You can obtain the list of all properties of an entity class (the real properties and the shadow properties) from the Properties object set of the EntityEntry<T> instance, which is obtained via the Entry() method of the context class (Listing 16-2).

***Listing 16-2.*** Printing a List of All Properties of an Entity Class Including the Shadow Properties

```
using (WWWingsContext ctx = new WWWingsContext())
{
 var flight = ctx.FlightSet.SingleOrDefault(x => x.FlightNo == flightNo);

 foreach (var p in ctx.Entry(flight).Properties)
 {
  Console.WriteLine(p.Metadata.Name + ": " + p.Metadata.IsShadowProperty);
 }
}
```

## Reading and Changing a Shadow Property

A shadow property cannot be used directly on the entity object since there is no real property for the corresponding database column. The use therefore takes place via an instance of EntityEntry<T>, which is obtained via the method Entry() of the context class. You can then call the method Property("ColumnName") and query the CurrentValue property from the supplied PropertyEntry object.

```
ctx.Entry(flight).Property("LastChange").CurrentValue
```

Incidentally, this way you can access any information from an entity object, including real properties.

```
ctx.Entry(flight).Property("FreeSeats").CurrentValue
```

However, this is usually not done because access through the entity object is easier.

```
flight.FreeSeats
```

---

**Note**   You can only access real properties and shadow properties via `Property` (`"Name"`). If the database table has additional columns, they are not accessible by Entity Framework. The call `ctx.Entry(Flight).Property("abc")`. `CurrentValue` causes the following runtime error: "The property 'abc' on entity type 'Flight' could not be found. Ensure that the property exists and has been included in the model."

---

You can also change the value via `CurrentValue`, as shown here:

```
ctx.Entry(Flight).Property("LastChange").CurrentValue = DateTime.Now;
```

Listing 16-3 shows the use of the `LastChange` shadow property on the `Flight` entity class.

***Listing 16-3.*** Using a Shadow Property

```
public static void ReadAndChangeShadowProperty()
  {
    int flightNo = 101;
    CUI.MainHeadline(nameof(ReadAndChangeShadowProperty));

    using (WWWingsContext ctx = new WWWingsContext())
    {
      var flight = ctx.FlightSet.SingleOrDefault(x => x.FlightNo == flightNo);

      CUI.Headline("List of all shadow property of type Flight");
      foreach (var p in ctx.Entry(flight).Properties)
```

```
  {
    Console.WriteLine(p.Metadata.Name + ": " + p.Metadata.IsShadowProperty);
  }

  CUI.Print("Before: " + flight.ToString() + " / " + ctx.Entry(flight).
  State, ConsoleColor.Cyan);
  Console.WriteLine("Free seats: " + ctx.Entry(flight).
  Property("FreeSeats").CurrentValue);
  Console.WriteLine("Last change: " + ctx.Entry(flight).
  Property("LastChange").CurrentValue);

  CUI.PrintWarning("Changing object...");
  flight.FreeSeats += 1;
  ctx.Entry(flight).Property("LastChange").CurrentValue = DateTime.Now;

  CUI.Print("After: " + flight.ToString() + " / " + ctx.Entry(flight).
  State, ConsoleColor.Cyan);
  Console.WriteLine("Free seats: " + ctx.Entry(flight).
  Property("FreeSeats").CurrentValue);
  Console.WriteLine("Last change: " + ctx.Entry(flight).
  Property("LastChange").CurrentValue);

  var count = ctx.SaveChanges();
  Console.WriteLine("Number of saved changes: " + count);
  }
}
```

# Writing LINQ Queries with Shadow Properties

You can also use shadow properties in Language Integrated Query (LINQ).
Here you cannot use Entry() and Property(); you must use the special construct
EF.Property<T>(). The class EF is a static class of Entity Framework Core.

This construct is not only allowed in conditions but also in sorting and projections.
Listing 16-4 identifies the last change that occurred in the last two days.

***Listing 16-4.*** LINQ Queries with Shadow Properties

```
CUI.Headline("LINQ query using a Shadow Property");
using (WWWingsContext ctx = new WWWingsContext())
{
 var date = ctx.FlightSet
  .Where(c => EF.Property<DateTime>(c, WWWingsContext.ShadowStateProp) >
  DateTime.Now.AddDays(-2))
  .OrderByDescending(c => EF.Property<DateTime>(c, WWWingsContext.
  ShadowStateProp))
  .Select(x => EF.Property<DateTime>(x, WWWingsContext.ShadowStateProp))
  .FirstOrDefault();

 Console.WriteLine("Last change: " + date);
}
```

# Practical Example: Automatically Updating the Shadow Property Every Time You Save

Shadow properties are especially useful if you want to hide information from the developer. Listing 16-5 shows how to override the SaveChanges() method in the context class, which will automatically update the LastChange shadow property to the current date and time each time a change to a Flight object is saved.

***Listing 16-5.*** Using a Shadow Property in the Overwritten SaveChanges() Method

```
public override int SaveChanges()
  {
   // Detect changes
   this.ChangeTracker.DetectChanges();

   // Search all new and changed flights
   var entries = this.ChangeTracker.Entries<Flight>()
       .Where(e => e.State == EntityState.Added || e.State == EntityState.
       Modified);
```

```
if (!String.IsNullOrEmpty(ShadowStateProp))
{
 // set the Shadow State column "LastChange" for all of them
 foreach (var entry in entries)
 {
  entry.Property(ShadowStateProp).CurrentValue = DateTime.Now;
 }
}

// Save changes (we do not need DetectChanges() to be called again!)
this.ChangeTracker.AutoDetectChangesEnabled = false;
var result = base.SaveChanges(); // Call base class now
this.ChangeTracker.AutoDetectChangesEnabled = true;
return result;
}
```

# Computed Columns

Computed columns are, from the point of view of Entity Framework Core, those database columns whose value the database management system assigns. These can be as follows:

- Auto-increment value columns (identity columns) such as PersonID in the Person class and the classes Employee, Pilot, and Passenger derived from Person

- Timestamp columns (see Chapter 11)

- Columns with default values

- Columns with a calculation formula

## Automatic SELECT

In all of these cases, Entity Framework Core responds with an automatic SELECT after any generated INSERT or UPDATE to read the new value of the calculated columns from the database management system into RAM. In the case of a default column, however, this makes sense only after an INSERT. Entity Framework Core therefore knows three strategies with corresponding methods in the Fluent API:

- ValueGeneratedOnAdd(): A SELECT is executed only after an INSERT.

- ValueGeneratedOnAddOrUpdate(): A SELECT is executed both after an INSERT and after an UPDATE.

- ValueGeneratedNever(): No SELECT is executed after INSERT or UPDATE.

ValueGeneratedNever() is the default for all columns, except in these three cases:

- Primary key columns that consist of a single integer column of type Int16 (short), Int32 (int), or Int64 (long) are created by default as auto-increment columns with ValueGeneratedOnAdd().

- Default columns automatically get ValueGeneratedOnAdd().

- Byte-array (byte[]) columns annotated with [Timestamp] are automatically given ValueGeneratedOnAddOrUpdate().

---

**Note**   While Microsoft SQL Server does allow identity columns of type tinyint (i.e., byte), Entity Framework Core always sets for the Byte data type, that the primary keys do not use auto-increment values.

---

# Practical Example: Creating Columns with a Calculation Formula

In contrast to its predecessor, Entity Framework Core supports the definition of calculation formulas in the program code for forward engineering that are to be executed in the database management system.

You create a property for the calculated column in the entity class. For example, the entity class Flight will get a property called Utilization, which will be the percentage of booked seats. It also gets Seats (total number of seats in the aircraft) and FreeSeats. It makes sense to declare the setter of the Utilization property for the calculated column as private because no user of the object should be able to set the value. You cannot omit the setter completely here because Entity Framework Core needs a way to set the value received from the database management system.

```
public class Flight
{
  public int FlightNo {get; set; }
  ...
  public short Seats {get; set; }
  public short? FreeSeats {get; set; }
  public decimal? Utilization {get; private set; }
}
```

Now define a formula in the Fluent API in OnModelCreating() with
HasComputedColumnSql(). In this case, the percentage utilization for each Flight is
calculated from the number of free seats and the total seats.

```
modelBuilder.Entity<Flight>().Property(p => p.Utilization)
            .HasComputedColumnSql("100.0-(([FreeSeats]*1.0)/[Seats])*100.0");
```

---

**Note**   It is not necessary to use ValueGeneratedOnAddOrUpdate() because
HasComputedColumnSql() implies this strategy.

---

When creating a schema migration, you can see the formula again in the
MigrationBuilder call.

```
    public partial class v2_FlightCostload: Migration
    {
        protected override void up (MigrationBuilder migrationBuilder)
        {
            migration builder.AddColumn<int>(
                name: "Utilization",
                table: "Flight",
                type: "decimal",
                nullable: true,
                computedColumnSql: "100.0-(([FreeSeats]*1.0)/
                [Seats])*100.0");
        }
    ...
    }
```

You can also see this formula in the database schema (Figure 16-1).

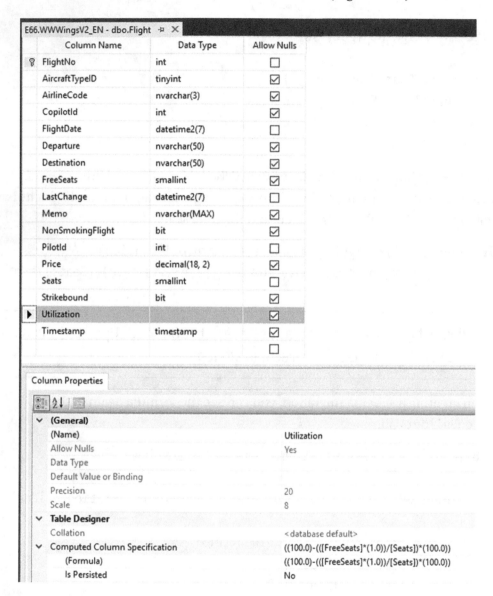

***Figure 16-1.*** *Calculation formula column in Microsoft SQL Server Management Studio*

# Using Columns with a Calculation Formula

The property of a column with a formula can be read like any other, as shown in
Listing 16-6.

---

**Note**   The value of a property of a column with a calculation formula column
will not change until you have executed SaveChanges(). In Listing 16-6, the
utilization initially shows the old value after the number of seats has been reduced.
Only after SaveChanges() do get the new value.

---

*Listing 16-6.*   The Utilization Column in the Flight Class Is Based on a Formula

```
public static void ComputedColumnWithFormula()
{
 CUI.MainHeadline(nameof(ComputedColumnWithFormula));

 int flightNo = 101;
 using (WWWingsContext ctx = new WWWingsContext())
 {
  ctx.Log();
  var flight = ctx.FlightSet.Find(flightNo);
  Console.WriteLine($"BEFORE: {flight}: Utilization={flight.
  Utilization:##0.00}%");

  flight.FreeSeats -= 10;
  //not possible: flight.Utilization = 100;

  // The change is not yet visible in the Utilization, since the
  Utilization is calculated in the DBMS
  Console.WriteLine($"After changes: {flight}: Utilization={flight.
  Utilization:##0.00}%");

  ctx.SaveChanges();
  // The change in Utilization is now visible
  Console.WriteLine($"After saving: {flight}: Utilization={flight.
  Utilization:##0.00}%");
```

```
CUI.Headline("Metadata of Flight properties");
foreach (PropertyEntry p in ctx.Entry(flight).Properties)
{
  Console.WriteLine(p.Metadata.Name + ": " + p.Metadata.ValueGenerated);
}
}
}
```

Figure 16-2 shows the output of Listing 16-6. Note the following:

- The utilization is updated only after SaveChanges().

- The utilization column has the strategy OnAddOrUpdate (what you get in a PropertyEntry object via Metadata.ValueGenerated).

- Therefore, after SaveChanges(), Entity Framework Core performs a SELECT on that value (and the timestamp).

```
H:\TFS\Demos\EFC\EFC_Samples\EFC_Console\bin\Debug\EFC_Console.exe                                    —   □   ×
ComputedColumnWithFormula
001:Debug #20100 Microsoft.EntityFrameworkCore.Database.Command.CommandExecuting:Executing DbCommand [Parameters=[@__get_Item_0='?'], CommandType='Tex
t', CommandTimeout='30']
SELECT TOP(1) [e].[FlightNo], [e].[AircraftTypeID], [e].[AirlineCode], [e].[CopilotId], [e].[FlightDate], [e].[Departure], [e].[Destination], [e].[Fre
eSeats], [e].[LastChange], [e].[Memo], [e].[NonSmokingFlight], [e].[PilotId], [e].[Price], [e].[Seats], [e].[Strikebound], [e].[Timestamp], [e].[Utili
zation]
FROM [Flight] AS [e]
WHERE [e].[FlightNo] = @__get_Item_0
BEFORE: Flight #101: from Berlin to Seattle on 24.04.18 13:31: 17 free Seats.: Utilization=93,20%
After changes: Flight #101: from Berlin to Seattle on 24.04.18 13:31: 7 free Seats.: Utilization=93,20%
002:Debug #20100 Microsoft.EntityFrameworkCore.Database.Command.CommandExecuting:Executing DbCommand [Parameters=[@p2='?', @p0='?', @p1='?', @p3='?' (
Size = 8)], CommandType='Text', CommandTimeout='30']
SET NOCOUNT ON;
UPDATE [Flight] SET [FreeSeats] = @p0, [LastChange] = @p1
WHERE [FlightNo] = @p2 AND [Timestamp] = @p3;
SELECT [Timestamp], [Utilization]
FROM [Flight]
WHERE @@ROWCOUNT = 1 AND [FlightNo] = @p2;
After saving: Flight #101: from Berlin to Seattle on 24.04.18 13:31: 7 free Seats.: Utilization=97,20%
Metadata of Flight properties
FlightNo: Never
AircraftTypeID: Never
AirlineCode: Never
CopilotId: Never
Date: OnAdd
Departure: OnAdd
Destination: OnAdd
FreeSeats: Never
LastChange: Never
Memo: Never
NonSmokingFlight: Never
PilotId: Never
Price: OnAdd
Seats: Never
Strikebound: Never
Timestamp: OnAddOrUpdate
Utilization: OnAddOrUpdate
```

***Figure 16-2.*** *Output of Listing 16-6*

# Using Columns with a Calculation Formula in Reverse Engineering

Entity Framework Core recognizes calculation formulas while reverse engineering with Scaffold-DbContext and creates a property accordingly. However, the property receives a public setter. The formula is stored by the code generator in the Fluent API, but this has no meaning in reverse engineering. It would only become meaningful if you later switched from reverse engineering to forward engineering.

```
public decimal? Utilization {get; set; }
```

The Fluent API contains the following:

```
entity.Property(e => e.Utilization)
.HasColumnType("numeric(20, 8)")
.HasComputedColumnSql("((100.0-(([FreeSeats]*1.0)/[Seats])*100.0))")
```

The basis for the reverse engineering was the database schema with the column Utilization, which was previously generated by forward engineering.

# Default Values

Entity Framework Core supports default values set by the database management system while creating records for columns for which no value has been passed. This support exists for both forward engineering and reverse engineering.

## Defining Default Values for Forward Engineering

Defining column default values to be assigned by the database management system when an explicit value is not supplied is done using the methods HasDefaultValue() and HasDefaultValueSql()in the Fluent API. The default value can be one of the following:

- A static value (e.g., number or string), using HasDefaultValue()

- A SQL expression (e.g., calling a function like getdate()), using HasDefaultValueSql()

Here are some examples:

```
f.Property(x => x.Price).HasDefaultValue(123.45m);
f.Property(x => x.Departure).HasDefaultValue("(not set)");
```

```
f.Property(x => x.Destination).HasDefaultValue("(not set)");
f.Property(x => x.Date).HasDefaultValueSql("getdate()");
```

Entity Framework Core considers these default values when creating the database (see Figure 16-3).

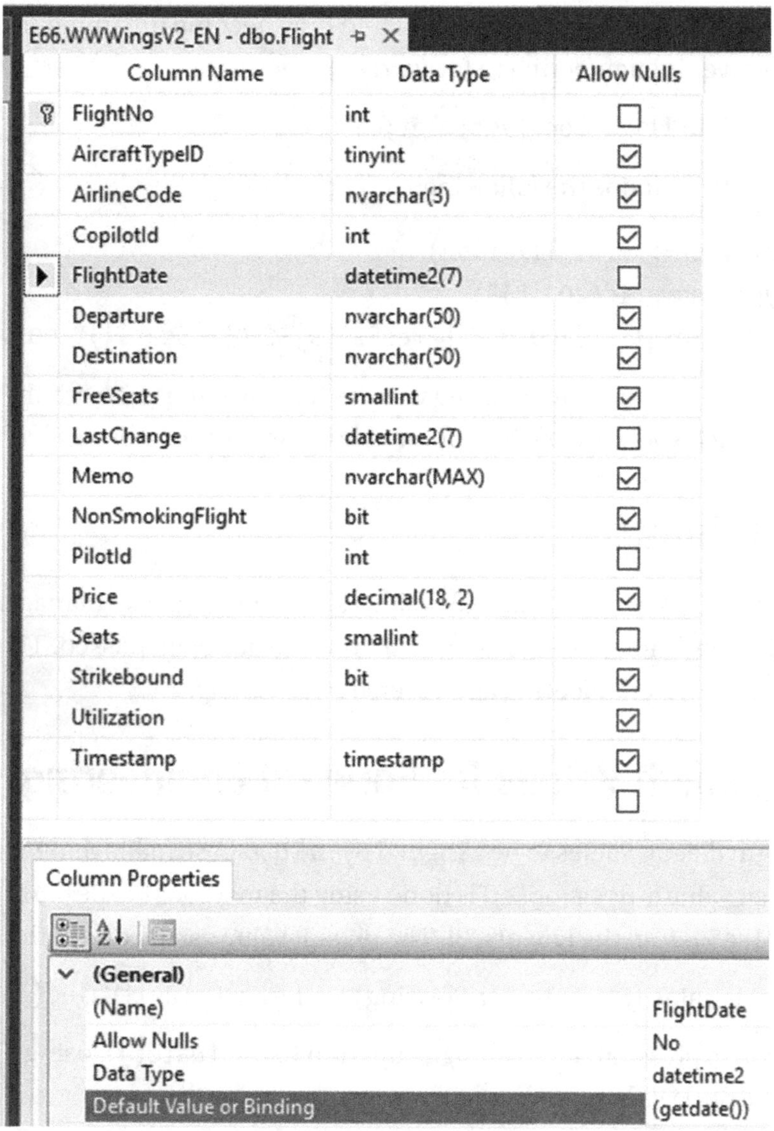

**Figure 16-3.** *Default value for the FlightDate column in Microsoft SQL Server Management Studio*

# Using Default Values

Defined default values are taken into account by Entity Framework Core. As part of the ValueOnAdd strategy, Entity Framework Core queries the default values after an INSERT, but only if no value has been passed before.

Listing 16-7 shows this very impressively.

- A value for the Departure is set in code. Therefore, Entity Framework Core does not ask for the departure location in SELECT.

- Destination is null. Entity Framework Core therefore does ask in SELECT for the Destination value given by the database management system.

- Flight Date and Price were not set in the new object. Entity Framework Core therefore queries in SELECT for the FlightDate and Price values assigned by the database management system.

***Listing 16-7.*** Using Default Values

```
public static void DefaultValues()
{
 CUI.MainHeadline(nameof(DefaultValues));
 using (WWWingsContext ctx = new WWWingsContext())
 {
  var pilot = ctx.PilotSet.FirstOrDefault();
  ctx.Log();
  var f = new Flight();
  f.FlightNo = ctx.FlightSet.Max(x => x.FlightNo) + 1;
  f.Departure = "Berlin";
  f.Destination = null;
  f.Pilot = pilot;
  f.Copilot = null;
  f.FreeSeats = 100;
  f.Seats = 100;
  CUI.Headline("Object has been created in RAM");
  Console.WriteLine($"{f} Price: {f.Price:###0.00} Euro.");
  ctx.FlightSet.Add(f);
```

```
CUI.Headline("Object has been connected to the ORM");
Console.WriteLine($"{f} Price: {f.Price:###0.00} Euro.");
ctx.SaveChanges();
CUI.Headline("Object has been saved");
Console.WriteLine($"{f} Price: {f.Price:###0.00} Euro.");

f.FreeSeats--;
CUI.Headline("Object has been changed in RAM");
Console.WriteLine($"{f} Price: {f.Price:###0.00} Euro.");
ctx.SaveChanges();
CUI.Headline("Object has been saved"); ;
Console.WriteLine($"{f} Price: {f.Price:###0.00} Euro.");

//if (f.Destination != "(not set)") Debugger.Break();
//if (f.Price != 123.45m) Debugger.Break();
 }
}
```

Figure 16-4 shows that the default values are set only after SaveChanges(). After updating with UPDATE, Entity Framework Core does not ask for the defaults again (only the column with the Utilization calculation formula and the timestamp column assigned by the database management system are part of the SELECT after UPDATE).

***Figure 16-4.*** *Output of Listing 16-7*

# Practical Example: Defaults Already Assigned When Creating the Object

If you want the defaults values to work in RAM immediately after an object has been created, then you do not have to assign these defaults in the database management system but in the constructor of the class (Listing 16-8).

***Listing 16-8.*** Assigning Default Values in the Constructor

```
public class Flight
 {

  /// <summary>
  /// Parameterless constructor
  /// </summary>
  public Flight()
  {
   // Default Values
   this.Departure = "(not set)";
   this.Destination = "(not set)";
   this.Price = 123.45m;
   this.Date = DateTime.Now;
  }
...
}
```

Figure 16-5 uses the example of the Date and Price properties to show that the default values apply immediately. However, the Departure and Destination default values set in the Flight constructor do not work because the program code overrides these values.

```
DefaultValues
003:Debug #20100 Microsoft.EntityFrameworkCore.Database.Command.CommandExecuting:Executing DbCommand [Parameters=[], CommandType='Text', CommandTimeou
t='30']
SELECT TOP(1) [e].[PersonID] [e].[Birthday], [e].[DetailID], [e].[Discriminator], [e].[EMail], [e].[GivenName], [e].[PassportNumber], [e].[Salary], [
e].[SupervisorPersonID], [e].[Surname], [e].[FlightHours], [e].[FlightSchool], [e].[LicenseDate], [e].[PilotLicenseType]
FROM [Employee] AS [e]
WHERE [e].[Discriminator] = N'Pilot'
004:Debug #20100 Microsoft.EntityFrameworkCore.Database.Command.CommandExecuting:Executing DbCommand [Parameters=[], CommandType='Text', CommandTimeou
t='30']
SELECT MAX([x].[FlightNo])
FROM [Flight] AS [x]
Object has been created in RAM
Flight #456801: from Berlin to  on 30.12.17 07:22: 100 free Seats. Price: 123,45 Euro.
Object has been connected to the ORM
Flight #456801: from Berlin to  on 30.12.17 07:22: 100 free Seats. Price: 123,45 Euro.
005:Debug #20100 Microsoft.EntityFrameworkCore.Database.Command.CommandExecuting:Executing DbCommand [Parameters=[@p0='?', @p1='?', @p2='?' (Size = 3)
, @p3='?', @p4='?', @p5='?' (Size = 50), @p6='?', @p7='?', @p8='?' (Size = 4000), @p9='?', @p10='?', @p11='?', @p12='?', @p13='?'], CommandType='Text'
, CommandTimeout='30']
SET NOCOUNT ON;
INSERT INTO [Flight] ([FlightNo], [AircraftTypeID], [AirlineCode], [CopilotId], [FlightDate], [Departure], [FreeSeats], [LastChange], [Memo], [NonSmok
ingFlight], [PilotId], [Price], [Seats], [Strikebound])
VALUES (@p0, @p1, @p2, @p3, @p4, @p5, @p6, @p7, @p8, @p9, @p10, @p11, @p12, @p13);
SELECT [Destination], [Timestamp], [Utilization]
FROM [Flight]
WHERE @@ROWCOUNT = 1 AND [FlightNo] = @p0;
Object has been saved
Flight #456801: from Berlin to (offen) on 30.12.17 07:22: 100 free Seats. Price: 123,45 Euro.
Object has been changed in RAM
Flight #456801: from Berlin to (offen) on 30.12.17 07:22: 99 free Seats. Price: 123,45 Euro.
006:Debug #20100 Microsoft.EntityFrameworkCore.Database.Command.CommandExecuting:Executing DbCommand [Parameters=[@p2='?', @p0='?', @p1='?', @p3='?' (
Size = 8)], CommandType='Text', CommandTimeout='30']
SET NOCOUNT ON;
UPDATE [Flight] SET [FreeSeats] = @p0, [LastChange] = @p1
WHERE [FlightNo] = @p2 AND [Timestamp] = @p3;
SELECT [Timestamp], [Utilization]
FROM [Flight]
WHERE @@ROWCOUNT = 1 AND [FlightNo] = @p2;
Object has been saved
Flight #456801: from Berlin to (offen) on 30.12.17 07:22: 99 free Seats. Price: 123,45 Euro.
```

***Figure 16-5.*** *Output of the Use Defaults listing when default values are specified in the constructor*

# Using Default Values for Reverse Engineering

Entity Framework Core recognizes default values on Scaffold-DbContext and places them in the Fluent API. All that matters here is that Entity Framework Core knows there is a default value. What the actual value is does not matter for reverse engineering. In fact, you could also enter an empty string here later. The generated code always uses HasDefaultValueSql(), even for static values.

```
entity.Property(e => e.FlightDate).HasDefaultValueSql("(getdate())")
entity.Property(e => e.Price).HasDefaultValueSql("((123.45))");
entity.Property(e => e.Departure)
                .HasMaxLength(50)
                .HasDefaultValueSql("(N'(not set)')");
entity.Property(e => e.Destination)
                .HasMaxLength(50)
                .HasDefaultValueSql("(N'(not set)')");
```

# Table Splitting

Since Entity Framework Core 2.0, the OR mapper allows you to spread a single database table across multiple entity classes. The table splitting is done as follows:

- You create an entity class (Master) and one or more dependent classes for the table.

- The entity class implements 1:1 navigation properties that reference the dependent classes.

- For these navigation properties you call the method OwnsOne() in OnModelCreating().

Listing 16-9 shows a Master class with three dependent Split classes.

***Listing 16-9.*** Project EFC_MappingTest, TableSplitting.cs

```
using ITVisions;
using Microsoft.EntityFrameworkCore;
using System;
using System.Collections.Generic;

namespace EFC_MappingScenarios.TableSplitting
{
 /// <summary>
 /// In this example, several classes are deliberately implemented in one
file, so that the example is clearer.
 /// </summary>
 class DEMO_TableSplitting
 {
  public static void Run()
  {
   CUI.MainHeadline(nameof(DEMO_TableSplitting));
   using (var ctx = new MyContext())
   {
    CUI.Print("Database: " + ctx.Database.GetDbConnection().ConnectionString);

    var e = ctx.Database.EnsureCreated();
```

```csharp
    if (e)
    {
     CUI.Print("Database has been created!");
    }
    else
    {
     CUI.Print("Database exists!");

    }


    CUI.Headline("Detail");
    var obj1 = new Detail();
    foreach (var p in ctx.Entry(obj1).Properties)
    {
     Console.WriteLine(p.Metadata.Name + ": " + p.Metadata.
     IsShadowProperty);
    }

    CUI.Headline("Master");
    var obj2 = new Master();
    foreach (var p in ctx.Entry(obj2).Properties)
    {
     Console.WriteLine(p.Metadata.Name + ": " + p.Metadata.
     IsShadowProperty);
    }
   }
  }
}

class MyContext : DbContext
{
 public DbSet<Master> MasterSet { get; set; }
 public DbSet<Detail> DetailSet { get; set; }
```

```csharp
protected override void OnConfiguring(DbContextOptionsBuilder builder)
{
 // Set provider and connectring string
 string connstring = @"Server=.;Database=EFC_MappingTest_
 TableSplitting;Trusted_Connection=True;MultipleActiveResultSets=True;";
 builder.UseSqlServer(connstring);
}

protected override void OnModelCreating(ModelBuilder modelBuilder)
{
 // Define a composite key
 modelBuilder.Entity<Master>().HasKey(b => new { b.MasterId1, b.MasterId2 });

 // Define table splitting
 modelBuilder.Entity<Master>().OwnsOne(c => c.Split1);
 modelBuilder.Entity<Master>().OwnsOne(c => c.Split2);
 modelBuilder.Entity<Master>().OwnsOne(c => c.Split3);
 }
}

public class Master
{
 public int MasterId1 { get; set; }
 public int MasterId2 { get; set; }
 public string Memo { get; set; }

 public List<Detail> DetailSet { get; set; }
 public Split1 Split1 { get; set; }
 public Split2 Split2 { get; set; }
 public Split3 Split3 { get; set; }
}

public class Detail
{
 public int DetailId { get; set; }
 public string DetailMemo { get; set; }

 public Master Master { get; set; }
}
```

```
public class Split1
{
 public string Memo1 { get; set; }
}
public class Split2
{
 public string Memo2 { get; set; }
}

public class Split3
{
 public string Memo3 { get; set; }
}
}
```

Figure 16-6 shows that the MasterSet database table includes all the properties of the Master class and the three classes named Split1, Split2, and Split3.

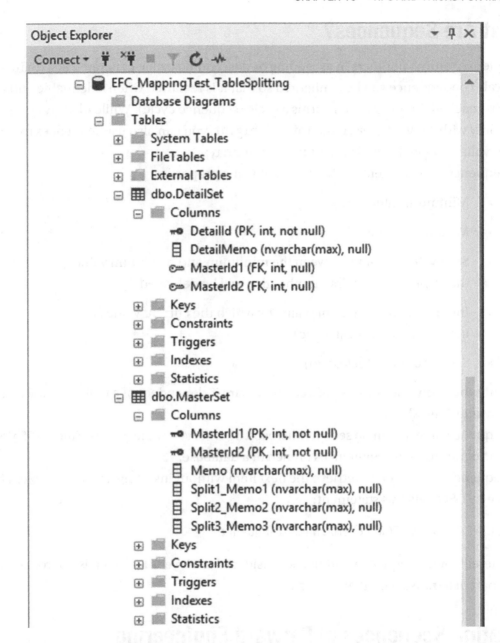

*Figure 16-6.* *Created database*

# Sequence Objects (Sequences)

A sequence object (aka a *sequence*) exists in some database management systems
(e.g., Oracle, Microsoft SQL Server) for the flexible allocation of auto-increment values.
In Microsoft SQL Server, sequences have existed since version 2012.

# What Are Sequences?

A sequence returns numbers in ascending or descending order, generated at a defined interval. The sequence can be configured so that it is restarted when a definable end value is reached, in other words, forms a cycle (sequence cycle). While identity columns work only with `tinyint`, `smallint`, `int`, and `bigint`, you can also use sequences to assign values to the decimal and numeric column types.

Sequence objects therefore have the following properties:

- Minimum value

- Maximum value

- Start value (a value between the minimum and maximum values) that represents the first number that will be delivered

- Increment or decrement (value by which the current value is increased for the next value)

- Sequence cycle (yes or no)

Unlike identity columns, sequence objects are independent of a table, so they can be used by multiple tables.

Sequence objects can be seen in Microsoft SQL Server Management Studio (SSMS) in the Programmability/Sequences branch of a database.

Sequences allow you to retrieve the next item without inserting a row in a table. This is done in T-SQL via this statement:

```
select NEXT VALUE FOR schema.NameofSequence
```

Thereafter, the sequence number is considered retrieved. The fetch is not rolled back if it is part of a transaction that is aborted.

# Creating Sequences at Forward Engineering

A sequence is defined in the Fluent API via `HasSequence()` and then used with `HasDefaultValueSql()` (e.g., for a primary key). Listing 16-10 shows the definition of a cyclic sequence between 1000 and 1300, with steps of 10 starting at 1100. This sequence is then used in three places (simple primary key, composite primary key, and other column). Figure 16-7 shows the output.

***Listing 16-10.*** Creating and Applying Sequences

```
// cyclic sequence between 1000 and 1300, step 10, starting at 1100
modelBuilder.HasSequence<int>("Setp10IDs", schema: "demo")
.StartsAt(1100).IncrementsBy(10).HasMin(1000).HasMax(1300).IsCyclic();

// Sequence used for primary key (Data type: short)
modelBuilder.Entity<EntityClass1>()
        .Property(o => o.EntityClass1Id)
        .HasDefaultValueSql("NEXT VALUE FOR demo.Setp10IDs");
// Sequence used for normal column (Data type: decimal)
modelBuilder.Entity<EntityClass2>()
      .Property(o => o.Value)
      .HasDefaultValueSql("NEXT VALUE FOR demo.Setp10IDs");

// Sequence used for part of a composite key (Data type: int)
modelBuilder.Entity<EntityClass3>().HasKey(b => new { b.EntityClass3Id1,
b.EntityClass3Id2 });
modelBuilder.Entity<EntityClass3>()
   .Property(o => o.EntityClass3Id1)
   .HasDefaultValueSql("NEXT VALUE FOR demo.Setp10IDs");
```

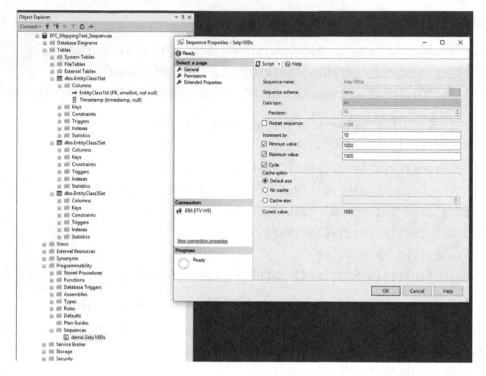

**Figure 16-7.** *Created sequence in Microsoft SQL Server Management Studio*

# Seeing Sequences in Action

Listing 16-11 shows a self-contained example:

1.  The context class references three entity classes.

2.  The context class creates a sequence.

3.  Entity class 1 uses the sequence for the primary key.

4.  Entity class 2 uses the sequence for the Value column (type decimal).

5.  Entity class 3 uses the sequence for part of the primary key.

6.  The program then creates and persists instances of all three classes.

***Listing 16-11.*** Project EFC_MappingTest, Sequences.cs

```
using ITVisions;
using Microsoft.EntityFrameworkCore;
using System;
using System.ComponentModel.DataAnnotations;

namespace EFC_MappingScenarios.Sequences
{
 /// <summary>
 /// In this example, several classes are deliberately implemented in one
file, so that the example is clearer.
 /// </summary>
 class DEMO_SequencesDemos
 {
  public static void Run()
  {
   CUI.MainHeadline(nameof(DEMO_SequencesDemos));
   using (var ctx = new Kontext())
   {
    CUI.Print("Database: " + ctx.Database.GetDbConnection().ConnectionString);
    var e = ctx.Database.EnsureCreated();
    if (e)
    {
     CUI.Print("Database has been created!");
    }
    else
    {
     CUI.Print("Database exists!");
    }
// This will fail, because we consume more IDs that the sequence defines!
    for (int i = 0; i < 30; i++)
    {
     var obj1 = new EntityClass1();
     ctx.EntityClass1Set.Add(obj1);
```

```csharp
      CUI.Headline("EntityClass1");
      Console.WriteLine($"BEFORE: PK: {obj1.EntityClass1Id}");
      var count1 = ctx.SaveChanges();
      Console.WriteLine($"Saved changes: {count1}.PK: {obj1.EntityClass1Id}");
      CUI.Headline("EntityClass2");
      var obj2 = new EntityClass2();
      ctx.EntityClass2Set.Add(obj2);
      Console.WriteLine($"BEFORE: PK: {obj2.EntityClass2Id} Value: {obj2.Value}");
      var count2 = ctx.SaveChanges();
      Console.WriteLine($"Saved changes: {count2}.PK: {obj2.EntityClass2Id}
      Value: {obj2.Value}");
      CUI.Headline("EntityClass3");
      var obj3 = new EntityClass3();
      ctx.EntityClass3Set.Add(obj3);
      Console.WriteLine($"BEFORE: PK: {obj3.EntityClass3Id1}/{obj3.
      EntityClass3Id2}");
      var count3 = ctx.SaveChanges();
      Console.WriteLine($"Saved changes: {count3}. PK: {obj3.
      EntityClass3Id1}/{obj3.EntityClass3Id2}");
    }
   }
  }
}
class Kontext : DbContext
{
 public DbSet<EntityClass1> EntityClass1Set { get; set; }
 public DbSet<EntityClass2> EntityClass2Set { get; set; }
 public DbSet<EntityClass3> EntityClass3Set { get; set; }
 protected override void OnConfiguring(DbContextOptionsBuilder builder)
 {
  // Set provider and connectring string
  string connstring = @"Server=.;Database=EFC_MappingTest_
  Sequences;Trusted_Connection=True;MultipleActiveResultSets=True;";
  builder.UseSqlServer(connstring);
  builder.EnableSensitiveDataLogging(true);
 }
```

```
protected override void OnModelCreating(ModelBuilder modelBuilder)
{
 // cyclic sequence between 1000 and 1300, step 10, starting at 1100
 modelBuilder.HasSequence<int>("Setp10IDs", schema: "demo")
 .StartsAt(1100).IncrementsBy(10).HasMin(1000).HasMax(1300).IsCyclic();

 // Sequence used for primary key (Data type: short)
 modelBuilder.Entity<EntityClass1>()
         .Property(o => o.EntityClass1Id)
         .HasDefaultValueSql("NEXT VALUE FOR demo.Setp10IDs");
 // Sequence used for normal column (Data type: decimal)
 modelBuilder.Entity<EntityClass2>()
       .Property(o => o.Value)
       .HasDefaultValueSql("NEXT VALUE FOR demo.Setp10IDs");

 // Sequence used for part of a composite key (Data type: int)
 modelBuilder.Entity<EntityClass3>().HasKey(b => new { b.EntityClass3Id1,
 b.EntityClass3Id2 });
 modelBuilder.Entity<EntityClass3>()
    .Property(o => o.EntityClass3Id1)
    .HasDefaultValueSql("NEXT VALUE FOR demo.Setp10IDs");
 }
}

public class EntityClass1
{
 public short EntityClass1Id { get; set; }
 [Timestamp]
 public byte[] Timestamp { get; set; }
}

public class EntityClass2
{
 public int EntityClass2Id { get; set; }
 public decimal Value { get; set; }
}
```

```
public class EntityClass3
{
/// Composite PK
public int EntityClass3Id1 { get; set; }
public int EntityClass3Id2 { get; set; }
}
}
```

Figure 16-8 doesn't show the first run starting at sequence number 1100, but one of the following runs. For entity class 1, the sequence number 1260 is assigned; for entity class 2, 1270 is assigned; and for entity class 3, 1280 is assigned. Then for entity class 1, 1290 is used, and for entity class 2, 1300 is assigned. Thus, the end of the value range of the cyclic sequence is reached. The next value is 1000.

```
H:\TFS\Demos\EFC\EFC_Samples\AdditionalSamples\EFC_MappingTest\bin\Debug\EFC_MappingTest.exe
EntityClass1
BEFORE: PK=-32641
Saved changes: 1 / PK=1260
EntityClass2
BEFORE: PK=-2147482621 Value=0
Saved changes: 1 / PK=27 Value=1270,00
EntityClass3
BEFORE: PK=-2147482621|0
Saved changes: 1 / PK=1280|0
EntityClass1
BEFORE: PK=-32640
Saved changes: 1 / PK=1290
EntityClass2
BEFORE: PK=-2147482620 Value=0
Saved changes: 1 / PK=28 Value=1300,00
EntityClass3
BEFORE: PK=-2147482620|0
Saved changes: 1 / PK=1000|0
EntityClass1
BEFORE: PK=-32639
Saved changes: 1 / PK=1010
EntityClass2
BEFORE: PK=-2147482619 Value=0
Saved changes: 1 / PK=29 Value=1020,00
EntityClass3
BEFORE: PK=-2147482619|0
Saved changes: 1 / PK=1030|0
EntityClass1
BEFORE: PK=-32638
Saved changes: 1 / PK=1040
```

***Figure 16-8.*** *Output of Listing 16-11*

**Note**   If it eventually happens that a value for a primary key that is already in use in this table is used in the sequence cycle, the following runtime error occurs: "Violation of PRIMARY KEY constraint 'PK_EntityClass1Set'. Can not insert duplicate key into object 'dbo.EntityClass1Set'. The duplicate key value is (…)."

# Alternative Keys

In addition to a primary key (which may consist of one or more columns), a table can have additional keys (which can also consist of one or more columns) that uniquely identify each row in the database table. For this purpose, database management systems have concepts such as unique indexes and unique constraints. Both concepts are similar. This is discussed at `https://technet.microsoft.com/en-us/library/aa224827(v=sql.80).aspx`.

*Table 16-2.* *Unique Index vs. Alternative Key*

| Entity Framework Concept | Unique Index | Alternative Key |
|---|---|---|
| Database concept | Unique index | Unique constraint |
| Supported in classic Entity Framework | Yes, from version 6.1 | No |
| Supported in Entity Framework Core | Yes, from version 1.0 | Yes, from version 1.0 |
| Use as foreign key in relationships | No | Yes |

The classic Entity Framework Core allows the creation of a unique index only. In Entity Framework Core, you can now also create unique constraints. Entity Framework Core calls this an *alternative key*.

# Defining Alternative Keys

An alternative key is defined in the Fluent API with HasAlternateKey(). Here the class Detail receives an alternative key for the column Guid.

```
modelBuilder.Entity<Detail>().HasAlternateKey(c => c.Guid);
```

An alternative key, like a primary key, can be composed of multiple columns. To do this, use an anonymous object as a parameter in HasAlternateKey():

```
modelBuilder.Entity<Detail>()
.HasAlternateKey(c => new { c.Guid, Bereich = c.Area });
```

Entity Framework Core automatically generates an alternate key if, in a relationship definition, you do not create the relationship between foreign key and primary key but use a different column of the parent class instead of the primary key.

```
modelBuilder.Entity<Detail>()
        .HasOne(p => p.Master)
        .WithMany(b => b.DetailSet)
        .HasForeignKey(p => p.MasterGuid)
        .HasPrincipalKey(b => b.Guid);
```

Figure 16-9 shows the created unique constraints in Microsoft SQL Server Management Studio. Entity Framework Core gives unique constraints names that begin with the letters AK (for "alternative key").

The MasterSet table has a unique constraint for the Guid column in addition to the primary key. In addition to the primary key and the foreign key leading to the Guid column of MasterSet, the DetailSet table still has two unique constraints: one on the Guid column and one on the Guid and Area columns together as a composite key. This proves that Entity Framework Core allows a column to be part of multiple alternative keys.

---

**Tip**   If possible, you should always use relationships to form integer columns, and the foreign key should refer to the primary key because this provides the greatest performance and the least memory overhead. The relationship shown in Figure 16-9 via a column of type GUID column is for illustrative purposes only.

---

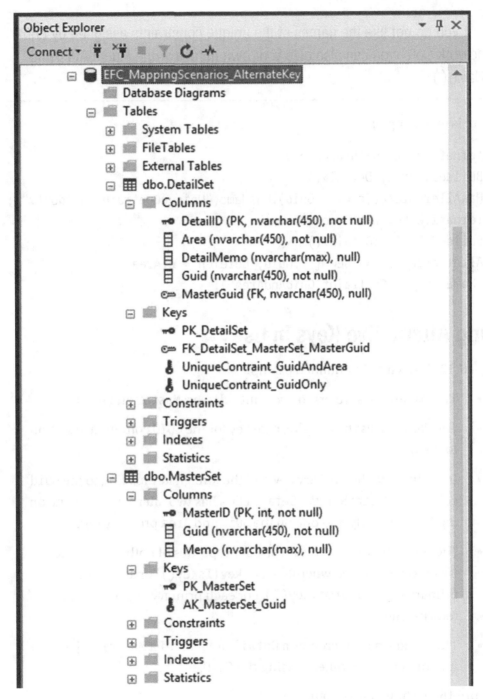

***Figure 16-9.*** *Unique constraints in Microsoft SQL Server Management Studio*

> **Tip**    If you do not like the names of the unique constraints assigned by Entity
> Framework Core, you can also give your own names in the Fluent API using
> HasName().

Here is an example:

```
// Alternative key with one column
modelBuilder.Entity<Detail>()
    .HasAlternateKey(c => c.Guid).HasName("UniqueContraint_GuidOnly"); ;
// Alternative key with two columns
modelBuilder.Entity<Detail>()
  .HasAlternateKey(c => new { c.Guid, Bereich = c.Area
}).HasName("UniqueContraint_GuidAndArea");
```

# Seeing Alternative Keys in Use

Listing 16-12 shows a self-contained example:

- The context class refers to two entity classes: Master and Detail.

- The Master class has an alternate key for the GUID column in addition
  to the primary key.

- In addition to the primary key and the foreign key leading to the Guid
  column of MasterSet, the Detail class has two alternate keys: one on
  the Guid property and one on the Guid and Area property together.

- The program first outputs the list of properties of both classes and
  writes on the screen whether it is a key (IsKey()) or additionally a
  primary key (IsPrimaryKey() ). For each primary key, IsKey() also
  returns true.

- The program then creates a Detail object and a Master object.
  It connects these objects using the GUID.

Figure 16-10 shows the output.

***Listing 16-12.*** Project EFC_MappingTest, AlternateKeys.cs

```csharp
using ITVisions;
using Microsoft.EntityFrameworkCore;
using System;
using System.Collections.Generic;

namespace EFC_MappingScenarios.AlternateKeys
{
 /// <summary>
 /// In this example, several classes are deliberately implemented in one
file, so that the example is clearer.
 /// </summary>
 class DEMO_AlternateKeys
 {
  public static void Run()
  {
   CUI.MainHeadline(nameof(DEMO_AlternateKeys));
   using (var ctx = new MyContext())
   {
    CUI.Print("Database: " + ctx.Database.GetDbConnection().ConnectionString);

    var e = ctx.Database.EnsureCreated();

    if (e)
    {
     CUI.Print("Database has been created!");
    }
    else
    {
     CUI.Print("Database exists!");

    }

    CUI.MainHeadline("Metadata");
    CUI.Headline("Detail");
    var obj1 = new Detail();
    foreach (var p in ctx.Entry(obj1).Properties)
```

```
  {
   Console.WriteLine(p.Metadata.Name + ": Key=" + p.Metadata.IsKey() + "
   PrimaryKey=" + p.Metadata.IsPrimaryKey());
  }

  CUI.Headline("Master");
  var obj2 = new Master();
  foreach (var p in ctx.Entry(obj2).Properties)
  {
   Console.WriteLine(p.Metadata.Name + ": Key=" + p.Metadata.IsKey() + "
   PrimaryKey=" + p.Metadata.IsPrimaryKey());
  }

  CUI.MainHeadline("Two new objects...");
  var h = new Master();
  h.Guid = Guid.NewGuid().ToString();
  var d = new Detail();
  d.Guid = Guid.NewGuid().ToString();
  d.Area = "AB";
  h.DetailSet.Add(d);
  ctx.MasterSet.Add(h);
  var count = ctx.SaveChanges();
  if (count > 0)
  {
   CUI.PrintSuccess(count + " Saved changes!");
   CUI.Headline("Master object");
   Console.WriteLine(h.ToNameValueString());
   CUI.Headline("Detail object");
   Console.WriteLine(d.ToNameValueString());
  }
 }
 }
 }
}
```

```
class MyContext : DbContext
{
 public DbSet<Master> MasterSet { get; set; }
 public DbSet<Detail> DetailSet { get; set; }

 protected override void OnConfiguring(DbContextOptionsBuilder builder)
 {
  // Set provider and connectring string
  string connstring = @"Server=.;Database=EFC_MappingScenarios_
  AlternateKey;Trusted_Connection=True;MultipleActiveResultSets=True;";
  builder.UseSqlServer(connstring);
 }

 protected override void OnModelCreating(ModelBuilder modelBuilder)
 {
  // Alternative key with one column
  modelBuilder.Entity<Detail>()
   .HasAlternateKey(c => c.Guid).HasName("UniqueContraint_GuidOnly"); ;
  // Alternative key with two columns
  modelBuilder.Entity<Detail>()
 .HasAlternateKey(c => new { c.Guid, Bereich = c.Area
}).HasName("UniqueContraint_GuidAndArea");

  // The Entity Framework Core automatically generates an alternate key
  // if, in a relationship definition, you do not create the relationship
  // between foreign key and primary key, but use a different column of the
  // parent class instead of the primary key.
  modelBuilder.Entity<Detail>()
        .HasOne(p => p.Master)
        .WithMany(b => b.DetailSet)
        .HasForeignKey(p => p.MasterGuid)
        .HasPrincipalKey(b => b.Guid);
 }
}
```

```
public class Master
{
 public int MasterID { get; set; }
 public string Guid { get; set; }
 public string Memo { get; set; }

 public List<Detail> DetailSet { get; set; } = new List<Detail>();

}

public class Detail
{
 public string DetailID { get; set; }
 public string DetailMemo { get; set; }
 public string Guid { get; set; }
 public string Area { get; set; }
 public string MasterGuid { get; set; }
 public Master Master { get; set; }
}

}
```

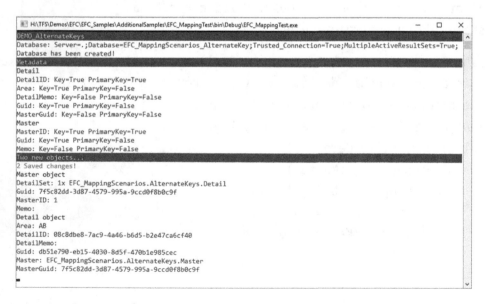

***Figure 16-10.*** *Output of Listing 16-12*

---

**Note**   Alternative keys (as primary keys) must not have NULL values. This is acknowledged by Entity Framework Core with the following runtime error, which also states that zero NULL values are allowed in a unique index: "Unable to track an entity of type 'DetailKlasse' because alternate key property 'Guid' is null. If the alternate key is not used in a relationship, then consider using a unique index instead. Unique indexes may contain nulls, while alternate keys must not."

---

# Cascading Delete

If a master record is deleted in an object relationship (1:1, 1:N), what happens to the one detail data record (1:1) or the many detailed data records (1:N)? *Cascading delete* is the answer.

## Delete Options in Entity Framework Core

Entity Framework Core offers a total of four options for the cascading deletion of dependent objects and thus more options than its predecessor ADO.NET Entity Framework.

If no related objects are materialized for the related records, Entity Framework Core will always send (no matter what setting) a single delete command for the master record to the database. What happens to the detailed data sets is up to the database management system.

However, if the related records are materialized for the related records, Entity Framework Core may send (depending on the setting) a DELETE or UPDATE record detail record command to the database management system before deleting the master record. In addition, depending on the setting, the foreign key properties in the RAM are also changed.

**Table 16-3.** *Cascading Delete Settings and Their Effects*

| Entity Framework Core Setting | Default at... | Resulting Setting of the Foreign Key in the Database | Runtime Behavior: Materialized Related Objects | Runtime Behavior: NOT Materialized Related Records |
|---|---|---|---|---|
| Cascade | Mandatory relations | Cascade | Entity Framework Core sends DELETE for related objects. Foreign key property in RAM is retained. | DBMS deletes related record automatically. |
| ClientSetNull | Optional relationships (since Entity Framework Core 2.0) | No Action | Entity Framework Core sends UPDATE SET = NULL for related objects. Foreign key property in RAM is set to NULL. | None. Error if records exist in database. |
| SetNull | | SetNull | Entity Framework Core sends UPDATE SET = NULL for related objects. Foreign key property in RAM is set to NULL. | DBMS sets foreign key column to NULL. |
| Restrict | Optional relationships (in Entity Framework Core 1.*x*) | No Action | Version 1.*x*: UPDATE SET = NULL. Version 2.*x*: Error when objects exist in RAM. | None. Error if records exist in database. |

The cascading delete settings for Entity Framework Core are made in the relationship declaration in the Fluent API using the OnDelete() method and using the DeleteBehavior enumeration.

```
modelBuilder.Entity<Detail1>()
.HasOne(f => f.Master)
.WithMany(t => t.Detail1Set)
.HasForeignKey(x => x.MasterId)
.OnDelete(DeleteBehavior.Cascade);

modelBuilder.Entity<Detail2>()
.HasOne(f => f.Master)
.WithMany(t => t.Detail2Set)
.HasForeignKey(x => x.MasterId)
.OnDelete(DeleteBehavior.ClientSetNull);

modelBuilder.Entity<Detail3>()
 .HasOne(f => f.Master)
 .WithMany(t => t.Detail3Set)
 .HasForeignKey(x => x.MasterId)
 .OnDelete(DeleteBehavior.SetNull);

modelBuilder.Entity<Detail4>()
 .HasOne(f => f.Master)
 .WithMany(t => t.Detail4Set)
 .HasForeignKey(x => x.MasterId)
 .OnDelete(DeleteBehavior.Restrict);
```

## Looking at an Example

Listing 16-13 sets up a test scenario for the deletion options. There is a Master class with four collections of four different Detail classes. One of the four cascading deletion options of Entity Framework Core is used for each of the four Detail classes.

*Listing 16-13.*  Project EFC_MappingScenarios, CascadingDelete.cs

```
using ITVisions;
using ITVisions.EFCore;
using Microsoft.EntityFrameworkCore;
using System;
using System.Collections.Generic;
using System.Linq;

namespace EFC_MappingScenarios_CascadingDelete
{
 /// <summary>
 /// In this example, several classes are deliberately implemented in one
file, so that the example is clearer.
 /// </summary>
 class DEMO_CascadingDelete
 {
  public static void Run()
  {
   CUI.MainHeadline(nameof(DEMO_CascasdingDelete));
   using (var ctx = new MyContext())
   {
    CUI.Print("Database: " + ctx.Database.GetDbConnection().ConnectionString);

    var e = ctx.Database.EnsureCreated();

    if (e)
    {
     CUI.Print("Database has been created!");
    }
    else
    {
     CUI.Print("Database exists!");
    }

    CUI.Headline("Metadata of Master");
    var obj2 = new Master();
```

```
foreach (var p in ctx.Entry(obj2).Properties)
{
 Console.WriteLine(p.Metadata.Name + ": ");
}
foreach (var p in ctx.Entry(obj2).Navigations)
{
 Console.WriteLine(p.Metadata.Name + ": " + p.Metadata);
}

CUI.Headline("Clean database");
ctx.Database.ExecuteSqlCommand("Delete from Detail1Set");
ctx.Database.ExecuteSqlCommand("Delete from Detail2Set");
ctx.Database.ExecuteSqlCommand("Delete from Detail3Set");
ctx.Database.ExecuteSqlCommand("Delete from Detail4Set");
ctx.Database.ExecuteSqlCommand("Delete from MasterSet");

CUI.Headline("Create one Master with three details");
var d1 = new Detail1();
var d2 = new Detail2();
var d3 = new Detail3();
var d4 = new Detail4();
var m = new Master();
m.Detail1Set.Add(d1);
m.Detail2Set.Add(d2);
m.Detail3Set.Add(d3);
//m.Detail4Set.Add(d4); // Code will fail with this
ctx.MasterSet.Add(m);
var count1 = ctx.SaveChanges();
Console.WriteLine("Saved changes: " + count1);

 PrintStatusDB();
}

CUI.Headline("Delete Master object...");
using (var ctx = new MyContext())
```

```
  {
    var m = ctx.MasterSet.Include(x => x.Detail1Set).Include(x =>
    x.Detail2Set).Include(x=>x.Detail3Set).FirstOrDefault();
    PrintStatusRAM(m);
    ctx.Log();
    ctx.Remove(m);
    var count2 = ctx.SaveChanges();
    DbContextExtensionLogging.DoLogging = false;
    Console.WriteLine("Saved changes: " + count2);
    PrintStatusDB();
    PrintStatusRAM(m);
  }
}

private static void PrintStatusRAM(Master m)
{
 Console.WriteLine("h.Detail1=" + m.Detail1Set.Count + " / Detail1.FK=" +
 (m.Detail1Set.Count > 0 ? m.Detail1Set.ElementAt(0)?.MasterId.ToString()
 : "--"));
 Console.WriteLine("h.Detail2=" + m.Detail2Set.Count + " / Detail2.FK=" +
 (m.Detail2Set.Count > 0 ? m.Detail2Set.ElementAt(0)?.MasterId.ToString()
 : "--"));
 Console.WriteLine("h.Detail3=" + m.Detail3Set.Count + " / Detail3.FK=" +
 (m.Detail3Set.Count > 0 ? m.Detail3Set.ElementAt(0)?.MasterId.ToString()
 : "--"));
 Console.WriteLine("h.Detail4=" + m.Detail4Set.Count + " / Detail4.FK=" +
 (m.Detail4Set.Count > 0 ? m.Detail4Set.ElementAt(0)?.MasterId.ToString()
 : "--"));
}

private static void PrintStatusDB()
{
 using (var ctx = new MyContext())
 {
  Console.WriteLine("DB Mastern: " + ctx.MasterSet.Count());
  Console.WriteLine("DB Detail1: " + ctx.Detail1Set.Count());
```

```
    Console.WriteLine("DB Detail2: " + ctx.Detail2Set.Count());
    Console.WriteLine("DB Detail3: " + ctx.Detail3Set.Count());
    Console.WriteLine("DB Detail4: " + ctx.Detail4Set.Count());
   }
  }
}

class MyContext : DbContext
{
 public DbSet<Master> MasterSet { get; set; }
 public DbSet<Detail1> Detail1Set { get; set; }
 public DbSet<Detail2> Detail2Set { get; set; }
 public DbSet<Detail3> Detail3Set { get; set; }
 public DbSet<Detail4> Detail4Set { get; set; }

 protected override void OnConfiguring(DbContextOptionsBuilder builder)
 {
  // Set provider and connectring string
  string connstring = @"Server=.;Database=EFC_MappingScenarios_
  CascadingDelete;Trusted_Connection=True;MultipleActiveResultSets=True;";
  builder.UseSqlServer(connstring);
 }

 protected override void OnModelCreating(ModelBuilder modelBuilder)
 {
  modelBuilder.Entity<Detail1>()
  .HasOne(f => f.Master)
  .WithMany(t => t.Detail1Set)
  .HasForeignKey(x => x.MasterId)
  .OnDelete(DeleteBehavior.Cascade);

  modelBuilder.Entity<Detail2>()
  .HasOne(f => f.Master)
  .WithMany(t => t.Detail2Set)
  .HasForeignKey(x => x.MasterId)
  .OnDelete(DeleteBehavior.ClientSetNull);
```

```
  modelBuilder.Entity<Detail3>()
   .HasOne(f => f.Master)
   .WithMany(t => t.Detail3Set)
   .HasForeignKey(x => x.MasterId)
   .OnDelete(DeleteBehavior.SetNull);

   modelBuilder.Entity<Detail4>()
   .HasOne(f => f.Master)
   .WithMany(t => t.Detail4Set)
   .HasForeignKey(x => x.MasterId)
   .OnDelete(DeleteBehavior.Restrict);
 }
}

public class Master
{
 public int MasterId { get; set; }
 public string Memo { get; set; }

 public List<Detail1> Detail1Set { get; set; } = new List<Detail1>();
 public List<Detail2> Detail2Set { get; set; } = new List<Detail2>();
 public List<Detail3> Detail3Set { get; set; } = new List<Detail3>();
 public List<Detail4> Detail4Set { get; set; } = new List<Detail4>();
}

public class Detail1
{
 public int Detail1Id { get; set; }
 public string DetailMemo { get; set; }
 public Master Master { get; set; }
 public int? MasterId { get; set; }
}

public class Detail2
{
 public int Detail2Id { get; set; }
 public string DetailMemo { get; set; }
```

```
 public Master Master { get; set; }
 public int? MasterId { get; set; }
}

public class Detail3
{
 public int Detail3Id { get; set; }
 public string DetailMemo { get; set; }
 public Master Master { get; set; }
 public int? MasterId { get; set; }
}

public class Detail4
{
 public int Detail4Id { get; set; }
 public string DetailMemo { get; set; }
 public Master Master { get; set; }
 public int? MasterId { get; set; }
 }
}
```

In this situation, assuming that a Master object in RAM is connected to one of each of the first three Detail objects, that these three Details objects are in RAM, and that the main class object has been deleted by calling Remove() and then SaveChanges(), the following reaction occurs:

- DELETE command for the record in table Detail1Set (in cascade mode). However, the foreign key value from Detail1 to the main class object in RAM is preserved.

- UPDATE command for record in table Detail2Set (in ClientSetNull mode). The foreign key value is set to NULL. The foreign key value from the Detail2 object to the Master object in RAM is also set to NULL.

- UPDATE command for the record in table Detail3Set (in SetNull mode). The foreign key value is set to NULL. The foreign key value from the Detail3 object to the Master class object in RAM is also set to NULL.

- DELETE command for the main class object.

Figure 16-11 shows the output.

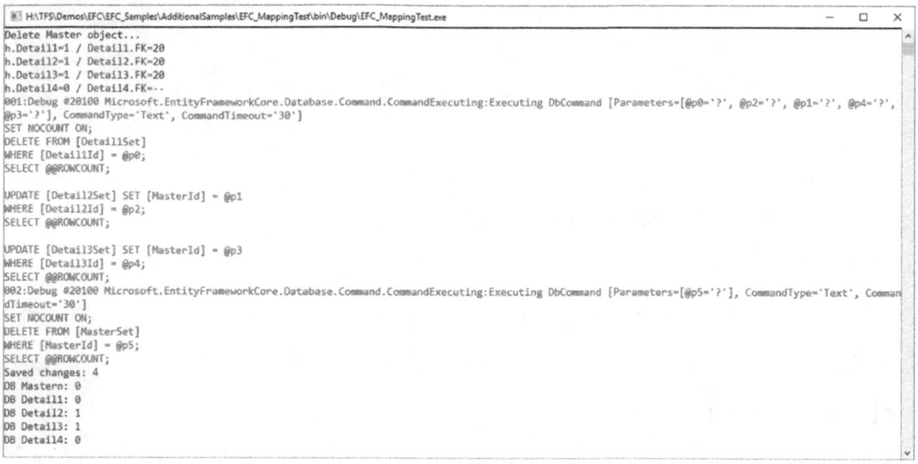

***Figure 16-11.*** *Entity Framework Core has disconnected the three DetailClass objects before deleting the main class object record*

If you uncomment the line m.Detail4Set.Add(d4), the code will fail with the following error: "The DELETE statement conflicted with the REFERENCE constraint "FK_Detail4Set_MasterSet_MasterId"." The conflict occurred in database EFC_MappingScenarios_CascadingDelete, table dbo.Detail4Set, and column MasterId because the relation to Detail4 is in Restrict mode. The same would happen with Detail2 (in ClientSetNull mode) if the object was not in RAM.

If you uncomment the line and also load Detail4Set to RAM with .Include (x => x.Detail4Set), that the code will fail with the following error: "The association between entity types 'Master' and 'Detail4' has been severed but the foreign key for this relationship cannot be set to null. If the dependent entity should be deleted, then setup the relationship to use cascade deletes."

# Mapping of Database Views

The mapping of database views is not yet officially supported in Entity Framework Core. In other words, it is not possible to create the reverse engineering program code for existing database views or to create forward engineering database views from the object model or the Fluent API.

However, it is possible to create database views manually in the database and treat them like tables in the program code. However, it is a bit tricky, as this section shows.

---

**Note**    Microsoft plans to provide better support for database views in version 2.1 of Entity Framework Core; see Appendix C.

---

# Creating a Database View

The database view must be created manually in the database (possibly with the help of a tool such as SQL Server Management Studio) via CREATE VIEW. The database view created by CREATE VIEW in Listing 16-14 provides the number of flights and the last flight for the Flight table per departure (from the World Wide Wings database).

***Listing 16-14.***  Creating a Database View via SQL Command

```
USE WWWingsV2_EN
GO
CREATE VIEW dbo.[V_DepartureStatistics]
AS
SELECT departure, COUNT(FlightNo) AS FlightCount
FROM dbo.Flight
GROUP BY departure
GO
```

# Creating an Entity Class for the Database View

You must create an entity class for the database view whose attributes correspond to the columns of the database view that are to be mapped. In this example, this class is named DepartureStatistics and will receive the data of the database view V_DepartureStatistics. The [Table ] annotation specifies the database view name in the database because the entity class differs by name.

Listing 16-15 deliberately ignores the LastFlight value of the view V_DepartureStatistics. It is important that the entity class requires a primary key to be specified with [Key] or the Fluent API method HasKey().

***Listing 16-15.*** Entity Class with Two Properties for the Two Columns of the Database View to Be Mapped

```
[Table("V_DepartureStatistics")]
 public class DepartureStatistics
 {
  [Key] // must have a PK
  public string Departure { get; set; }
  public int FlightCount { get; set; }
 }
```

# Including the Entity Class in the Context Class

The entity class for the database view is now included as an entity class for a table in the context class via DbSet<T>, as shown in Listing 16-16.

***Listing 16-16.*** Including the Entity Class for the Database View in the Context Class

```
public class WWWingsContext: DbContext
{
  #region Entities for tables
  public DbSet<Airline> AirlineSet { get; set; }
  public DbSet<Flight> FlightSet { get; set; }
  public DbSet<Pilot> PilotSet { get; set; }
  public DbSet<Passenger> PassengerSet { get; set; }
  public DbSet<Booking> BookingSet { get; set; }
  public DbSet<AircraftType> AircraftTypeSet { get; set; }
  #endregion

  #region Pseudo-entities for views
  public DbSet<DepartureStatistics> DepartureStatisticsSet { get; set; }
  // for view
  #endregion
...
}
```

# Using the Database View

You can now use the entity class for the database view, such as for a table in LINQ queries or in FromSql() for direct SQL queries. If the database view is writable, then you could also use the API of Entity Framework Core via SaveChanges() to change, add, or delete records (Listing 16-17).

*Listing 16-17.* Using the Entity Class for the Database View

```
public static void DatabaseViewWithPseudoEntity()
{
 CUI.MainHeadline(nameof(DatabaseViewWithPseudoEntity));

 using (var ctx = new WWWingsContext())
 {
  var query = ctx.DepartureStatisticsSet.Where(x => x.FlightCount > 0);
  var liste = query.ToList();
  foreach (var stat in liste)
  {
   Console.WriteLine($"{stat.FlightCount:000} Flights departing from
   {stat.Departure}.");
  }
 }
}
```

# Challenge: Migrations

Although the previous sections showed some manual work, integrating database views did not seem to be all that challenging beyond some typing required. Unfortunately, on closer inspection, this is not the whole story.

If you create a schema migration in the context class after creating the database view, you detect that Entity Framework Core now undesirably wants to create a table for the database view in the database (Listing 16-18).

**Listing 16-18.**   Entity Framework Core Creates a CreateTable() for the Database View in the Schema Migration Class, Which Is Not Desirable

```
using Microsoft.EntityFrameworkCore.Migrations;
using System;
using System.Collections.Generic;

namespace DA.Migrations
{
    public partial class v8 : Migration
    {
        protected override void Up(MigrationBuilder migrationBuilder)
        {

            migrationBuilder.CreateTable(
                name: "V_DepartureStatistics",
                columns: table => new
                {
                    Departure = table.Column<string>(nullable: false),
                    FlightCount = table.Column<int>(nullable: false)
                },
                constraints: table =>
                {
                    table.PrimaryKey("PK_V_DepartureStatistics", x =>
                    x.Departure);
                });
        }

        protected override void Down(MigrationBuilder migrationBuilder)
        {

            migrationBuilder.DropTable(
                name: "V_DepartureStatistics");
        }
    }
}
```

This is correct from the point of view of Entity Framework Core, as the code told the OR mapper that V_DepartureStatistics would be a table and Entity Framework Core in the current version 2.0 has absolutely no understanding of database views.

Unfortunately, this schema migration could not be executed because there can be only one object named V_DepartureStatistics in the database.

There are two possible solutions to this situation, explained here:

- You manually delete CreateTable() in the Up() method and the corresponding DropTable() in Down() from the migration class.

- You trick Entity Framework Core so that the OR mapper ignores the entity class DepartureStatistics when creating the migration step at development time but not at runtime.

---

**Tip**   The trick is realized in Listing 16-19. Entity Framework Core instantiates the context class as part of creating or deleting a schema migration and calls OnModelCreating(). However, this does not happen at development time via the actual starting point of the application (in which case the application would start). This happens when hosting the DLL with the context class in the command-line tool ef.exe. In OnModelCreating(), you therefore check whether the current process has the name ef. If so, then you are not in the runtime of the application but in the development environment and want to ignore the database view with Ignore(). At runtime of the application, however, Ignore() will not be executed, and thus using the database view through the entity class is possible.

---

*Listing 16-19.*   Entity Framework Core Should Only Ignore the Entity Class for the Database View at Development Time

```
// Trick: hide the view or grouping pseudo entities from the EF
migration tool so it does not want to create a new table for it
if (System.Diagnostics.Process.GetCurrentProcess().ProcessName.ToLower()
== "ef")
{
modelBuilder.Ignore<DepartureStatistics>();
...
}
```

---

**Note**    If the query of the process name is too uncertain because Microsoft could change this name, you can instead use a switch in the context class in the form of a static attribute (e.g., `bool IsRuntime {get; set; } = false`). By default, this `IsRuntime` is false and ignores the entity class for the database view. At runtime, however, the application sets `IsRuntime` to true before the first instantiation of the context class.

---

# Global Query Filters

Global query filters are a nice new feature in Entity Framework Core 2.0. This allows you to define filter conditions centrally in `OnModelCreating()`, which Entity Framework Core then appends to any LINQ query, any direct SQL query, any call to a table-valued function, and any explicit load operation. This feature is well-suited to the following scenarios:

- *Multitenancy*: A column in a record expresses which tenant a record belongs to. The global filter ensures that each tenant only sees their data. Without a global filter, you would have to remember to consider the tenant condition in each query.

- *Soft delete*: Records that are deleted should not really be deleted; they should only be marked. The global filter ensures that the user does not see any "deleted" data. Without a global filter, you would have to remember to include the `deleted = false` condition in each query.

## Defining a Filter

You set a global filter in `OnModelCreating()` per an entity class using the method `HasQueryFilter()`.

The following is a global filter where only the flights of an specific airline (i.e., one tenant) and only flights that are not fully booked are returned for all inquiries:

```
modelBuilder.Entity<Flight>().HasQueryFilter(x => x.FreeSeats > 0 &&
x.AirlineCode == "WWW");
```

---

**Note**   You can only define at most one filter per entity class. If you call
HasQueryFilter() multiple times, only the conditions of the last filter apply. To
link multiple conditions, use the AND operator (in C#, &&), as shown earlier.

---

## Using Filters in LINQ

The previous filter forces this LINQ query in the database management system:

```
List<Flight> flightSet = (from f in ctx.FlightSet //
                         where f.Departure == "Berlin"
                         select f).ToList();
```

It executes the following SQL with additional conditions from the global filter:

```
SELECT [f].[FlightNo], [f].[AircraftTypeID], [f].[AirlineCode], [f].
[CopilotId], [f].[FlightDate], [f].[Departure], [f].[Destination], [f].
[FreeSeats], [f].[LastChange], [f].[Memo], [f].[NonSmokingFlight], [f].
[PilotId], [f].[Price], [f].[Seats], [f].[Strikebound], [f].[Timestamp],
[f].[Utilization]
FROM [Flight] AS [f]
WHERE ((([f].[FreeSeats] > 0) AND ([f].[AirlineCode] = N'WWW')) AND ([f].
[Departure] = N'Berlin')
```

Entity Framework Core considers the global filters also during eager loading and
explicit loading, as shown in Listing 16-20 and Listing 16-21.

***Listing 16-20.***  Explicit Load Example with Load()

```
CUI.Headline("Pilots (Eager Loading)");

var pilotWithFlights= ctx.PilotSet.Include(x => x.FlightAsPilotSet).
ToList();

foreach (var p in pilotWithFlights)
{
 Console.WriteLine(p);
 foreach (var f in p.FlightAsPilotSet.ToList())
```

```
 {
  Console.WriteLine(" - " + f.ToString());
 }
}
```

*Listing 16-21.* Explicit Load Example with Load()

```
CUI.Headline("Pilots (Explicit Loading)");

var pilotenSet = ctx.PilotSet.ToList();

foreach (var p in pilotenSet)
{
 Console.WriteLine(p);
 ctx.Entry(p).Collection(x => x.FlightAsPilotSet).Load();
 foreach (var f in p.FlightAsPilotSet.ToList())
 {
  Console.WriteLine(" - " + f.ToString());
 }
}
```

# Practical Example: Ignoring a Filter

You can decide in each individual query to ignore the global filter. This is done with
IgnoreQueryFilters(), as shown here:

```
List<Flight> FlightAllSet = (from f in ctx.FlightSet.IgnoreQueryFilters()
                     where f.Departure == "Berlin"
                     select f).ToList();
```

---

**Note**    However, it is not possible to ignore only individual parts of a filter. Although
this is desired by users, an implementation has not yet been provided by Microsoft.

---

## Global Query Filters for SQL Queries

The global query filters (see chapter 14) also work when querying directly via SQL with FromSql(). For a global filter that returns only flights from a specific airline and only flights that are not fully booked, this is the SQL query in code:

```
List<Flight> flightSet2 = ctx.FlightSet.FromSql("select * from Flight where
Departure = 'Berlin'").ToList();
```

Entity Framework Core will embed your SQL query in the global filter query and send it to the database management system.

```
SELECT [f].[FlightNo], [f].[AircraftTypeID], [f].[AirlineCode], [f].
[CopilotId], [f].[FlightDate], [f].[Departure], [f].[Destination], [f].
[FreeSeats], [f].[LastChange], [f].[Memo], [f].[NonSmokingFlight], [f].
[PilotId], [f].[Price], [f].[Seats], [f].[Strikebound], [f].[Timestamp],
[f].[Utilization]
FROM (
    select * from Flight where Departure = 'Berlin'
) AS [f]
WHERE ([f].[FreeSeats] > 0) AND ([f].[AirlineCode] = N'WWW')
```

## Global Query Filters for Stored Procedures and Table-Valued Functions

The global query filters also work with FromSql() when using table-valued functions (TVFs).

With the previous global filter, where only the flights of an airline and only nonbooked flights are returned, this SQL query

```
List<Flight> flightSet3 = ctx.FlightSet.FromSql("Select * from
GetFlightsFromTVF({0})", "Berlin").Where(f=>f.NonSmokingFlight == true).
ToList();
```

will result in the following:

```
SELECT [f].[FlightNo], [f].[AircraftTypeID], [f].[AirlineCode],
[f].[CopilotId], [f].[FlightDate], [f].[Departure], [f].[Destination],
[f].[FreeSeats], [f].[LastChange], [f].[Memo], [f].[NonSmokingFlight],
```

```
[f].[PilotId], [f].[Price], [f].[Seats], [f].[Strikebound],
[f].[Timestamp], [f].[Utilization]
FROM (
    Select * from GetFlightsFromTVF(@p0)
) AS [f]
WHERE ((([f].[FreeSeats] > 0) AND ([f].[AirlineCode] = N'WWW')) AND
([f].[NonSmokingFlight] = 1)
```

---

**Warning**    When a stored procedure is called, the global filters in RAM do not work!

From the following query: `List<Flight> flightSet4 = ctx.FlightSet.FromSql("EXEC GetFlightsFromSP {0}", "Berlin").ToList();`

Entity Framework Core will execute only the following: `EXEC GetFlightsFromSP @p0`

---

# Future Queries

Entity Framework Plus (see Chapter 20) implements an additional feature called *future queries*. This allows you to define a series of queries, by calling the `Future()` extension method, that are not executed immediately. These queries can then be executed individually later, but this would also work with the standard functionality of Entity Framework Core. The special feature of future queries is that all the defined queries are executed together when the data of a query is needed because a conversion or aggregate operator is applied.

Listing 16-22 first defines three future queries and then uses `ToList()` to retrieve the data from two queries (all pilots and flights from Berlin). As Figure 16-12 shows, all three queries are executed the first time the data is accessed. Then two more queries are defined (flights from London and flights from Paris). Then you get access to the data of the third query (flights from Rome) of the first action. There is no round-trip to the database management system since this data has already been loaded. Only then, on a call to `ToList()` on the flights from London, does Entity Framework Core execute the queries of the flights from London and Paris.

**_Listing 16-22._**  Future Queries with EFPlus

```
public static void EFPlus_FutureQuery()
{
 CUI.MainHeadline(nameof(EFPlus_FutureQuery));
 using (var ctx = new DA.WWWingsContext())
 {
  ctx.Log();
  CUI.Headline("Define three future queries ... nothing happens in the
  database");
  QueryFutureEnumerable<Pilot> qAllePilots = ctx.PilotSet.Future();
  QueryFutureEnumerable<Flight> qflightSetRome = ctx.FlightSet.Where(x =>
  x.Departure == "Rome").Future();
  QueryFutureEnumerable<Flight> qFlightSetBerlin = ctx.FlightSet.Where(x
  => x.Departure == "Berlin").Future();
  CUI.Headline("Access the pilots:");
  var allePilots = qAllePilots.ToList();
  Console.WriteLine(allePilots.Count + " Pilots are loaded!");
  CUI.Headline("Access the flights from Rome:");
  var flightSetRom = qflightSetRome.ToList();
  Console.WriteLine(flightSetRom.Count + " flights from Berlin are
  loaded!");
  CUI.Headline("Define another two future queries ... nothing happens in
  the database");
  QueryFutureEnumerable<Flight> qFugSetLondon = ctx.FlightSet.Where(x =>
  x.Departure == "London").Future();
  QueryFutureEnumerable<Flight> qflightSetParis = ctx.FlightSet.Where(x
  => x.Departure == "Paris").Future();
  CUI.Headline("Access the flights from Berlin:");
  var flightSetBerlin = qFlightSetBerlin.ToList();
  Console.WriteLine(flightSetBerlin.Count + " flights from Rome are
  loaded!");
  CUI.Headline("Access the flights from London:");
  var flightSetLondon = qFugSetLondon.ToList();
  Console.WriteLine(flightSetLondon.Count + " flights from London are
  loaded!");
```

```
CUI.Headline("Access the flights from Paris:");
var flightSetParis = qflightSetParis.ToList();
Console.WriteLine(flightSetParis.Count + " flights from Paris are
loaded!");
}
}
```

*Figure 16-12.* *Output of Listing 16-22*

# CHAPTER 17

# Performance Tuning

This chapter provides guidance on accelerating database access with Entity Framework Core.

## Process Model for Performance Optimization in Entity Framework Core

Like the classic Entity Framework, the following process model for optimizing the performance of database accesses has proven itself in Entity Framework Core:

- The first step is to implement all data accesses using Entity Framework Core and LINQ, except for UPDATE, DELETE, and INSERT bulk operations. Bulk operations use SQL or a bulk insert directly.

- The application is then tested with realistic data sets. Where the speed is too slow, it is optimized in three stages.

  - In stage 1, tricks within Entity Framework Core are used, such as no tracking, caching, paging, changing the loading strategy (for example, eager loading or preloading instead of explicit loading), and reducing the number of round-trips.

  - Where this is not enough, in stage 2 the LINQ commands are checked and replaced with better-optimized SQL commands or other database constructs such as views, stored procedures, or table-valued functions (TVFs). Access to them continues via Entity Framework Core.

  - Only in stage 3 is Entity Framework Core replaced by DataReader and Command objects for access to SQL commands, views, stored procedures, and TVFs.

© Holger Schwichtenberg 2018
H. Schwichtenberg, *Modern Data Access with Entity Framework Core*,
https://doi.org/10.1007/978-1-4842-3552-2_17

# Best Practices for Your Own Performance Tests

If you are performing performance testing to check the speed of different alternatives, consider the following:

- Do not run the performance test in Visual Studio. The debugger and the possibly activated IntelliTrace feature will slow down your program code in different ways, depending on the procedure, and you will not receive either absolutely or proportionally correct results.

- Do not run the performance test in a graphical user interface (GUI) application or do any console output. Write a console application for the performance test, but do print anything to the console and do not include the time of the console output in your measures.

- Repeat each test several times, at least ten times, and calculate the average. There are many factors that can influence the results (for example, .NET garbage collection and the Windows paging file).

- Do not include the first run (cold start) in the average. Additional tasks of Entity Framework Core and the database management system have to be done on the first run (for example, starting up the database, establishing a connection, generating the mapping code, and so on) that falsify your result. Do eleven runs if you want to get ten valid results.

- Test against a remote database management system on a different system (unless your solution really uses a local database).

- Make sure that the test machine does not perform any other significant processes during the test and the network is not significantly active (run the tests when everyone else is at home or set up your own network!).

# Performance Comparison of Various Data Access Techniques in .NET

When developing Entity Framework Core, Microsoft's goals were platform independence and increased performance over the classic ADO.NET Entity Framework. This chapter covers one of many performance test scenarios.

Figure 17-1 shows that Entity Framework Core 1.1 on .NET Framework 4.7 in no-tracking mode is almost as fast as a `DataReader` with manual mapping (in other words, copying from the `DataReader` to a .NET object: `obj.x = Convert (dr ["x"])`).

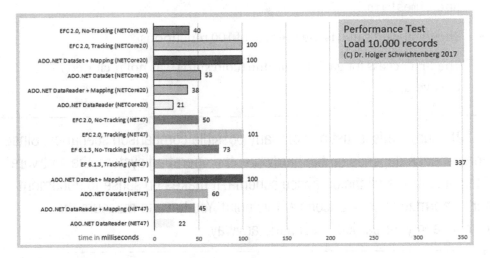

***Figure 17-1.*** *Performance comparison of various data access techniques in .NET*

With .NET Core 2.0, both manual mapping and Entity Framework Core are faster than .NET Framework 4.7. Interestingly, Entity Framework Core's no-tracking mode benefits greatly from .NET Core 2.0 and is equivalent to manual mapping in .NET Core 2.0.

Also in tracking mode, Entity Framework Core is faster than ADO.NET Entity Framework 6.*x*.

The fastest measurement (22 ms in .NET Framework 4.7 or 21 ms in .NET Core 2.0) comes from a `DataReader`, but the data sets were not mapped to objects.

Note the following about this measurement scenario:

- The database server is running Microsoft SQL Server 2016 (virtualized) on Windows Server 2016. The performance of the database server would be even better if it were not virtualized or at least could use nonvirtualized hard disk space.

- The client is a Windows 10 machine.

- The two computers are connected via a 1GB Ethernet network.

- The executed command is a simple `SELECT` without joins or aggregate operators.

- 10,000 records were loaded.

- A result record consists of 13 columns from a single table.

- Data types are int, smallint, nvarchar(30), nvarchar(max), bit, and timestamp.

- The displayed value is the average of 100 repetitions.

- The first access for each technique (cold start) was not included in the average.

---

**Note**    Of course, this is just one of many possible comparison scenarios. Since the performance depends on the hardware, the operating system, the software versions, and above all the database schema, it makes no sense to document further performance comparisons at this point. You have to measure the performance in your particular scenario anyway.

---

# Optimizing Object Assignment

To assign related objects, Entity Framework Core gives you two options.

- Assignment via an object reference (Listing 17-1)

- Assignment via a foreign key property (Listing 17-2)

***Listing 17-1.***  Assignment via an Object Reference

```
public static void ChangePilotUsingObjectAssignment()
{
 CUI.Headline(nameof(ChangePilotUsingObjectAssignment));
 var flightNo = 102;
 var newPilotID = 123;

 using (var ctx = new WWWingsContext())
 {
  ctx.Log();
  Flight flight = ctx.FlightSet.Find(flightNo);
  Pilot newPilot = ctx.PilotSet.Find(newPilotID);
```

```
  flight.Pilot = newPilot;
  var count = ctx.SaveChanges();
  Console.WriteLine("Number of saved changes: " + count);
 }
}
```

***Listing 17-2.*** Assignment via a Foreign Key Property

```
public static void ChangePilotUsingFK()
{
 CUI.Headline(nameof(ChangePilotUsingFK));
 var flightNo = 102;
 var newPilotID = 123;

 using (var ctx = new WWWingsContext())
 {
  ctx.Log();
  Flight flight = ctx.FlightSet.Find(flightNo);
  flight.PilotId = newPilotID;
  var count = ctx.SaveChanges();
  Console.WriteLine("Number of saved changes: " + count);
 }

}
```

Using a foreign key property is a bit more efficient because the object does not have to be explicitly loaded. The UPDATE command sent by Entity Framework Core to the database looks the same in both cases (see Figure 17-2).

**Figure 17-2.** *Output of Listing 17-2*

If the object to be assigned is already in RAM, you have a choice between both syntax forms. But if you have only the primary key value of the object to be allocated in RAM, then you should use the foreign key property. This situation is common in web applications and web services because the client gets only the primary key value.

---

**Tip**    Foreign key properties are optional in Entity Framework Core. However, the optimization option in the assignment is a good reason why you should implement foreign key properties in addition to the navigation properties in the entity classes!

Although there is no foreign key property, you can bypass loading the object by using the foreign key column's shadow property in Entity Framework Core (see Listing 17-3). The disadvantage of this procedure, however, is that the name of the foreign key column must then be used as a string in the program code, which is more cumbersome at input time and more error-prone.

---

***Listing 17-3.*** Assignment via the Shadow Key Property of the Foreign Key
Column

```
public static void ChangePilotUsingFKShadowProperty()
{
 CUI.Headline(nameof(ChangePilotUsingFKShadowProperty));
 var flightNo = 102;
 var neuerPilotNr = 123;
 using (var ctx = new WWWingsContext())
 {
  ctx.Log();
  Flight flight = ctx.FlightSet.Find(flightNo);
  ctx.Entry(flight).Property("PilotId").CurrentValue = neuerPilotNr;
  var count = ctx.SaveChanges();
  Console.WriteLine("Number of saved changes: " + count);
 }
}
```

# Bulk Operations

Entity Framework Core does not support bulk operations on more than one record in its
default configuration but handles each record individually for deletion and modification.
This chapter discusses the topic using a delete command (DELETE) for 1,000 records. The
information also applies to bulk data changes with UPDATE.

## Single Delete

Listing 17-4 shows an inefficient way of deleting all records from the Flight table where
the primary key has a value greater than 20,000. All records must first be loaded and
materialized into .NET objects. Each .NET object is then marked for deletion using the
Remove() method, and the deletion is eventually translated into a DELETE command by
the Entity Framework Core context when the SaveChanges() method is executed for
each object.

So, if 1,000 flights are to be deleted, you need 1,001 commands (one SELECT and 1,000 DELETE commands). In the implementation, the 1,000 DELETE commands, one at a time, are sent to the database management system because the SaveChanges() method occurs after each individual Remove() within the loop.

*Listing 17-4.* Bulk Clear Without Batching with the Entity Framework Core API

```
public static void BulkDeleteEFCAPIwithoutBatching()
{
 CUI.Headline(nameof(BulkDeleteEFCAPIwithoutBatching));
 var sw = new Stopwatch();
 sw.Start();
 int total = 0;
 using (var ctx = new WWWingsContext())
 {
  var min = 20000;
  var flightSet = ctx.FlightSet.Where(x => x.FlightNo >= min).ToList();
  foreach (Flight f in flightSet)
  {
   ctx.FlightSet.Remove(f);
   var count = ctx.SaveChanges();
   total += count;
  }
 }
 sw.Stop();
 Console.WriteLine("Number of DELETE statements: " + total);
 Console.WriteLine("Duration: " + sw.ElapsedMilliseconds);
}
```

# Optimization with Batching

In the classic Entity Framework, it does not matter whether SaveChanges() is in the loop after each Remove() or out of the loop. The old OR mapper always transfers each DELETE command to the database management system one at a time. In Entity Framework Core, there is batching (see Chapter 10), which mitigates the problem.

Listing 17-5, which executes SaveChanges() only once at the end, therefore does not lead to 1,000 DELETE round-trips to the database management system, but to solely one (see Figure 17-3). In total, there are only two round-trips left (one for loading with SELECT and one for the 1,000 DELETE commands). The execution time is reduced significantly (see Table 17-1).

***Listing 17-5.*** Bulk Deleting Batching with the Entity Framework Core API

```
public static void BulkDeleteEFCAPIwithBatching()
{
 CUI.Headline(nameof(BulkDeleteEFCAPIwithBatching));
 int total = 0;
 var sw = new Stopwatch();
 sw.Start();
 using (var ctx = new WWWingsContext())
 {
  var min = 20000;
  var flightSet = ctx.FlightSet.Where(x => x.FlightNo >= min).ToList();
  foreach (Flight f in flightSet)
  {
   ctx.FlightSet.Remove(f);
  }
  total = ctx.SaveChanges();
 }
 sw.Stop();
 Console.WriteLine("Number of DELETE statements: " + total);
 Console.WriteLine("Duration: " + sw.ElapsedMilliseconds);
}
```

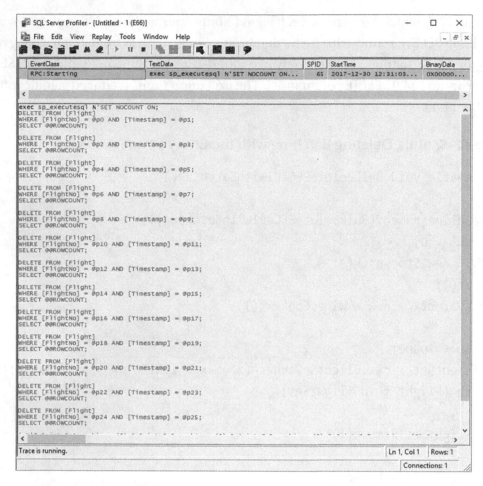

***Figure 17-3.*** *Here, 1,000 delete commands are transferred in one round-trip to the database management system*

***Table 17-1.*** *Execution Time*

| Method | Number of Round-Trips | Execution Time |
| --- | --- | --- |
| Bulk delete 1,000 Flight records without batching with the Entity Framework Core API | 1,001 | 11,110 seconds |
| Bulk deletion of 1,000 Flight records batched with the Entity Framework Core API | 2 | 3,395 seconds |

# Delete Without Loading with Pseudo-Objects

Even with the use of batching, this operation remains inefficient because initially all records are loaded only to then delete them. When it is running Entity Framework Core expects an entity Object.

Here is the trick to manually construct such an entity object as a pseudo-object in RAM and pass this as a parameter to Remove() (see Listing 17-6). This significantly increases performance (see Table 17-2). However, this is possible only under these two conditions:

- The primary keys of the objects to be deleted are known.

- There is no concurrency check by [ConcurrencyCheck] running, no IsConcurrencyToken() method, and no timestamp column.

***Listing 17-6.*** Bulk Deleting Batching with Entity Framework Core API Using Pseudo-Objects

```
public static void BulkDeleteEFCAPIusingPseudoObject()
{
 CUI.Headline(nameof(BulkDeleteEFCAPIusingPseudoObject));
 int total = 0;
 var sw = new Stopwatch();
 sw.Start();
 using (var ctx = new WWWingsContext())
 {
  for (int i = 20001; i < 21000; i++)
  {
   var f = new Flight() { FlightNo = i };
   ctx.FlightSet.Attach(f);
   ctx.FlightSet.Remove(f);
  }
 total = ctx.SaveChanges();
 }
 sw.Stop();
 Console.WriteLine("Number of DELETE statements: " + total);
 Console.WriteLine("Duration: " + sw.ElapsedMilliseconds);
}
```

**Table 17-2.** *Execution Time*

| Method | Number of Round-Trips | Execution Time |
| --- | --- | --- |
| Bulk delete 1,000 Flight records without batching with the Entity Framework Core API | 1001 | 11,110 seconds |
| Bulk deletion of 1,000 Flight records batched with the Entity Framework Core API | 2 | 3,395 seconds |
| Bulk deletion of 1,000 Flight records batched with the Entity Framework Core API using pseudo-objects | 1 | 0,157 seconds |

# Using Classic SQL Instead of the Entity Framework Core API

Through the measures taken so far, the duration of implementation has already been significantly reduced, but the duration is still much longer than necessary. Rather than using the Entity Framework Core API here, in a situation where a consecutive set of records is to be dropped, it is much more efficient to issue a single classic SQL command.

```
DELETE dbo.Flight where FlightNo >= 20000
```

You set this SQL command either in the classical way via an ADO.NET command with a `Parameter` object (see Listing 17-7) or more succinctly via the direct SQL support in Entity Framework Core (see Listing 17-8) with the `ExecuteSqlCommand()` method in the `Database` subobject of the Entity Framework Core context. It should be emphasized that the `string.Format()`-based wildcard syntax protects against SQL injection attacks, as shown in the parameterization in Listing 17-7. Here the strings are not simply put together, as the syntax suggests, but SQL parameter objects are generated internally.

The execution time drops in both cases to less than 40 ms, which is not surprising because the program has now nothing more to do than build a database connection and transfer a few characters. The performance difference between the transmission via the Entity Framework Core context or a `Command` object cannot be measured.

The disadvantage of this approach, of course, is that again SQL strings are used, for which there is no compiler check, and thus there is a risk of syntax and type errors that are not noticeable until runtime.

| Method | Number of Round-Trips | Execution Time |
|---|---|---|
| Bulk delete 1,000 Flight records without batching with the Entity Framework Core API | 1,001 | 11,110 seconds |
| Bulk deletion of 1,000 Flight records batched with the Entity Framework Core API | 2 | 3,395 seconds |
| Bulk deletion of 1,000 Flight records batched with the Entity Framework Core API using pseudo-objects | 1 | 0,157 seconds |
| Bulk delete 1,000 Flight records using SQL through the Entity Framework context | 1 | 0,034 seconds |
| Bulk delete 1,000 Flight records with SQL via the ADO.NET command object with a parameter | 1 | 0,034 seconds |

***Listing 17-7.*** Bulk Delete with SQL via ADO.NET Command Object with a Parameter

```
public static void BulkDeleteADONETCommand()
  {
   CUI.Headline(nameof(BulkDeleteADONETCommand));
   int total = 0;
   var min = 20000;
   var sw = new Stopwatch();
   sw.Start();
   using (SqlConnection connection = new SqlConnection(Program.CONNSTRING))
   {
    connection.Open();
    SqlCommand command = new SqlCommand("DELETE dbo.Flight where FlightNo
    >= @min", connection);
```

```
    command.Parameters.Add(new SqlParameter("@min", min));
    total = command.ExecuteNonQuery();
  }
  sw.Stop();
  Console.WriteLine("Number of DELETE statements: " + total);
  Console.WriteLine("Duration: " + sw.ElapsedMilliseconds);
 }
```

***Listing 17-8.*** Bulk Erase with SQL via Entity Framework Core Context

```
public static void BulkDeleteEFCSQL()
{
 CUI.Headline(nameof(BulkDeleteEFCSQL));
 int total = 0;
 var min = 20000;
 var sw = new Stopwatch();
 sw.Start();
 using (var ctx = new WWWingsContext())
 {
  total = ctx.Database.ExecuteSqlCommand("DELETE dbo.Flight where FlightNo
  >= {0}", min);
 }
 sw.Stop();
 Console.WriteLine("Number of DELETE statements: " + total);
 Console.WriteLine("Duration: " + sw.ElapsedMilliseconds);
}
```

# Lambda Expressions for Mass Deletion with EFPlus

The extension component EFPlus (see Chapter 20) allows to formulate UPDATE and
DELETE commands in lambda expressions based on LINQ commands so you can avoid
the error-prone SQL.

The EFPlus component implements extension methods named Update() and
Delete(). To enable this, using Z.EntityFramework.Plus is necessary.

Listing 17-9 shows how to use Delete() on a LINQ command. Unfortunately, as Figure 17-4 shows, the SQL commands generated by EFPlus are suboptimal. They always work with a nested SELECT, although this would not be necessary. From the point of view of the authors of EFPlus, this was the simplest implementation because it is easy to use the existing SELECT command generation of Entity Framework Core. The execution results in the same performance as using a direct SQL command (see Table 17-3).

***Listing 17-9.*** Mass Deletion with EFPlus

```
public static void BulkDeleteEFPlus()
{
 CUI.Headline(nameof(BulkDeleteEFPlus));
 int min = 20000;
 int total = 0;
 var sw = new Stopwatch();
 sw.Start();
 using (var ctx = new WWWingsContext())
 {
  var count = ctx.FlightSet.Where(x => x.FlightNo >= min).Delete();
  Console.WriteLine("Number of DELETE statements: " + count);
 }
 sw.Stop();
 Console.WriteLine("Duration: " + sw.ElapsedMilliseconds);
 Timer_BulkDeleteEFPlus += sw.ElapsedMilliseconds;
}
```

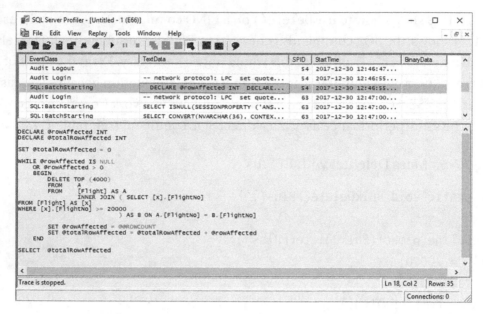

**Figure 17-4.** *SQL DELETE command that EFPlus sends to the database management system*

**Table 17-3.** *Execution Time*

| Method | Number of Round-Trips | Execution Time |
|---|---|---|
| Bulk delete 1,000 Flight records without batching with the Entity Framework Core API | 1001 | 11,110 seconds |
| Bulk deletion of 1,000 Flight records batched with the Entity Framework Core API | 2 | 3,395 seconds |
| Bulk deletion of 1,000 Flight records batched with the Entity Framework Core API using pseudo-objects | 1 | 157 seconds |
| Bulk delete 1,000 Flight records using SQL through the Entity Framework context | 1 | 34 seconds |
| Bulk delete 1,000 Flight records with SQL via the ADO.NET command object with a parameter | 1 | 34 seconds |
| Bulk deletion of 1,000 Flight records with EFPlus | 1 | 45 seconds |

# Bulk Update with EFPlus

Listing 17-10 shows the formulation of an UPDATE command with the additional component EFPlus (see Chapter 20) with the extension method Update(), which reduces the number of free seats by one on each future flight from Berlin. Figure 17-5 shows the SQL UPDATE command sent to the database management system, which works in the same way as Delete().

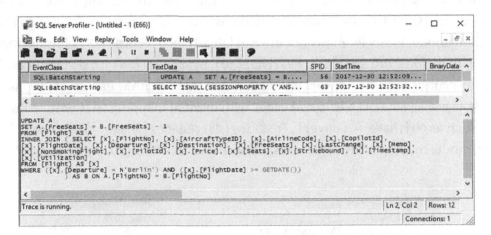

***Figure 17-5.*** *SQL UPDATE command that EFPlus sends to the database management system*

***Listing 17-10.*** Bulk Update with EFPlus

```
public static void BulkUpdateEFPlus()
{
 CUI.Headline(nameof(BulkUpdateEFPlus));
 using (var ctx = new WWWingsContext())
 {
  var count = ctx.FlightSet.Where(x => x.Departure == "Berlin" && x.Date
  >= DateTime.Now).Update(x => new Flight() { FreeSeats = (short)
  (x.FreeSeats - 1) });
  Console.WriteLine("Changed records: " + count);
 }
}
```

# Performance Optimization Through No-Tracking

As with its predecessor ADO.NET Entity Framework, Entity Framework Core has a no-tracking mode, which significantly speeds up the loading of data records. With the new implementation, Microsoft has improved the practical application of this function by adding a context option.

The performance measurement in Figure 17-6 shows that the optional no-tracking mode provides significant speed advantages over the standard tracking mode—both in the legacy ADO.NET Entity Framework and the new Entity Framework Core. In the no-tracking mode, Entity Framework Core can fetch 10,000 records (13 columns from a table, with no join, int, smallint, nvarchar(30), nvarchar (max), bit, timestamp) in 46 milliseconds over the network and materialize them in RAM into objects. This is almost as fast as an ADO.NET DataReader with manual mapping (self-written lines of code such as obj.Name = Convert.ToString(dataReader["name"])). In normal tracking mode, reading the records takes just over twice as long (100 milliseconds).

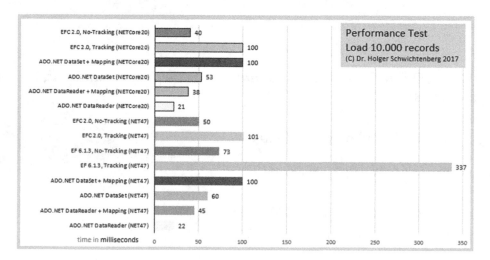

***Figure 17-6.***  *Performance comparison*

In comparison, Figure 17-6 also shows Entity Framework 6.*x*. Here it takes 263 ms in tracking mode. At 53 milliseconds, there is only a marginal difference compared to Entity Framework Core in no-tracking mode. Microsoft has therefore accelerated Entity Framework Core compared to Entity Framework 6.1.3, especially the tracking mode. Nevertheless, the no-tracking mode in Entity Framework Core also pays off.

# Activating No-Tracking Mode

In the first versions of the classic Entity Framework, you had to set no-tracking mode using the property MergeOption for each entity class or every query with an additional line of code. Since Entity Framework 4.1, you can set the mode at the level of a query with the much more elegant AsNoTracking() extension method (see Listing 17-11). In Entity Framework Core, only AsNoTracking() is used for this.

***Listing 17-11.*** Activation of No-Tracking Mode with AsNoTracking( ) in Entity Framework 6.x and Entity Framework Core

```
CUI.Headline("No-Tracking mode");
using (WWWingsContext ctx = new WWWingsContext())
{
 var flightSet = ctx.FlightSet.AsNoTracking().ToList();
 var flight = flightSet[0];
 Console.WriteLine(flight + " object state: " + ctx.Entry(flight).
 State); // Detached
 flight.FreeSeats--;
 Console.WriteLine(flight + " object state: " + ctx.Entry(flight).
 State); // Detached
 int count = ctx.SaveChanges();
 Console.WriteLine($"Saved changes: {count}"); // 0
}
```

The consequence of no-tracking mode is shown in Figure 17-7, with the output of Listing 17-11. The change tracking feature of Entity Framework Core no longer works if you activate no-tracking mode. In the default case, objects are in the state Unchanged after loading, and they change to the state Modified after a change. When loading in no-tracking mode, they are Detached after loading and remain so even after the change. Execution of the SaveChanges() method then sends no change to the database management system because Entity Framework Core does not notice the change.

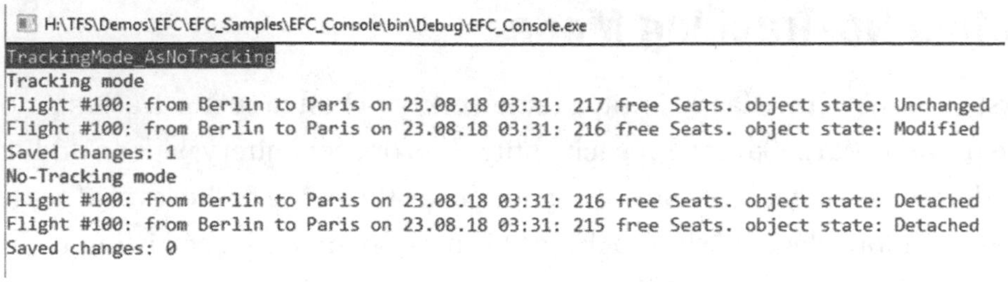

```
H:\TFS\Demos\EFC\EFC_Samples\EFC_Console\bin\Debug\EFC_Console.exe

TrackingMode AsNoTracking
Tracking mode
Flight #100: from Berlin to Paris on 23.08.18 03:31: 217 free Seats. object state: Unchanged
Flight #100: from Berlin to Paris on 23.08.18 03:31: 216 free Seats. object state: Modified
Saved changes: 1
No-Tracking mode
Flight #100: from Berlin to Paris on 23.08.18 03:31: 216 free Seats. object state: Detached
Flight #100: from Berlin to Paris on 23.08.18 03:31: 215 free Seats. object state: Detached
Saved changes: 0
```

***Figure 17-7.*** *Screen output from Listing 17-11*

# No-Tracking Mode Is Almost Always Possible

In any case, no-tracking mode should always be used for objects that only display data and should not be modified at all. But even if you want to modify individual objects, you can first load the objects in no-tracking mode and then later attach them to the context class. Therefore you have to change the objects only—in the best case before the change—and add them to the context using the Attach() method. This method exists both in the DbContext class and in the dbSet<T> class.

The Attach() method adds an object to Entity Framework Core change tracking. The object is thereby transferred from the state Detached to the state Unchanged. Of course, only instances of the entity class may be passed to Attach(). If you pass instances of classes that the Entity Framework Core context does not know, you will get the following error message: "The entity type xy was not found. Ensure that the entity type has been added to the model."

Listing 17-12 (and the accompanying screen output in Figure 17-8) shows the use of the Attach() method in these three scenarios:

- Attach() is executed before the actual change. In this case, there is nothing else to do because Entity Framework Core recognizes all changes after Attach() and transfers the object independently from the state Unchanged to the state Modified.

- If a change occurs before Attach() is executed, Entity Framework Core knows nothing about the change that occurred. Therefore, you must subsequently register the change with ctx.Entry(obj).Property (f => f.Property).IsModified = true.

- If you do not know the changed properties of the object (for
  example, because the changes have occurred in the calling program
  code or another process) or it is too annoying to set the individual
  properties to IsModified, you can use ctx.Entry(Flight).State =
  EntityState.Modified to set the state of the whole object.

***Listing 17-12.*** Using the Attach( ) Method

```
public static void TrackingMode_NoTracking_Attach()
{

 CUI.MainHeadline(nameof(TrackingMode_NoTracking_Attach));

 CUI.Headline("Attach() before change");

 using (WWWingsContext ctx = new WWWingsContext())
 {
  var flightSet = ctx.FlightSet.AsNoTracking().ToList();
  var flight = flightSet[0];
  Console.WriteLine(flight + " object state: " + ctx.Entry(flight).
  State); // Detached
  ctx.Attach(flight);
  Console.WriteLine(flight + " object state: " + ctx.Entry(flight).
  State); // Unchanged
  flight.FreeSeats--;
  Console.WriteLine(flight + " object state: " + ctx.Entry(flight).
  State); // Modified
  int count = ctx.SaveChanges();
  Console.WriteLine($"Saved changes: {count}"); // 0
 }

 CUI.Headline("Attach() after change (change state per property)");
 using (WWWingsContext ctx = new WWWingsContext())
 {
  var flightSet = ctx.FlightSet.AsNoTracking().ToList();
  var flight = flightSet[0];
  Console.WriteLine(flight + " object state: " + ctx.Entry(flight).
  State); // Detached
```

```
  flight.FreeSeats--;
  Console.WriteLine(flight + " object state: " + ctx.Entry(flight).
  State); // Detached
  ctx.Attach(flight);
  Console.WriteLine(flight + " object state: " + ctx.Entry(flight).
  State); // Unchanged
  // Register changed property at EFC
  ctx.Entry(flight).Property(f => f.FreeSeats).IsModified = true;
  Console.WriteLine(flight + " object state: " + ctx.Entry(flight).
  State); // Modified
  int count = ctx.SaveChanges();
  Console.WriteLine($"Saved changes: {count}"); // 1
}

CUI.Headline("Attach() after change (change state per object)");
using (WWWingsContext ctx = new WWWingsContext())
{
 var flightSet = ctx.FlightSet.AsNoTracking().ToList();
 var flight = flightSet[0];
 Console.WriteLine(flight + " object state: " + ctx.Entry(flight).
 State); // Detached
 flight.FreeSeats--;
 Console.WriteLine(flight + " object state: " + ctx.Entry(flight).
 State); // Detached
 ctx.Attach(flight);
 Console.WriteLine(flight + " object state: " + ctx.Entry(flight).
 State); // Unchanged
 ctx.Entry(flight).State = EntityState.Modified;
 Console.WriteLine(flight + " object state: " + ctx.Entry(flight).
 State); // Modified
 int count = ctx.SaveChanges();
 Console.WriteLine($"Saved changes: {count}"); // 1
 }

}
```

As Figure 17-8 shows, the change is saved by `SaveChanges()` in all three cases. Behind the scenes, however, there is a difference between these three scenarios. In the first two scenarios, Entity Framework Core sends a SQL `UPDATE` command to the database, which updates only the actual `Free Spend` column.

```
exec sp_executesql N'SET NOCOUNT ON;
UPDATE [Flight] SET [FreeSeats] = @p0
WHERE [FlightNo] = @p1;
```

```
H:\TFS\Demos\EFC\EFC_Samples\EFC_Console\bin\Debug\EFC_Console.exe
-------------------- START -------------
TrackingMode_NoTracking_Attach
Attach() before change
Flight #100: from Berlin to Paris on 23.08.18 03:31: 216 free Seats. object state: Detached
Flight #100: from Berlin to Paris on 23.08.18 03:31: 216 free Seats. object state: Unchanged
Flight #100: from Berlin to Paris on 23.08.18 03:31: 215 free Seats. object state: Modified
Saved changes: 1
Attach() after change (change state per property)
Flight #100: from Berlin to Paris on 23.08.18 03:31: 215 free Seats. object state: Detached
Flight #100: from Berlin to Paris on 23.08.18 03:31: 214 free Seats. object state: Detached
Flight #100: from Berlin to Paris on 23.08.18 03:31: 214 free Seats. object state: Unchanged
Flight #100: from Berlin to Paris on 23.08.18 03:31: 214 free Seats. object state: Modified
Saved changes: 1
Attach() after change (change state per object)
Flight #100: from Berlin to Paris on 23.08.18 03:31: 214 free Seats. object state: Detached
Flight #100: from Berlin to Paris on 23.08.18 03:31: 213 free Seats. object state: Detached
Flight #100: from Berlin to Paris on 23.08.18 03:31: 213 free Seats. object state: Unchanged
Flight #100: from Berlin to Paris on 23.08.18 03:31: 213 free Seats. object state: Modified
Saved changes: 1
```

***Figure 17-8.** Output from Listing 17-12*

However, in the third scenario, the developer failed to give Core Entity Framework information about which properties actually changed. Entity Framework Core cannot help but send the values of all properties back to the database, even if they are already known there.

```
UPDATE [Flight] SET [AircraftTypeID] = @p0, [AirlineCode] = @p1, [CopilotId]
= @p2, [FlightDate] = @p3, [Departure] = @p4, [Destination] = @p5,
[FreeSeats] = @p6, [LastChange] = @p7, [Memo] = @p8, [NonSmokingFlight] =
@p9, [PilotId] = @p10, [Price] = @p11, [Seats] = @p12, [Strikebound] = @p13
WHERE [FlightNo] = @p14 AND [Timestamp] = @p15;
SELECT [Timestamp], [Utilization]
FROM [Flight]
WHERE @@ROWCOUNT = 1 AND [FlightNo] = @p14;
```

**Note**    In addition to the fact that unnecessary data is being sent over the line, updating all columns poses a potential data change conflict. If other processes have already changed parts of the record, the changes will mercilessly overwrite those other processes. So, you should always ensure that Entity Framework Core knows the changed columns. If the modifications of the object took place before the `Attach()` method by the caller, then the caller must be obligated to supply corresponding meta-information about the changed properties.

## No-Tracking Mode in an Editable Data Grid

When using the method `Attach()`, you can load almost all records in no-tracking mode. Figure 17-9 and Figure 17-10 show the usual data grid scenario. The user can load a (larger) amount of data and change any data sets. The changes are then persisted by clicking Save. (see Figure 17-12).

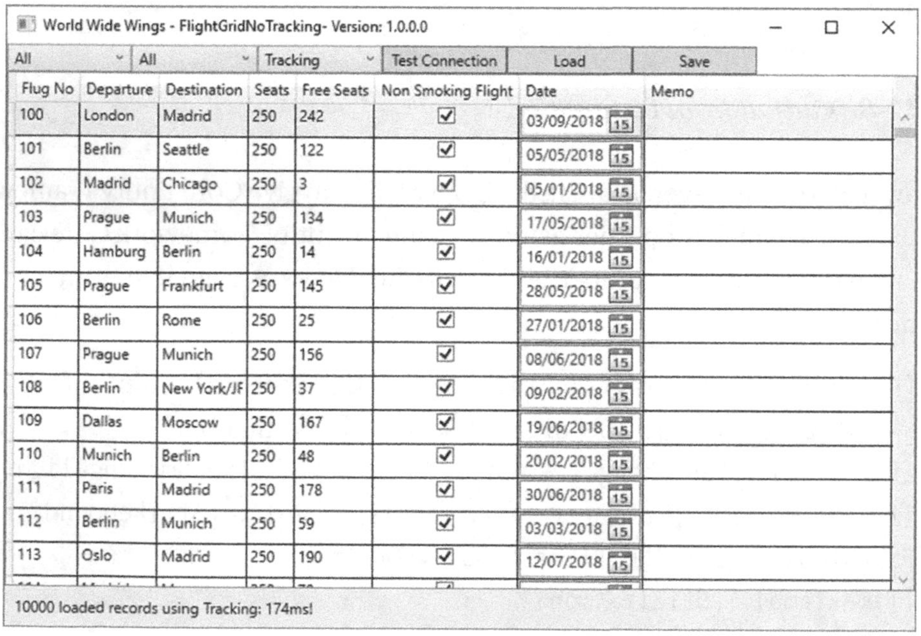

*Figure 17-9.* *10,000 records loaded in tracking mode in 174 milliseconds*

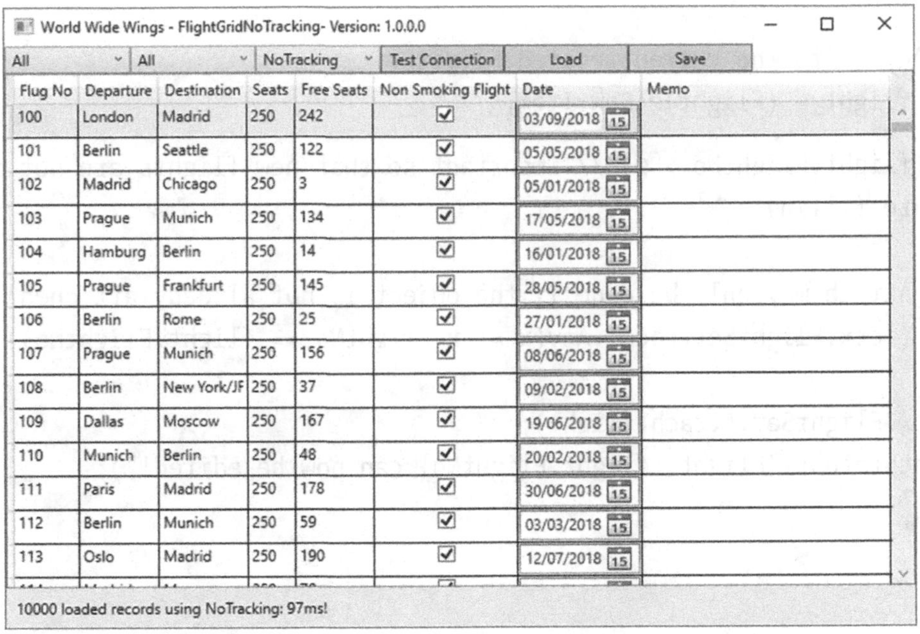

**Figure 17-10.** *10,000 records loaded in no-tracking mode in 96 milliseconds*

In this case, it is not necessary at all to waste the extra time for tracking mode when loading. It is quite sufficient to use `Attach()` to log the individual records that the user is working on at the Entity Framework context (see Listing 17-13). The DataGrid control supplied by Microsoft for the Windows Presentation Foundation (WPF) uses the `BeginningEdit()` event. In the event handler, `Attach()` transfers the detached object to an attached object (see Figure 17-11) and thus registers the object as part of the change tracking of the Entity Framework context.

However, after loading with `AsNoTracking()`, it is not a good idea to attach all objects in a loop with `Attach()`. `Attach()` takes less than a millisecond per object. This is not noticeable when you attach individual objects to it. But in total, such a loop is slower than loading all the objects in tracking mode directly. So if you know for sure that all objects have to be changed, you should use tracking mode when loading.

**Listing 17-13.** Attaching an Object to the Context When the User Starts Editing

```
/// <summary>
/// Called when starting to editing a flight in the grid
/// </summary>
private void C_flightDataGrid_BeginningEdit(object sender,
DataGridBeginningEditEventArgs e)
```

```
{
  // Access to the current edited Flight
  var flight = (Flight)e.Row.Item;

  if (flight.FlightNo > 0) // important so that new flights are not added
  before filling
  {
    // Attach may only be done if the object is not already attached!
    if (!ctx.FlightSet.Local.Any(x => x.FlightNo == flight.FlightNo))
    {
      ctx.FlightSet.Attach(flight);
      SetStatus($"Flight {flight.FlightNo} can now be edited!");
    }
  }
}
```

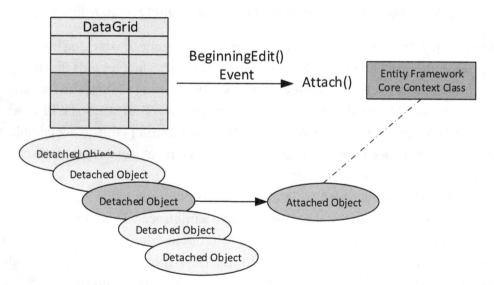

***Figure 17-11.*** *The developer logs the object to be changed in the DataGrid to Attach( ) at the Entity Framework context when the change begins*

**Note**  Attach() may be executed only if the object is not already connected to the context instance. Otherwise, a runtime error will occur. Whether the object is already connected to the context, you cannot ask the object itself. However, the DbSet<T> class has a property called Local that includes all objects in the local cache of Entity Framework Core. To query this cache, use ctx.FlightSet. Local.Any (x => x.FlightNo == flight.FlightNo).

**Warning**  The property Local has a method Clear(). Not only does this cause—as you would expect—an emptying of the cache of the Entity Framework Core context, but it also puts all objects contained in it in the state Deleted, which would delete them at the next SaveChanges()! To actually remove the objects from the cache only, you must set the objects individually to the state Detached, as shown here:

```
foreach (var f in ctx.FlightSet.Local.ToList ())
{
  ctx.Entry (f) .State = EntityState.Detached;
}
```

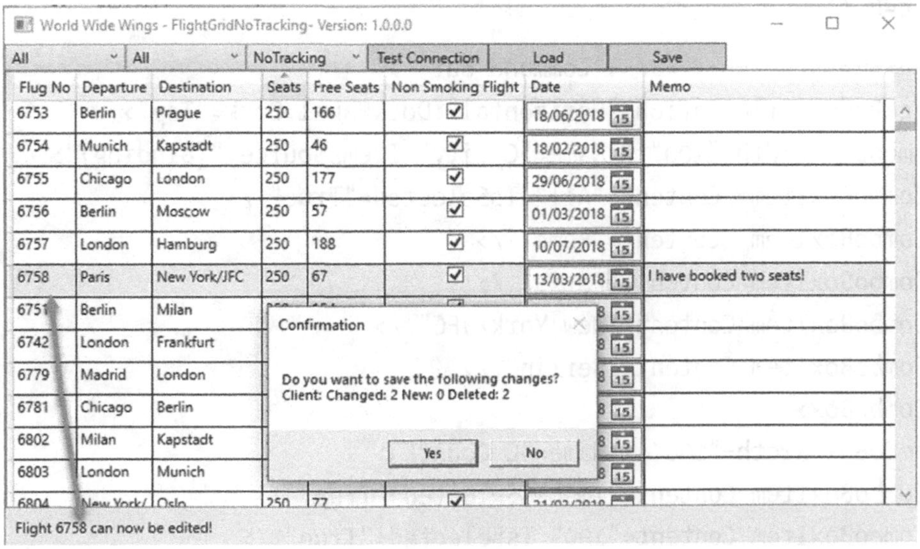

*Figure 17-12.* *Saving changes although loaded in no-tracking mode*

Listing 17-14 and Listing 17-15 show the XAML code and the complete code-behind class.

***Listing 17-14.*** XAML Code FlightGridNoTracking.xaml (Project EFC_GUI)

```xaml
<Window x:Class="GUI.WPF.FlightGridNoTracking"
        xmlns="http://schemas.microsoft.com/winfx/2006/xaml/presentation"
        xmlns:x="http://schemas.microsoft.com/winfx/2006/xaml"
        xmlns:d="http://schemas.microsoft.com/expression/blend/2008"
        xmlns:mc="http://schemas.openxmlformats.org/markup-
        compatibility/2006"
        xmlns:local="clr-namespace:GUI.WPF"
        xmlns:wpf="clr-namespace:ITVisions.WPF;assembly=ITV_DemoUtil"
        mc:Ignorable="d"
        Title="World Wide Wings - FlightGridNoTracking" Height="455.233"
        Width="634.884">

<Window.Resources>
 <wpf:InverseBooleanConverter x:Key="InverseBooleanConverter">
 </wpf:InverseBooleanConverter>
</Window.Resources>

<Grid x:Name="LayoutRoot" Background="White">
 <DockPanel>

  <!--===================== Command Bar-->
  <StackPanel Orientation="Horizontal" DockPanel.Dock="Top">
   <ComboBox Width="100" x:Name="C_City" ItemsSource="{Binding}">
    <ComboBoxItem  Content="All" IsSelected="True" />
    <ComboBoxItem  Content="Rome" />
    <ComboBoxItem Content="Paris" />
    <ComboBoxItem Content="New York/JFC" />
    <ComboBoxItem Content="Berlin" />
   </ComboBox>
   <ComboBox Width="100" x:Name="C_Count" >
    <ComboBoxItem Content="10" IsSelected="True" />
    <ComboBoxItem Content="100" IsSelected="True" />
```

```xml
  <ComboBoxItem Content="1000" IsSelected="True" />
  <ComboBoxItem Content="All" IsSelected="True" />
</ComboBox>
<ComboBox Width="100" x:Name="C_Mode" >
  <ComboBoxItem Content="Tracking" IsSelected="True" />
  <ComboBoxItem Content="NoTracking" IsSelected="False" />
</ComboBox>
<Button Width="100" x:Name="C_Test" Content="Test Connection"
Click="C_Test_Click" ></Button>
<Button Width="100" x:Name="C_Load" Content="Load" Click="C_Load_Click">
</Button>
<Button Width="100" x:Name="C_Save" Content="Save" Click="C_Save_Click">
</Button>
</StackPanel>
<!--===================== Status Bar-->
<StatusBar DockPanel.Dock="Bottom">
  <Label x:Name="C_Status"></Label>
</StatusBar>
<!--===================== Datagrid-->
<DataGrid Name="C_flightDataGrid" AutoGenerateColumns="False"
EnableRowVirtualization="True"  IsSynchronizedWithCurrentItem="True"
SelectedIndex="0" Height="Auto" BeginningEdit="C_flightDataGrid_
BeginningEdit"  PreviewKeyDown="C_flightDataGrid_PreviewKeyDown"
RowEditEnding="C_flightDataGrid_RowEditEnding">
  <DataGrid.Columns>
   <DataGridTextColumn Binding="{Binding Path=FlightNo}" Header="Flight
   No" Width="SizeToHeader" />
   <DataGridTextColumn Binding="{Binding Path=Departure}"
   Header="Departure" Width="SizeToHeader" />
   <DataGridTextColumn Binding="{Binding Path=Destination}"
   Header="Destination" Width="SizeToHeader" />
   <DataGridTextColumn Binding="{Binding Path=Seats}" Header="Seats"
   Width="SizeToHeader" />
   <DataGridTextColumn Binding="{Binding Path=FreeSeats}" Header="Free
   Seats" Width="SizeToHeader" />
```

```
      <DataGridCheckBoxColumn Binding="{Binding Path=NonSmokingFlight,
      Converter={StaticResource InverseBooleanConverter}}" Header="Non
      Smoking Flight" Width="SizeToHeader" />
      <DataGridTemplateColumn Header="Date" Width="100">
       <DataGridTemplateColumn.CellTemplate>
        <DataTemplate>
         <DatePicker SelectedDate="{Binding Path=Date}" />
        </DataTemplate>
       </DataGridTemplateColumn.CellTemplate>
      </DataGridTemplateColumn>
      <DataGridTextColumn Binding="{Binding Path=Memo}" Width="200"
      Header="Memo"  />
     </DataGrid.Columns>
    </DataGrid>

  </DockPanel>
 </Grid>
</Window>
```

***Listing 17-15.*** Code-Behind Class FlightGridNoTracking.cs (Project EFC_GUI)

```
using BO;
using DA;
using Microsoft.EntityFrameworkCore;
using System;
using System.Diagnostics;
using System.Linq;
using System.Reflection;
using System.Windows;
using System.Windows.Controls;
using System.Windows.Input;

namespace GUI.WPF
{

 public partial class FlightGridNoTracking : Window
 {
```

```csharp
public FlightGridNoTracking()
{
 InitializeComponent();
 this.Title = this.Title + "- Version: " + Assembly.
 GetExecutingAssembly().GetName().Version.ToString();
}

private void SetStatus(string s)
{
 this.C_Status.Content = s;
}

WWWingsContext ctx;

/// <summary>
/// Load flights
/// </summary>
private void C_Load_Click(object sender, RoutedEventArgs e)
{
 ctx = new WWWingsContext();
 // Clear grid
 this.C_flightDataGrid.ItemsSource = null;
 // Get departure
 string Ort = this.C_City.Text.ToString();
 // Show status
 SetStatus("Loading with " + this.C_Mode.Text + "...");

 // Prepare query
 var q = ctx.FlightSet.AsQueryable();
 if (this.C_Mode.Text == "NoTracking") q = q.AsNoTracking();
 if (Ort != "All") q = (from f in q where f.Departure == Ort select f);

 if (Int32.TryParse(this.C_Count.Text, out int count))
 {
  if (count>0) q = q.Take(count);
 }
```

```
var sw = new Stopwatch();
sw.Start();
// Execute query
var fluege = q.ToList();
sw.Stop();

// Databinding to grid
this.C_flightDataGrid.ItemsSource = fluege; // Local is empty at
NoTracking;

// set state
SetStatus(fluege.Count() + " loaded records using " + this.C_Mode.Text +
": " + sw.ElapsedMilliseconds + "ms!");
}

/// <summary>
/// Save the changed flights
/// </summary>
private void C_Save_Click(object sender, RoutedEventArgs e)
{
// Get changes and ask
var added = from x in ctx.ChangeTracker.Entries() where x.State ==
EntityState.Added select x;
var del = from x in ctx.ChangeTracker.Entries() where x.State ==
EntityState.Deleted select x;
var mod = from x in ctx.ChangeTracker.Entries() where x.State ==
EntityState.Modified select x;

if (MessageBox.Show("Do you want to save the following changes?\n" +
String.Format("Client: Changed: {0} New: {1} Deleted: {2}", mod.Count(),
added.Count(), del.Count()), "Confirmation", MessageBoxButton.YesNo) ==
MessageBoxResult.No) return;

string Ergebnis = "";

// Save
Ergebnis = ctx.SaveChanges().ToString();
```

```csharp
 // Show status
 SetStatus("Number of saved changes: " + Ergebnis);
}

/// <summary>
/// Called when starting to editing a flight in the grid
/// </summary>
private void C_flightDataGrid_BeginningEdit(object sender,
DataGridBeginningEditEventArgs e)
{
 // Access to the current edited Flight
 var flight = (Flight)e.Row.Item;

 if (flight.FlightNo > 0) // important so that new flights are not added
 before filling
 {
  // Attach may only be done if the object is not already attached!
  if (!ctx.FlightSet.Local.Any(x => x.FlightNo == flight.FlightNo))
  {
   ctx.FlightSet.Attach(flight);
   SetStatus($"Flight {flight.FlightNo} can now be edited!");
  }
 }
}

/// <summary>
/// Called when deleting a flight in the grid
/// </summary>
private void C_flightDataGrid_PreviewKeyDown(object sender, KeyEventArgs e)
{
 var flight = (Flight)((DataGrid)sender).CurrentItem;

 if (e.Key == Key.Delete)
 {
  // Attach may only be done if the object is not already attached!
  if (!ctx.FlightSet.Local.Any(x => x.FlightNo == flight.FlightNo))
```

```
   {
    ctx.FlightSet.Attach(flight);
   }

   ctx.FlightSet.Remove(flight);
   SetStatus($"Flight {flight.FlightNo} can be deleted!");
  }
}

/// <summary>
/// Called when adding a flight in the grid
/// </summary>
private void C_flightDataGrid_RowEditEnding(object sender,
DataGridRowEditEndingEventArgs e)
{
 var flight = (Flight)e.Row.Item;
 if (!ctx.FlightSet.Local.Any(x => x.FlightNo == flight.FlightNo))
 {
  ctx.FlightSet.Add(flight);
  SetStatus($"Flight {flight.FlightNo} has bee added!");
 }
}

private void C_Test_Click(object sender, RoutedEventArgs e)
{
 try
 {
  ctx = new WWWingsContext();
  var flight = ctx.FlightSet.FirstOrDefault();
  if (flight == null) MessageBox.Show("No flights :-(", "Test
  Connection", MessageBoxButton.OK, MessageBoxImage.Warning);
  else MessageBox.Show("OK!", "Test Connection", MessageBoxButton.OK,
  MessageBoxImage.Information);
 }
 catch (Exception ex)
 {
```

```
    MessageBox.Show("Error: " + ex.ToString(), "Test Connection",
    MessageBoxButton.OK, MessageBoxImage.Error);
  }

 }
 }
}
```

# QueryTrackingBehavior and AsTracking()

No-tracking mode dramatically improves performance when reading data with Entity Framework and Entity Framework Core data sets. You have seen in earlier sections that the no-tracking mode should almost always be used. Unfortunately, in the classic Entity Framework, you always had to remember to use AsNoTracking() in each of your queries. This is not only annoying but is also easy to forget. In the classical Entity Framwork, you need additional solutions, for example an abstraction for accessing DbSet<T>, which automatically enabled no-tracking mode each time.

In Entity Framework Core, Microsoft has introduced a more elegant solution: you can put the entire Entity Framework core context into no-tracking mode. There is the enumeration property QueryTrackingBehavior in the class Microsoft.EntityFrameworkCore.DbContext in the subobject ChangeTracker. By default, it is set to QueryTrackingBehavior.TrackAll; in other words, the tracking is activated. However, if you change it to QueryTrackingBehavior.NoTracking, all queries are executed in no-tracking mode, even without the AsNoTracking()extension method. To execute individual queries in tracking mode, there is a new extension method AsTracking() for a no-tracking basic mode (see Listing 17-16). Figure 17-13 shows the output.

***Listing 17-16.*** Setting QueryTrackingBehavior and Using AsTracking()

```
  public static void TrackingMode_QueryTrackingBehavior()
  {

  CUI.MainHeadline("Default setting: TrackAll. Use AsNoTracking()");
  using (WWWingsContext ctx = new WWWingsContext())
  {
```

```
 ctx.ChangeTracker.QueryTrackingBehavior = QueryTrackingBehavior.
 TrackAll; // Standard
 var flightSet = ctx.FlightSet.AsNoTracking().ToList();
 var flight = flightSet[0];
 Console.WriteLine(flight + " object state: " + ctx.Entry(flight).
 State); // Detached
 flight.FreeSeats--;
 Console.WriteLine(flight + " object state: " + ctx.Entry(flight).
 State); // Modified
 int count = ctx.SaveChanges();
 Console.WriteLine($"Saved changes: {count}"); // 0
}

CUI.MainHeadline("Default setting: NoTracking.");
using (WWWingsContext ctx = new WWWingsContext())
{
 ctx.ChangeTracker.QueryTrackingBehavior = QueryTrackingBehavior.
 NoTracking; // NoTracking
 var flightSet = ctx.FlightSet.ToList();
 var flight = flightSet[0];
 Console.WriteLine(flight + " object state: " + ctx.Entry(flight).
 State); // Unchanged
 flight.FreeSeats--;
 Console.WriteLine(flight + " object state: " + ctx.Entry(flight).
 State); // Modified
 int count = ctx.SaveChanges();
 Console.WriteLine($"Saved changes: {count}"); // 0
}

CUI.MainHeadline("Default setting: NoTracking. Use AsTracking()");
using (WWWingsContext ctx = new WWWingsContext())
{
 ctx.ChangeTracker.QueryTrackingBehavior = QueryTrackingBehavior.
 NoTracking; // NoTracking
 var flightSet = ctx.FlightSet.AsTracking().ToList();
 var flight = flightSet[0];
```

```
Console.WriteLine(flight + " object state: " + ctx.Entry(flight).
State); // Unchanged
flight.FreeSeats--;
Console.WriteLine(flight + " object state: " + ctx.Entry(flight).
State); // Modified
int count = ctx.SaveChanges();
Console.WriteLine($"Saved changes: {count}"); // 1
 }

}
```

```
Default setting: TrackAll. Use AsNoTracking()
Flight #100: from Berlin to Paris on 23.08.18 03:31: 204 free Seats. object state: Detached
Flight #100: from Berlin to Paris on 23.08.18 03:31: 203 free Seats. object state: Detached
Saved changes: 0
Default setting: NoTracking.
Flight #100: from Berlin to Paris on 23.08.18 03:31: 204 free Seats. object state: Detached
Flight #100: from Berlin to Paris on 23.08.18 03:31: 203 free Seats. object state: Detached
Saved changes: 0
Default setting: NoTracking. Use AsTracking()
Flight #100: from Berlin to Paris on 23.08.18 03:31: 204 free Seats. object state: Unchanged
Flight #100: from Berlin to Paris on 23.08.18 03:31: 203 free Seats. object state: Modified
Saved changes: 1
```

***Figure 17-13.*** *Output of Listing 17-16*

# Consequences of No-Tracking Mode

No-tracking mode has further consequences besides the missing change tracking, listed here:

- The objects do not load in the first-level cache of Entity Framework Core. When accessing the object again (for example, with DbSet<T>. Find()), it is always loaded by the database management system.

- There is no relationship fixup. Relationship fixup is a feature of Entity Framework Core that connects two independently loaded objects in RAM if they belong together according to the database. For example, say Pilot 123 has been loaded. Flight 101 is now loaded and has a value of 123 for Pilot in its foreign key relationship. Entity Framework Core will connect Flight 101 and Pilot 123 in RAM so you can navigate from Flight to Pilot and, for bidirectional navigation, from Pilot to Flight.

- Lazy loading does not work with no-tracking, but lazy loading is not currently available in Entity Framework Core anyway.

## Best Practices

The new default `QueryTrackingBehavior.NoTracking` and the new extension method `AsTracking()` are meaningful additions in Entity Framework Core. But having seen many poorly performing applications of Entity Framework/Entity Framework Core in practice, for me this does not go far enough. `QueryTrackingBehavior.NoTracking` should be standard so that all developers get a high-performance query execution. Currently, in Entity Framework Core with `QueryTrackingBehavior.TrackAll` as the standard setting, every developer still has to remember to set `QueryTrackingBehavior. NoTracking`. It's best to do this in the constructor of the context class itself, and as a result, you will have no more overhead for tracking queries!

## Selecting the Best Loading Strategy

The loading strategies available for Entity Framework Core for related master or detail data (explicit reload, eager loading, and preloading) were discussed in Chapter 9. Unfortunately, I cannot say in general what the best loading strategy is. It always depends on the situation, and the best loading strategy for your situation can be determined only on a case-by-case basis through performance tests. However, some blanket statements are still possible.

Basically, it is advisable not to load connected data records as a whole, if they are not absolutely necessary, but to load connected data sets only when they are actually needed. It depends on the number of potentially connected and connected data sets to determine whether eager loading is worthwhile.

If you know that connected data is needed (for example, in the context of a data export), you should choose eager loading or preloading. The preloading trick shown can significantly increase the performance in many cases compared to eager loading with `Include()`.

If you do not know exactly whether the linked data is needed, then the choice between delayed loading and eager loading is often a choice between the plague and cholera. Reloading slows everything down through the additional round-trips to the server, but eager loading slows everything down through the larger set of results. In most cases, however, an increased number of round-trips is worse for performance than the bigger result sets.

If you're not sure, do not wire delayed loading or eager loading into the code, but instead allow it to be controlled via a configuration at runtime. Thus, the operator of the application can tune the application with increasing amounts of data and according to the typical user behavior of the application.

# Caching

Both web and desktop applications have data that is constantly in use but rarely updated in the data store. In these cases, time-based caching of the data in RAM is useful. The classic .NET has had the component `System.Runtime.Caching.dll` since version 4.0. A precursor to `System.Runtime.Caching` has been around since .NET 1.0 within ASP.NET in the namespace `System.WebCaching` in `System.Web.dll`. The component `System.Runtime.Caching.dll` introduced in .NET 4.0, on the other hand, can be used in all types of applications. `System.Runtime.Caching` inherently offers only one kind of cache: `MemoryCache` for RAM caching. You can develop other cache methods (for example, on dedicated cache servers or in the file system) by deriving from the base class `ObjectCache`. The caching feature of Windows Server's `AppFabric` is another caching option, but it is not `System.Runtime.Caching`-based.

---

**Note**   .NET Core replaces `System.Runtime.Caching` with the NuGet package `Microsoft.Extensions.Caching.Memory`. However, `System. Runtime.Caching` is now part of the Windows Compatibility Pack for .NET Core (`https://blogs.msdn.microsoft.com/dotnet/2017/11/16/ announcing-the-windows-compatibility-pack -for-net-core`), which is also available for .NET Core.

---

# MemoryCache

Listing 17-17 shows an example of using `MemoryCache` in conjunction with Entity Framework Core. First, `GetFlights1()` checks whether the list of flights from a departure is already in the cache. If the list does not exist, all relevant flights are loaded with a new Entity Framework context instance. For this data set, GetFlights1() creates a cache entry called FlightSet. The program code with policy.AbsoluteExpiration = DateTime.Now.AddSeconds (5) determines that the cache entry should expire after five seconds. It then automatically disappears from the RAM.

Of course, it would also be possible to create a separate cache entry for each departure location to cache all flights in one entry and then to filter them from the RAM. Then the number of database accesses would be even lower, but you would also have data in RAM that you might not need. This can be considered only if the data volumes are not too large. How much RAM is used for caching can be set in the application configuration file (app.config/web.config), either in absolute megabytes (cacheMemoryLimitMegabytes) or in a percent of physical memory (physicalMemoryLimitPercentage). The check interval for these limits (pollingInterval) can also be defined. As an alternative to defining these parameters through the application configuration file, it is possible to pass them to the constructor of MemoryCache as a NameValueCollection.

The Demo_MemoryCache() method in Listing 17-17 tests the operation of GetFlights1() by making two calls per second within 15 seconds. Listing 17-17 shows that the caching solution works as expected and reloads the flights only every five seconds. Figure 17-14 shows the output.

***Listing 17-17.*** Timed Caching of Data Loaded with Entity Framework Using MemoryCache

```
public static void Demo_MemoryCache()
{
 CUI.MainHeadline(nameof(Demo_MemoryCache));
 DateTime Start = DateTime.Now;
 do
 {
  var flightSet = GetFlight1("Rome");
  // you can process the flights here...
  Console.WriteLine("Processing " + flightSet.Count + " flights...");
  System.Threading.Thread.Sleep(500);
 } while ((DateTime.Now - Start).TotalSeconds < 60);

 CUI.Print("done!");
}
```

```csharp
/// <summary>
/// GetFlight with MemoryCache (5 sek)
/// </summary>
private static List<Flight> GetFlight1(string departure)
{
 string cacheItemName = "FlightSet_" + departure;

 // Access to the cache entry
 System.Runtime.Caching.MemoryCache cache = System.Runtime.Caching.
 MemoryCache.Default;
 List<Flight> flightSet = cache[cacheItemName] as List<Flight>;
 if (flightSet == null) // Element is NOT in the cache
 {
  CUI.Print($"{DateTime.Now.ToLongTimeString()}: Cache missed",
  ConsoleColor.Red);
  using (var ctx = new WWWingsContext())
  {
   ctx.Log();
   // Load flights
   flightSet = ctx.FlightSet.Where(x => x.Departure == departure).ToList();
  }
  // Store flights in cache
  CacheItemPolicy policy = new CacheItemPolicy();
  policy.AbsoluteExpiration = DateTime.Now.AddSeconds(5);
  //or: policy.SlidingExpiration = new TimeSpan(0,0,0,5);
  cache.Set(cacheItemName, flightSet, policy);
 }
 else // Data is already in cache
 {
  CUI.Print($"{DateTime.Now.ToLongTimeString()}: Cache hit",
  ConsoleColor.Green);
 }
 return flightSet;
}
```

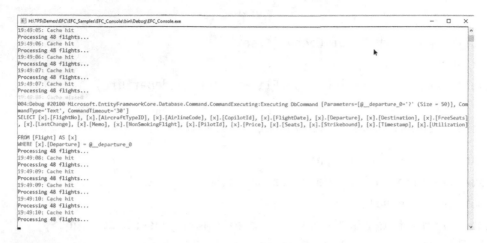

***Figure 17-14.*** *Output to Listing 17-17*

---

**Note** Incidentally, `System.Runtime.Caching` can do even more, especially so-called cache invalidation based on resource changes. For example, a cache entry can be removed immediately (even before the expiration of the set cache period) if a file changes (`HostFileChangeMonitor`) or the contents of a database table (`SqlChangeMonitor`) change.

---

# CacheManager

Data access methods such as `GetFlights1()` appear hundreds or even thousands of times in each application. The same program code appearing repeatedly to check for the existence of the cache entry and possibly creating a new entry is certainly not a good solution.

In Listing 17-18, the task is much more concise and tidier. `GetFlights2()` consists only of calling the generic `Get()` method of an instance of the `CacheManager` class. The `CacheManager` receives the cache duration in seconds during instantiation. In addition to the type parameter describing the return type, the `Get()` method expects the cache entry name and a reference to a load method for the data. The third and any subsequent parameters are passed to the load method by `Get()`. The load method `GetFlights2Internal()` is then completely free of cache aspects and is responsible only for loading the data with Entity Framework. It could also be called directly, but this is usually undesirable. Therefore, it is also "private" here.

*Listing 17-18.* Simplified Implementation of the Task Now with the CacheManager

```
public static void Demo_CacheManager()
  {
   CUI.MainHeadline(nameof(Demo_CacheManager));
   DateTime Start = DateTime.Now;
   do
   {
    var flightSet = GetFlight2("Rome");
    // you can process the flights here...
    Console.WriteLine("Processing " + flightSet.Count + " flights...");
    System.Threading.Thread.Sleep(500);
   } while ((DateTime.Now - Start).TotalSeconds < 60);
  }

  /// <summary>
  /// GetFlight with CacheManager (5 sek)
  /// </summary>
  private static List<Flight> GetFlight2(string departure)
  {
   string cacheItemName = "FlightSet_" + departure;
   var cm = new CacheManager<List<Flight>>(5);
   cm.CacheHitEvent += (text) => { CUI.Print($"{DateTime.Now.
   ToLongTimeString()}: Cache hit: " + text, ConsoleColor.Green); };
   cm.CacheMissEvent += (text) => { CUI.Print($"{DateTime.Now.
   ToLongTimeString()}: Cache missed: " + text, ConsoleColor.Red); };
   return cm.Get(cacheItemName, GetFlight2Internal, departure);
  }

  private static List<Flight> GetFlight2Internal(object[] param)
  {
   using (var ctx = new WWWingsContext())
   {
```

```
    ctx.Log();
    string departure = param[0] as string;
    // Load flights
    return ctx.FlightSet.Where(x => x.Departure == departure).ToList();
    }
}
```

However, this elegant CacheManager class is not a .NET Framework class but a self-implementation. The full source code for this class is shown in Listing 17-20. In addition to the generic Get() method used in Listing 17-18, which expects a type parameter and a load method, you can also directly retrieve data from the cache with another overload of Get(). If the data does not exist, you get zero back here. With Save() you can also save directly. The user of the CacheManager class does not see anything of the underlying library System.Runtime.Caching.

By its very nature, using the property SlidingExpiration instead of AbsoluteExpiration sounds seductive for the given task. However, the policy leads.SlidingExpiration = new TimeSpan (0,0,0,5) says that the data will never be reloaded after the first load because the time span of 5 seconds set by TimeSpan (0,0,0,5) refers to SlidingExpiration. The last access, i.e. only five seconds after the last read access the cache entry is removed. To force a reload, you would have to set the duration for Sleep() to 5000 or higher in the method Demo_CacheManager().

If you want it to be a bit more succinct, you should look at Listing 17-20, which shows a variant with an anonymous function. It is no longer necessary to write a separate load method; the necessary code is completely embedded in GetFlights4(). Thanks to the Closure technique, Get() no longer needs to get Departure as a parameter, because the anonymous method embedded in GetFlights3() can directly access all the variables of the GetFlights3() method.

***Listing 17-19.*** Variant for Using CacheManager with an Anonymous Function

```
public static void Demo_CacheManagerLambda()
{
CUI.MainHeadline(nameof(Demo_CacheManagerLambda));
 DateTime Start = DateTime.Now;
 do
```

```csharp
{
  var flightSet = GetFlight3("Rome");
  // you can process the flights here...
  Console.WriteLine("Processing " + flightSet.Count + " flights...");
  System.Threading.Thread.Sleep(500);

} while ((DateTime.Now - Start).TotalSeconds < 60);
}

public static List<Flight> GetFlight3(string departure)
{
  string cacheItemName = "FlightSet_" + departure;
  Func<string[], List<Flight>> getData = (a) =>
  {
    using (var ctx = new WWWingsContext())
    {
      // Load flights
      return ctx.FlightSet.Where(x => x.Departure == departure).ToList();
    }
  };

  var cm = new CacheManager<List<Flight>>(5);
  cm.CacheHitEvent += (text) => { CUI.Print($"{DateTime.Now.
ToLongTimeString()}: Cache Hit: " + text, ConsoleColor.Green); };
  cm.CacheMissEvent += (text) => { CUI.Print($"{DateTime.Now.
ToLongTimeString()}: Cache Miss: " + text, ConsoleColor.Red); };

  return cm.Get(cacheItemName, getData);
}
```

***Listing 17-20.*** The Auxiliary Class CacheManager Simplifies the Use of System.
Runtime.Caching

```
using System;
using System.Collections.Generic;
using System.Runtime.Caching;

namespace ITVisions.Caching
{
 /// <summary>
 /// CacheManager for simplified caching with System.Runtime.Caching
 /// (C) Dr. Holger Schwichtenberg 2013-2017
 /// </summary>
 public class CacheManager
 {
  public static List<MemoryCache> AllCaches = new List<MemoryCache>();

  public static bool IsDebug = false;
  /// <summary>
  /// Default cache duration
  /// </summary>
  public static int DefaultCacheSeconds = 60 * 60; // 60 minutes

  /// <summary>
  /// Reduced cache duration in debug mode
  /// </summary>
  public static int DefaultCacheSeconds_DEBUG = 10; // 10 seconds
  /// <summary>
  /// Removes all entries from all caches
  /// </summary>
  public static void Clear()
  {
   MemoryCache.Default.Dispose();
   foreach (var c in AllCaches)
```

```
  {
   c.Dispose();
  }
 }

 /// <summary>
 /// Removes all entries with name part from all caches
 /// </summary>
 /// <param name="name"></param>
 public static void RemoveLike(string namepart)
 {
  foreach (var x in MemoryCache.Default)
  {
   if (x.Key.Contains(namepart)) MemoryCache.Default.Remove(x.Key);
  }
  foreach (var c in AllCaches)
  {
   foreach (var x in MemoryCache.Default)
   {
    if (x.Key.Contains(namepart)) MemoryCache.Default.Remove(x.Key);
   }
  }
 }
}

/// <summary>
/// CacheManager for simplified caching with System.Runtime.Caching
/// (C) Dr. Holger Schwichtenberg 2013-2017
/// </summary>
/// <typeparam name="T">type of cached data</typeparam>
/// <example>
/// public List<Datentyp> GetAll()
/// {
/// var cm = new CacheManager<List<Datentyp>>();
/// return cm.Get("Name", GetAllInternal, "parameter");
/// }
```

```csharp
/// public List<Datentyp> GetAllInternal(string[] value)
/// {
/// var q = (from x in Context.MyDbSet where x.Name == value select x);
/// return q.ToList();
/// }
/// </example>
public class CacheManager<T> where T : class
{

 /// <summary>
 /// CacheHit or CassMiss
 /// </summary>
 public event Action<string> CacheEvent;
 /// <summary>
 /// triggered when requested data is in the cache
 /// </summary>
 public event Action<string> CacheHitEvent;
 /// <summary>
 /// triggered when requested data is not in the cache
 /// </summary>
 public event Action<string> CacheMissEvent;

 private readonly int _seconds = CacheManager.DefaultCacheSeconds;

 public MemoryCache Cache { get; set; } = MemoryCache.Default;

 /// <summary>
 /// Created CacheManager with MemoryCache.Default
 /// </summary>
 public CacheManager()
 {
  if (CacheManager.IsDebug || System.Diagnostics.Debugger.IsAttached)
  {
   this._seconds = CacheManager.DefaultCacheSeconds_DEBUG;
  }
```

```
 else
 {
  this._seconds = CacheManager.DefaultCacheSeconds;
 }
}

public CacheManager(int seconds) : this()
{
 this._seconds = seconds;
}

/// <summary>
/// Generated CacheManager with its own MemoryCache instance
/// </summary>
/// <param name="seconds">Gets or sets the maximum memory size, in
///     megabytes, that an instance of a MemoryCache object can grow to.</param>
/// <param name="cacheMemoryLimitMegabytes"></param>
/// <param name="physicalMemoryLimitPercentage">Gets or sets the
///     percentage of memory that can be used by the cache.</param>
/// <param name="pollingInterval">Gets or sets a value that indicates
///     the time interval after which the cache implementation compares the
///     current memory load against the absolute and percentage-based memory
///     limits that are set for the cache instance.</param>
public CacheManager(int seconds, int cacheMemoryLimitMegabytes,
int physicalMemoryLimitPercentage, TimeSpan pollingInterval)
{
 var config = new System.Collections.Specialized.NameValueCollection();
 config.Add("CacheMemoryLimitMegabytes", cacheMemoryLimitMegabytes.
 ToString());
 config.Add("PhysicalMemoryLimitPercentage",
 physicalMemoryLimitPercentage.ToString());
 config.Add("PollingInterval", pollingInterval.ToString());
 Cache = new MemoryCache("CustomMemoryCache_" + Guid.NewGuid().ToString(),
 config);
```

```
    Console.WriteLine(Cache.PhysicalMemoryLimit);
    Console.WriteLine(Cache.DefaultCacheCapabilities);
    this._seconds = seconds;
}

/// <summary>
/// Get element from cache. It will not load if it is not there!
/// </summary>
public T Get(string name)
{
  object objAlt = Cache[name];
  return objAlt as T;
}

/// <summary>
/// Get element from cache or data source. Name becomes the name of the
generic type
/// </summary>
public T Get(Func<string[], T> loadDataCallback, params string[] args)
{
  return Get(typeof(T).FullName, loadDataCallback, args);
}

/// <summary>
/// Retrieves item from cache or data source using the load method.
/// </summary>
public T Get(string name, Func<string[], T> loadDataCallback, params
string[] args)
{
  string cacheInfo = name + " (" + Cache.GetCount() + " elements in cache.
  Duration: " + _seconds + "sec)";
  string action = "";
  object obj = Cache.Get(name);
  if (obj == null) // not in cache
```

```csharp
{
 action = "Cache miss";
 CacheMissEvent?.Invoke(cacheInfo);
 CUI.PrintVerboseWarning(action + ": " + cacheInfo);

 #region DiagnoseTemp
 string s = DateTime.Now + "################ CACHE MISS for: " + cacheInfo
 + ": " + loadDataCallback.ToString() + System.Environment.NewLine;
 int a = 0;
 var x = Cache.DefaultCacheCapabilities;
 foreach (var c in Cache)
 {
  a++;
  s += $"{a:00}: LIMIT: {Cache.PhysicalMemoryLimit}:" + c.Key + ": " +
  c.Value.ToString().Truncate(100) + System.Environment.NewLine;

 }

 Console.WriteLine(s);
 #endregion

 // load data now
 obj = loadDataCallback(args);
 // and store it in cache
 Save(name, obj as T);

}
else // found in cache
{
 action = "Cache hit";
 CUI.PrintVerboseSuccess(action + ": " + cacheInfo);
 CacheHitEvent?.Invoke(cacheInfo);
}

// return data
CacheEvent?.Invoke(action + " for " + cacheInfo);
return obj as T;
}
```

```csharp
/// <summary>
/// Saves an object in the cache
/// </summary>
public void Save(string name, T obj)
{
 if (obj == null) return;
 object objAlt = Cache[name];
 if (objAlt == null)
 {
  CacheItemPolicy policy = new CacheItemPolicy();
  policy.AbsoluteExpiration = DateTime.Now.AddSeconds(_seconds);
  policy.RemovedCallback = new CacheEntryRemovedCallback(this.
  RemovedCallback);
  Cache.Set(name, obj, policy);
 }
}

public void RemovedCallback(CacheEntryRemovedArguments arguments)
{

}


/// <summary>
/// Removes an entry with specific names from this cache
/// </summary>
/// <param name="name"></param>
public void Remove(string name)
{
 if (Cache.Contains(name)) Cache.Remove(name);
}
/// <summary>
/// Removes all entries with specific name part from this cache
/// </summary>
public void RemoveLike(string namepart)
```

```
 {
  foreach (var x in Cache)
  {
   if (x.Key.Contains(namepart)) Cache.Remove(x.Key);
  }
 }
}
}
```

# Second-Level Caching with EFPlus

The CacheManager shown in Listing 17-18 is a general solution that allows you to cache not just Entity Framework objects but any form of data. Caching on Entity Framework Core can be even more elegant! For Entity Framework Core there is a special caching solution within the additional libraries Entity Framework Plus (EFPlus) and EFSecondLevelCache.Core (see Chapter 20).

These components implement a context-independent query result cache based on System.MemoryCache.Runtime.Caching. Such a cache is called a *second-level cache*. These extra components can manipulate queries so that Entity Framework Core materializes their results into objects and stores them not only in the first-level cache of the context instance but also in a second-level cache at the process level. Another context instance can then look up the same query in this second-level cache and deliver the objects stored there instead of a new query from the database management system (see Figure 17-15).

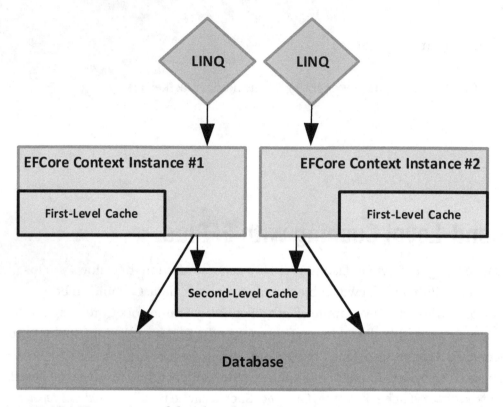

***Figure 17-15.*** *How a second-level cache works*

---

**Note**    This section discusses the second-level cache in
EFPlus. EFSecondLevelCache.Core is much more complex in configuration
but also more flexible because in addition to the main memory cache
(MemoryCache), Redis as a cache is possible.

---

## Setting Up a Second-Level Cache

Setting up a second-level cache in the context class is not necessary with EFPlus.

# Using the Second-Level Cache

Listing 17-21 shows the application of the second-level cache in EFPlus. In GetFlights4(), the FromCache() method is used in the LINQ query, specifying a caching duration (here: five seconds), in the form of an object of type MemoryCacheEntryOptions from the NuGet package Microsoft.Extensions.Caching.Abstraction.

Alternatively, you can set the cache duration centrally and then omit the FromCache() parameter.

```
var options = new MemoryCacheEntryOptions() { AbsoluteExpiration =
DateTime.Now.AddSeconds(5) };
QueryCacheManager.DefaultMemoryCacheEntryOptions = options;
```

Note that GetFlights4() creates a new context instance each time it is called, but caching still works, as Figure 17-16 demonstrates.

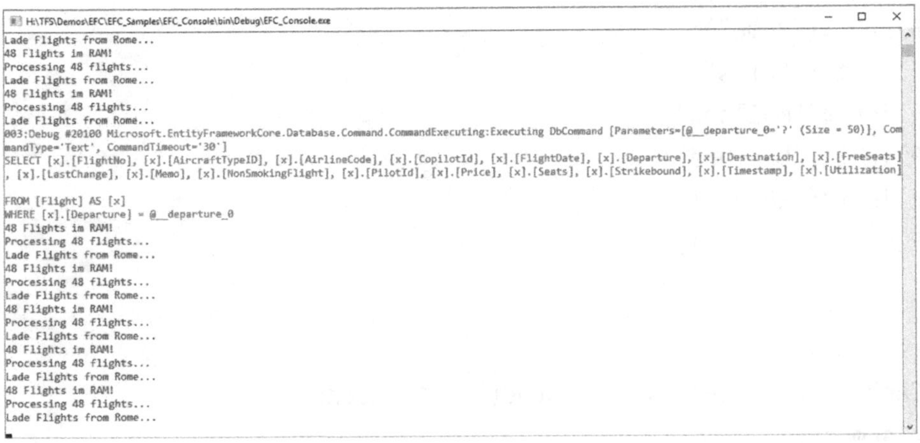

***Figure 17-16.***  *Output of Listing 17-21*

Unfortunately, unlike the solutions that directly use the MemoryCache object, you have no way of getting information whether the objects came from the cache or at which time the database was consulted because unfortunately the cache manager of EFPlus in this cases fires no results. So, you can derive the cache behavior from the database accesses, which you do through Entity Framework logging (ctx.Log(); see Chapter 12) or via an external profiler (for example, Entity Framework Profiler or SQL Server Profiler).

**_Listing 17-21._** Second-Level Caching with EFPlus

```csharp
public static void Demo_SecondLevelCache()
{
 CUI.MainHeadline(nameof(Demo_SecondLevelCache));
 DateTime Start = DateTime.Now;
 do
 {
  var flightSet = GetFlight4("Rome");
  // you can process the flights here...
  Console.WriteLine("Processing " + flightSet.Count + " flights...");
  System.Threading.Thread.Sleep(500);
 } while ((DateTime.Now - Start).TotalSeconds < 30);

 GetFlight4("Rome");
 GetFlight4("Rome");
 GetFlight4("Rome");
 GetFlight4("Paris");
 GetFlight4("Mailand");
 GetFlight4("Mailand");
 GetFlight4("Rome");
 GetFlight4("Paris");
}

/// <summary>
/// Caching with EFPlus FromCache() / 5 seconds
/// </summary>
/// <param name="departure"></param>
/// <returns></returns>
public static List<Flight> GetFlight4(string departure)
{
 using (var ctx = new WWWingsContext())
 {
  ctx.Log();

  var options = new MemoryCacheEntryOptions() { AbsoluteExpiration =
  DateTime.Now.AddSeconds(5) };
```

```
// optional: QueryCacheManager.DefaultMemoryCacheEntryOptions = options;

Console.WriteLine("Load flights from " + departure + "...");

var flightSet = ctx.FlightSet.Where(x => x.Departure == departure).
FromCache(options).ToList();
Console.WriteLine(flightSet.Count + " Flights im RAM!");
return flightSet;
 }
}
```

# CHAPTER 18

# Software Architecture with Entity Framework Core

Entity Framework Core undoubtedly belongs to the data access layer. But what does the layer model look like overall when using Entity Framework Core? In this chapter, I briefly discuss several architectural alternatives.

## Monolithic Model

Entity Framework Core can be used in a monolithic software model. In other words, the instantiation of the Entity Framework Core context and the execution of the commands (LINQ, stored procedures, SQL) are in the presentation layer (see Figure 18-1). However, this makes sense only in very small applications (see the app MiracleList Light in Appendix A).

© Holger Schwichtenberg 2018
H. Schwichtenberg, *Modern Data Access with Entity Framework Core*,
https://doi.org/10.1007/978-1-4842-3552-2_18

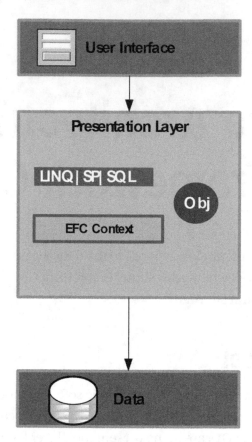

**Figure 18-1.** *Entity Framework Core in the monolithic software architecture model*

# Entity Framework Core as a Data Access Layer

Figure 18-2 shows the general structure of a multilayer application on the left and shows a simple multilayered software architecture model on the right, using Entity Framework Core for data access. This pragmatic software architecture model dispenses with a dedicated data access layer. Rather, the Entity Framework context is the complete data layer. The overlying layer is the business logic layer, which controls data access through Language Integrated Query (LINQ) commands and the invocation of stored procedures, including direct SQL commands as needed. According to statements in the business logic layer, the Entity Framework context fills the entity classes. Entity classes are passed down through all layers to the presentation layer.

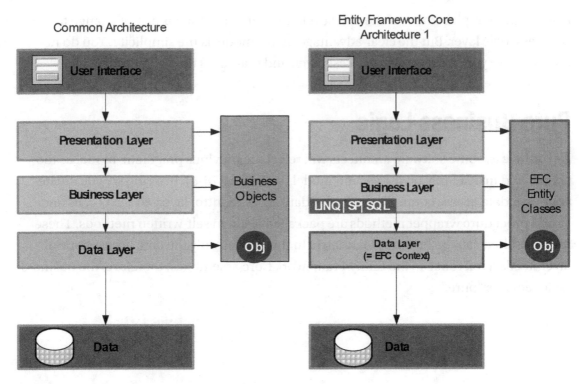

***Figure 18-2.*** *The pragmatic Entity Framework Core–based software architecture model*

Some software architects criticize this simplified model because the business logic layer is contaminated with data access commands. The business logic layer should not actually contain any database access commands. You can see this if you really equate LINQ with SQL. But you can also understand LINQ as a true abstraction of SQL. After all, LINQ commands are just a sequence of database-neutral method calls; the SQL-like syntax in C# and Visual Basic is just syntactic sugar for the software developer. The C# or Visual Basic compiler immediately makes the LINQ commands a method call string again. You can also use this method call string itself, that is, `collection.Where(x => x.CatID > 4).OrderBy(x => x.Name)` instead of `x in collection where x.CatID > 4 orderby x.Name`. But method calls are exactly the form in which business logic and data access control normally communicate with each other; that is, the necessary use of LINQ in the business logic layer does nothing here other than what is common practice between the business logic and the data layer. LINQ is only more generic than most APIs of data layers.

What is actually some contamination of the business logic layer is the use of the Entity Framework context instance in the business logic layer. This means that the business logic layer must have a reference to the Entity Framework Core assemblies.

A subsequent replacement of the object-relational mapper then means a change to the business logic layer. But the clear advantage of this model is the simplicity. You do not have to write your own database access layer, and that saves time and money.

# Pure Business Logic

Nevertheless, some software architects will reject the previous pragmatic model as too simple and instead rely on the second model (see Figure 18-3). In doing so, you create your own data access control layer. In this data access control layer, all LINQ calls and stored procedure wrapper methods are packaged again in self-written methods. These methods then call the business logic layer. In this model, only the data access control layer needs a reference to the Entity Framework Core assemblies; therefore, the business logic remains "pure."

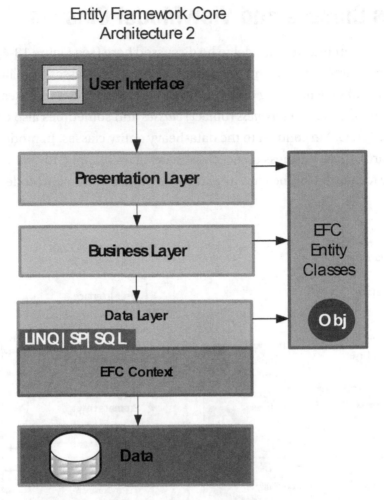

*Figure 18-3.* *The textbook Entity Framework Core–based software architecture model without distribution*

This second software architecture model corresponds to the "pure" doctrine, but in practice it also requires much more implementation effort. Especially in cases of "form-over-data" applications with little business logic in the strict sense, the developer has to implement many "annoying" wrapper routines. For LINQ, GetCustomers() in the database access layer contains the LINQ command, and GetCustomers() in the business logic forwards to GetCustomers() in the database access layer. With stored procedure usage, both layers only pass on.

# Business Classes and ViewModel Classes

The third software architecture model to be discussed here (see Figure 18-4) goes even further with an abstraction step and also prohibits the entity classes from being passed on to all layers. Rather, a mapping of the entity classes to other classes takes place. These other classes are often called *business* (object) *classes* and sometimes also called *data transfer objects* (DTOs) in contrast to the data-heavy entity classes. In model 3b (the right side of the figure), these business object classes are again mapped to classes that have been specially formatted for the view as part of the Model-View-ViewModel (MVVM) pattern.

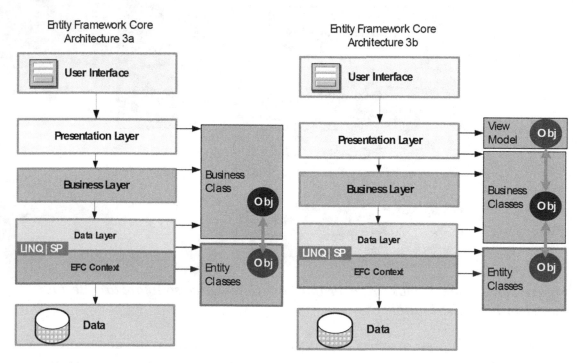

***Figure 18-4.***  *Business objects and ViewModel classes*

The business class–based software architecture model would be mandatory if the created entity classes had a relationship to Entity Framework Core (something like the `EntityObject` base class in the first version of the classic ADO.NET Entity Framework). But that's not the case in Entity Framework Core. A good reason for using the business class–based model is if the Design of the entity classes match the needs of the presentation layer, for example, because it is a "historically grown" database schema.

However, this business class–based model implies a considerable implementation overhead because all data must be transferred from the entity classes to the business classes. And of course this transfer must be implemented in the opposite direction for new and changed objects. Such an object-to-object mapping (OOM) does not work with an object-relational mapper like Entity Framework Core. However, there are other frameworks for object-to-object mapping, such as AutoMapper (`http://automapper.org`) and ValueInjecter (`http://valueinjecter.codeplex.com`). But even with such frameworks, the implementation effort is significant, especially since there is no graphical designer for object-to-object mapping.

Also, the effort is higher not only at development time but also at runtime since the additional mapping requires computation time.

# Distributed Systems

Figure 18-5, Figure 18-6, and Figure 18-7 show six software architecture models for distributed systems with Entity Framework Core for data access. There is now no direct access to the database from the client, but there is a service façade on the application server, as well as proxy classes in the client (which call the service façade). In terms of the division of the business logic layer and the data access layer, you have the same options as in architecture 1 and architecture 2. These options are not shown here. It's more about the entity classes. If you use the same classes on the client side as on the server side, this is called *shared contracts*. This is possible whenever the server and client are written in .NET, and therefore the client can reference the assembly with the classes from the server. The case of shared contracts is shown in architecture 4 on the left of Figure 18-5; here the client also uses the Entity Framework Core entity classes.

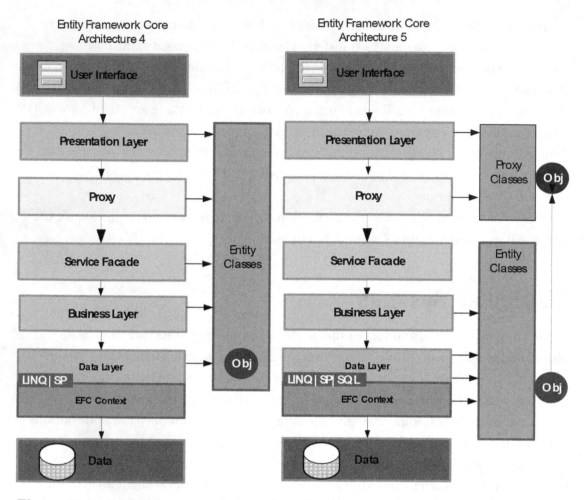

***Figure 18-5.*** *Entity Framework Core–based software architecture models in a distribution system*

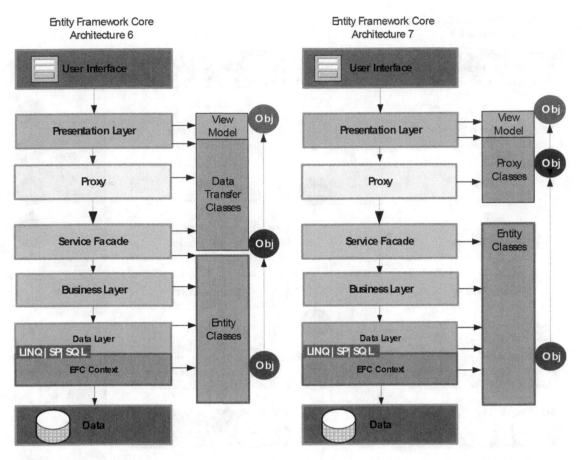

**Figure 18-6.** *Entity Framework Core–based software architecture models in a distribution system*

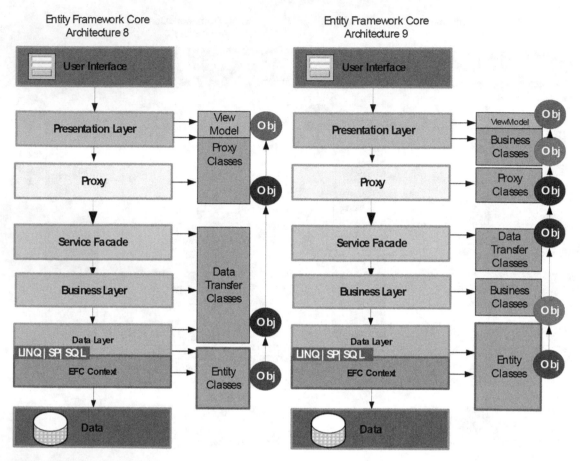

***Figure 18-7.*** *More Entity Framework Core–based software architecture models with distribution*

If the client has a different platform, then you have to create proxy class for the entity classes. In architecture 5 shown in Figure 18-5, explicit proxy classes are desired or necessary because the client is not .NET.

The architecture models 6 to 9 differ only in terms of the mapping of the entity classes.

- Although architecture 6 works with shared contracts, it maps the entity classes to DTO classes that are optimized for online transmission. In the client, there is another OO mapping to ViewModel classes.

- Architecture 7 assumes proxy classes and an OO mapping to ViewModel classes.

- Architecture 8 uses DTO, proxy, and ViewModel classes.

- The most elaborate model, architecture 9, also uses business object classes in the client.

You might be wondering who uses architecture 9. In fact, I see in my job as a consultant that many software architectures are elaborately designed like this. These are projects in which larger teams work, and nevertheless every small user request needs a long time in the implementation.

# Conclusion

Software architects have many architectural options when using Entity Framework Core. The bandwidth starts with a simple, pragmatic model (with a few compromises), where developers have to implement only three assemblies. On the other side of the architectural models presented here, you need at least 12 assemblies.

Which of the architectural models to choose depends on various factors. Of course, this includes the specific requirements, the system environment, and the know-how of the available software developers. But also the budget is an important factor. I experience in my everyday life as a consultant with companies again and again that software architects choose a too-complex architecture because of the "pure" doctrine that is not adapted to the business conditions. In such systems, even the smallest user request ("We still need a field on the left") is typically extremely time-consuming and costly to implement. Many projects fail because of unnecessarily complex software architecture.

---

**Tip**    Use as few layers as possible. Think twice before adding another abstraction to your software architecture model.

---

# CHAPTER 19

# Commercial Tools

This chapter introduces commercial tools that you can use for Entity Framework Core. I am in no way involved in the development or distribution of these tools.

## Entity Framework Core Power Tools

Microsoft offered Power Tools for the classic Entity Framework, but the reissue for Entity Framework Core is now implemented by an external developer. Entity Framework Core Power Tools is a free extension to Visual Studio 2017.

## EF Core Power Tools

| | |
|---|---|
| Tool name | Entity Framework Core Power Tools |
| Web site | `https://www.visualstudiogallery.msdn.microsoft.com/9674e1bb-d942-446a-9059-a8b4bd18dde2` |
| Manufacturer | Erik Ejlskov Jensen (MVP), `https://github.com/ErikEJ` |
| Free version | Yes |
| Commercial version | No |

## Features

Once installed, Entity Framework Core Power Tools (Figure 19-1) can be accessed via the context menu of a project in the Visual Studio Solution Explorer. The add-on offers the following functions:

- Graphical user interface for the reverse engineering of existing databases in SQL Server, SQL Compact, and SQLite

447

H. Schwichtenberg, *Modern Data Access with Entity Framework Core*,
https://doi.org/10.1007/978-1-4842-3552-2_19

- The ability to create a diagram for a given Entity Framework Core context

- The ability to create a diagram for a database schema

- The ability to display the SQL DDL commands for creating a database schema for an Entity Framework Core context with its entity classes

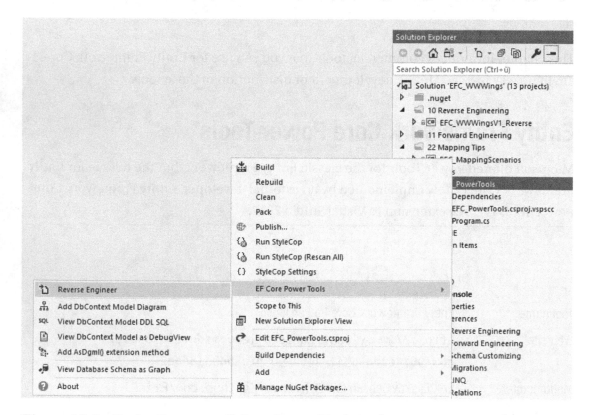

***Figure 19-1.*** *Entity Framework Core Power Tools in the context menu of a project in Solution Explorer*

# Reverse Engineering with Entity Framework Core Power Tools

Reverse engineering with Entity Framework Core Power Tools consists of three steps. In the first step, you select the database via the standard dialog of Visual Studio (Figure 19-2). In the second step, you select the tables (Figure 19-3). You can save the table selection as a text file and reload it for a new call (Figure 19-4). In the third step, you set the options, that are also allowed by the Scaffold-DbContext cmdlet (Figure 19-5). After that, the code generation is the same as in Scaffold-DbContext (Figure 19-6).

---

**Note**   Just like with Scaffold-DbContext, updating the program code after changes in the database (Update Model from Database) isn't implemented by Power Tools.

---

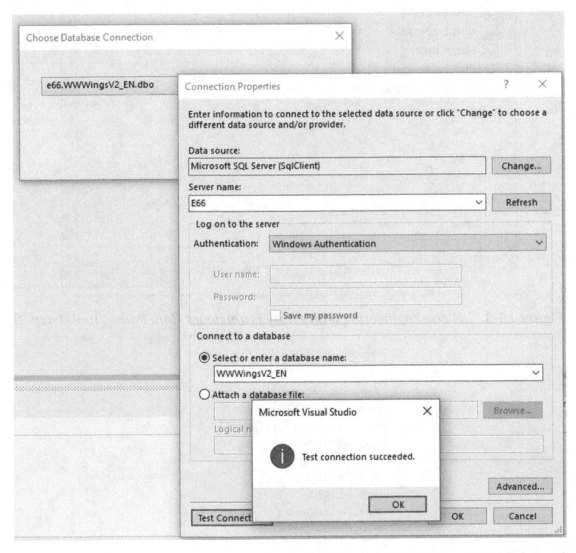

***Figure 19-2.***  *Reverse engineering with Entity Framework Core Power Tools (step 1)*

**Figure 19-3.** *Reverse engineering with Entity Framework Core Power Tools (step 2)*

**Figure 19-4.** *Storage of the table selection in a text file*

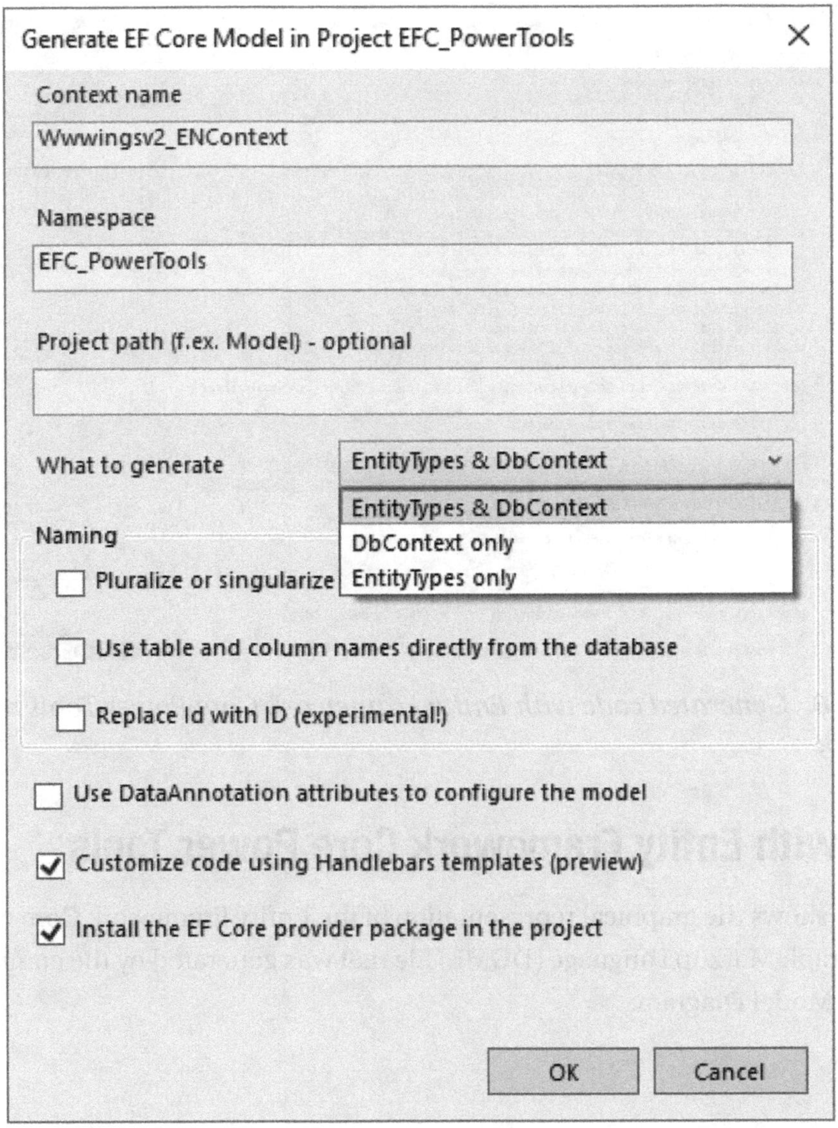

***Figure 19-5.*** *Reverse engineering with Entity Framework Core Power Tools (step 3)*

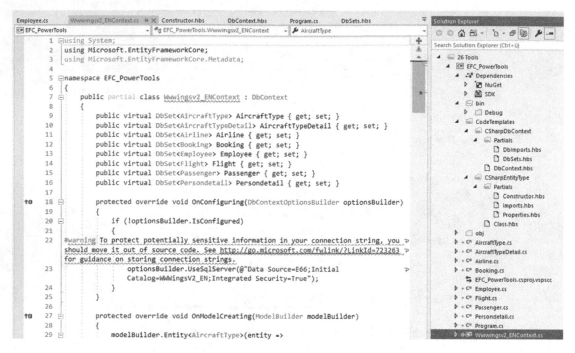

***Figure 19-6.*** *Generated code with Entity Framework Core Power Tools reverse engineering*

# Charts with Entity Framework Core Power Tools

Figure 19-7 shows the graphical representation of the Entity Framework Core model as a Directed Graph Markup Language (DGML) file that was generated by the command Add DbContext Model Diagram.

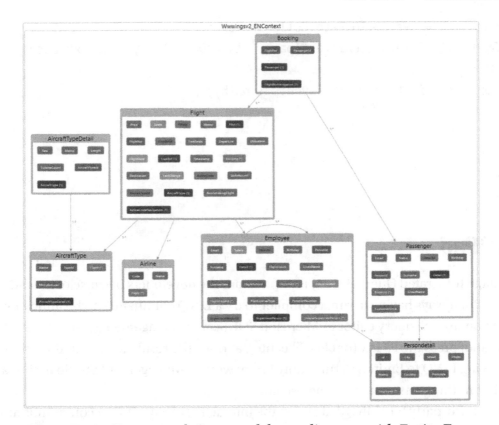

*Figure 19-7.* *Entity Framework Core model as a diagram with Entity Framework Core Power Tools*

You can also generate this diagram at runtime with the extension method AsDgml(), which is available in the NuGet package ErikEJ.EntityFrameworkCore.DgmlBuilder (Listing 19-1).

*Listing 19-1.* Using AsDgml( )

```
using System;

namespace EFC_PowerTools
{
 class Program
 {
  static void Main(string[] args)
  {
   using (var ctx = new Wwwingsv2_ENContext())
   {
```

```
    var path = System.IO.Path.GetTempFileName() + ".dgml";
    System.IO.File.WriteAllText(path, ctx.AsDgml(), System.Text.Encoding.
    UTF8);
    Console.WriteLine("file saved:" + path);
  }

 }
 }
}
```

# LINQPad

Language Integrated Query (LINQ) is popular among developers because of its static typing. But always having to run a compiler to try a LINQ command can be annoying. When you use the query editor in Microsoft SQL Server Management, you enter a SQL command, press the F5 key (or click Execute), and see the result. Microsoft once thought of allowing LINQ to Entities to run Entity Framework in Management Studio in the same way, but nothing has been published so far.

The third-party tool LINQPad allows the interactive input of LINQ commands and the direct execution in an editor. You can execute LINQ commands against objects in RAM (LINQ to Objects), Entity Framework/Entity Framework Core, and various other LINQ providers.

| Tool name | LINQPad |
|---|---|
| Website | www.linqpad.net |
| Manufacturer | Joseph Albahari, Australia |
| Free version | Yes |
| Commercial version | From $45 |

LINQPad is available as a free freeware version. But if you want to be pampered by IntelliSense input support in the style of Visual Studio, you have to buy the Professional or Premium version. In the Premium version, there are also numerous included program code snippets. Likewise, in the Premium version, you can define queries using several databases. The system requirement for the current version 5 is .NET Framework 4.6. At only 5MB in size, the application is very lightweight. The author of the tool says, "It does not slow down your computer when you install it!"

## Using LINQPad

LINQPad presents a window for the connections in the top-left corner (see Figure 19-8). Underneath you can choose from a supplied sample collection (from the book *C# 6.0 in a Nutshell*) or save your own commands (under My Queries). In the main area you will find the editor at the top and the output area at the bottom (see the center/right of Figure 19-8).

LINQPad supports the syntax of C#, Visual Basic .NET, and F# as well as SQL and Entity SQL (the latter only for the classic Entity Framework).

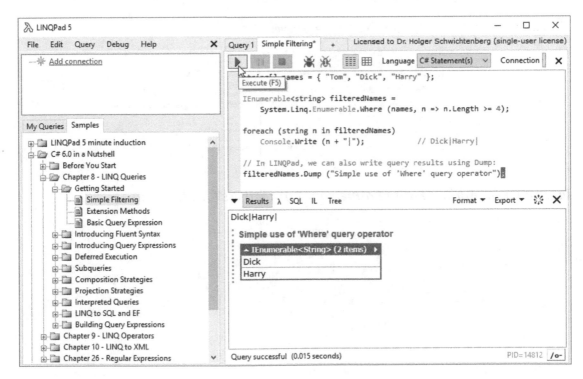

***Figure 19-8.*** *LINQPad in action with LINQ to Objects*

# Including Data Sources

To run LINQ commands against an Entity Framework Core context, you must add a connection using Add Connection. However, the dialog currently only shows drivers for LINQ to SQL and the classic Entity Framework. With View More Drivers, you can download a driver for Entity Framework Core (Figure 19-9).

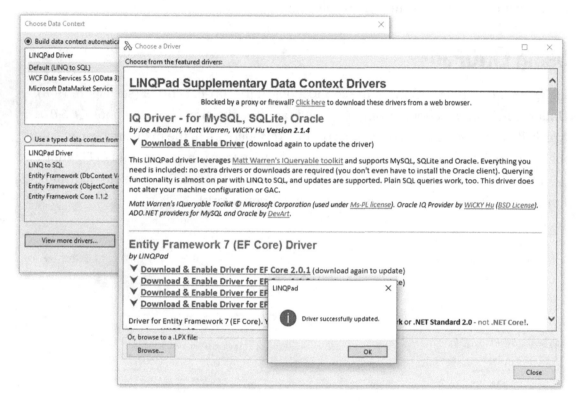

***Figure 19-9.*** *Adding Entity Framework Core drivers for LINQPad*

After adding the driver, you should be able to select Entity Framework Core (Figure 19-10).

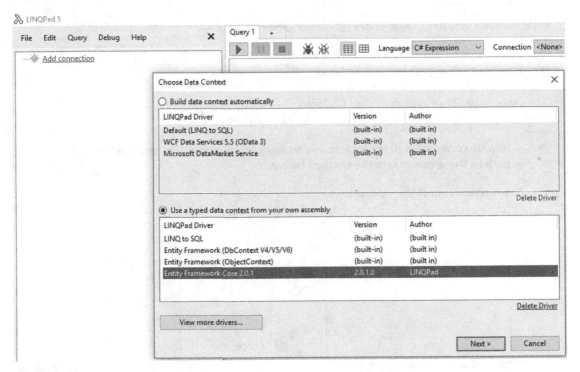

***Figure 19-10.*** *Selecting Entity Framework Core as the LINQPad driver*

After selecting the provider, you have to integrate an Entity Framework Core context. To do this, use Browse (see Figure 19-11) to select a .NET assembly that implements such a context.

---

**Note**   LINQPad itself does not create context classes for Entity Framework and Entity Framework Core. You always have to create and compile such a class using Visual Studio or another tool.

---

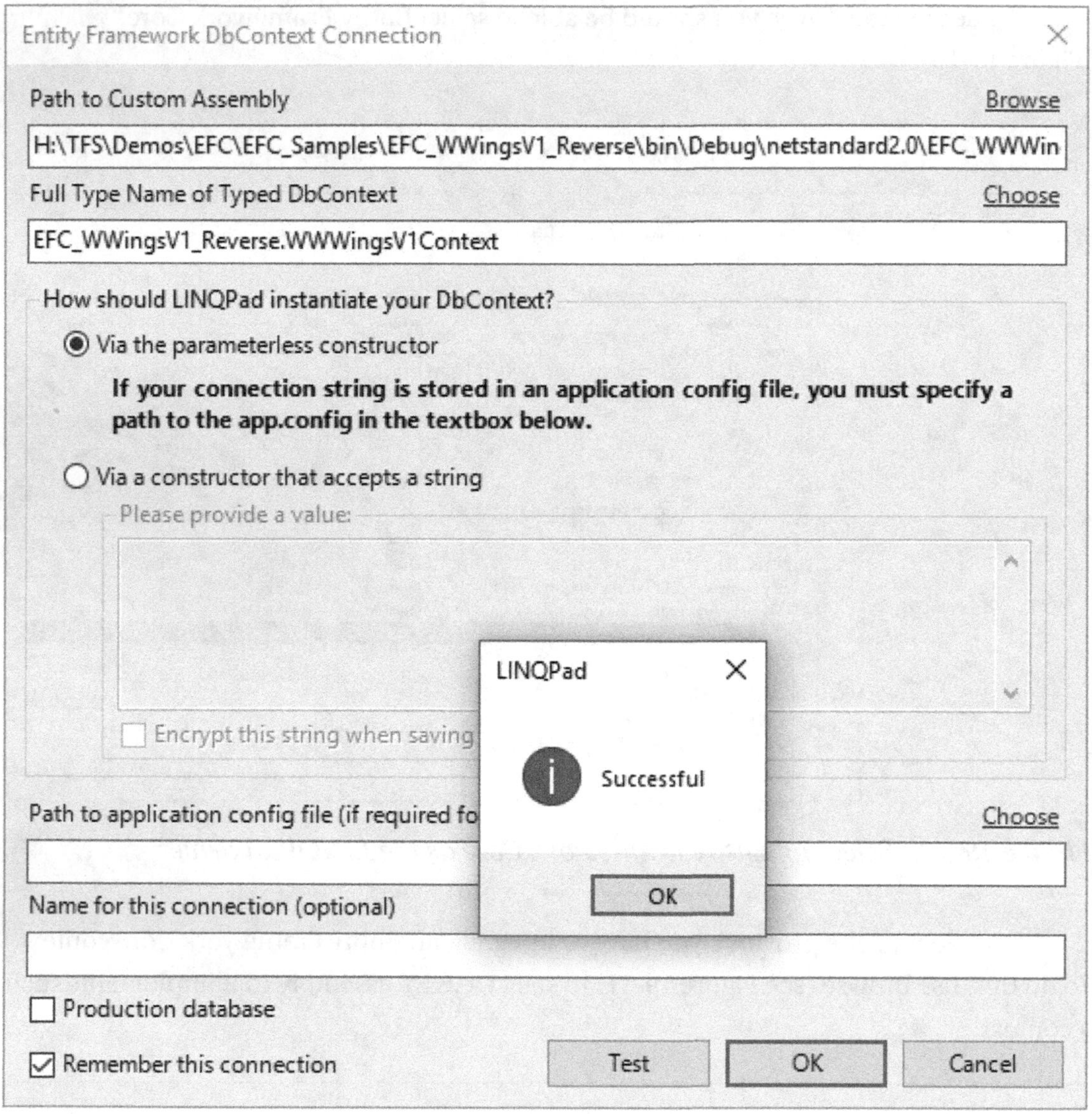

*Figure 19-11.* *An Entity Framework Core context class has been selected*

After incorporating the context, you can see the existing entity classes on the left in LINQPad (Figure 19-12).

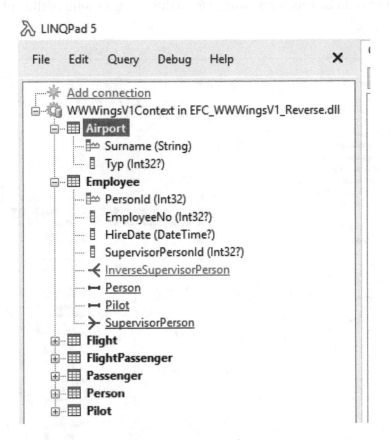

**Figure 19-12.**  *After incorporating the context class*

# Executing LINQ Commands

Some commands can be executed directly from the context menu of the entity classes (see Figure 19-13).

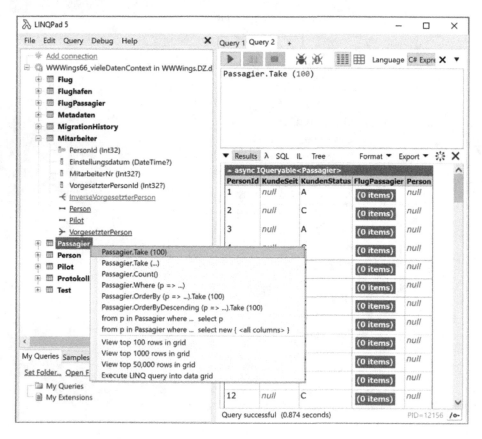

***Figure 19-13.*** *Predefined commands in the context menu of the entity class*

In the query area you can even enter commands (in the commercial version with input support). Figure 19-14 shows a LINQ command with conditions, projection, and eager loading. The result view is hierarchical in the case of eager loading.

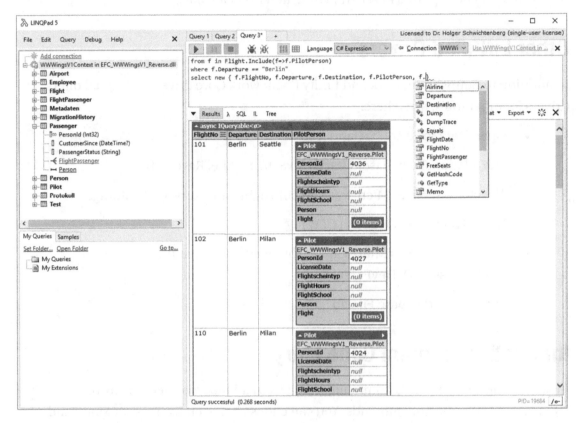

***Figure 19-14.** Execution of a custom LINQ command in LINQPad*

In addition to the result view, you can display the LINQ to Entities command in the other tabs of the output area in the following formats:

- LINQ command in lambda syntax

- LINQ command in SQL form

- LINQ command in Microsoft Intermediate Language (IL)

- LINQ command as expression tree

461

# Saving

Queries can be saved as text files with the file extension `.linq`.

Results can be exported in HTML, Word, and Excel formats.

# Other LINQPad Drivers

In addition to Entity Framework and Entity Framework Core drivers, LINQPad offers drivers for other tools, listed here:

- Open Data Protocol (OData) feeds

- The relational databases Oracle, MySQL, SQLite, RavenDB

- The cloud services Microsoft StreamInsight and Azure Table Storage

- Event Traces for Windows (ETW)

- The ORM mappers Mindscape LightSpeed, LLBLGen Pro, DevExpress XPO, DevForce

- The NoSQL database FileDb

# Interactive Program Code Entry

In addition to running LINQ commands, the LINQPad tool can execute any other C#, F#, and Visual Basic commands. You can choose between Expression Mode and Statement Mode under Language. Expression Mode captures individual expressions, the result of which is then printed, as in `System.DateTime.Now.ToString(new System.Globalization.CultureInfo("ya-JP"))`. These expressions should be terminated by a semicolon. You can do only one expression at a time. If you have multiple expressions in the editor, you must first mark the expression to be executed.

In Statement Mode, on the other hand, complete program code snippets are recorded, with each command terminated by a semicolon. You issue a command with `Console.WriteLine()`. Listing 19-2 shows a small test program.

***Listing 19-2.*** Small Test Program for LINQPad

```
<Query Kind="Statements" />

for (int i = 0; i < 10; i++)
```

```
{
    Console.WriteLine(i);
}
```

Also the definition of your own types (z.Classes; see Listing 19-3) is possible. However, note that LINQPad embeds the captured code in its own default code. Therefore, the following rules apply:

- The main program code to be executed must be at the top.

- It has to be closed with an additional curly brace before the following type definitions.

- The type definitions must be at the end, and the last type definition must not have a closing curly bracket.

So, internally, LINQPad obviously complements a type definition with a main() at the top and a curly bracket at the bottom.

***Listing 19-3.*** Small Test Program for LINQPad with Class Definition

```
<Query Kind="Statements" />

var e = new Result() { a = 1, b = 20 };

for (int i = 0; i < e.b; i++)
{
    e.a += i;
    Console.WriteLine(i + ";" + e.a);
}
} // This extra parenthesis is required!

// The type definition must be after the main program!
class Result
{
    public int a { get; set; }
    public int b { get; set; }
// // here you have to omit the parenthesis!
```

## Conclusion to LINQPad

LINQPad is a useful tool to learn LINQ, to test LINQ commands and, in general, to test commands in C#, Visual Basic, and F# without having to start a heavyweight program like Visual Studio or install a work-around routine into an existing project. Because of the practical export function, LINQPad can be used not only for development but also in everyday practice as a tool for ad hoc database queries.

## Entity Developer

Microsoft does not yet offer GUI development tools for Entity Framework Core. This gap has been closed by DevArt with the product Entity Developer.

In the past, DevArt has offered more tooling capabilities for the classic Entity Framework than Microsoft itself. Now it is again leading with tools for Entity Framework Core. Entity Developer supports both reverse engineering and forward engineering in Entity Framework Core with a graphical designer.

| | |
|---|---|
| Tool name | Entity Developer |
| Web site | www.devart.com/entitydeveloper |
| Manufacturer | DevArt, Czech Republic |
| Free version | Yes |
| Commercial version | From $99.95 |

Figure 19-15 shows the available product variants. The free Express edition can manage models with a maximum of ten tables.

| Feature | Professional | NHibernate | Entity Framework | LINQ to SQL | Express |
|---|---|---|---|---|---|
| NHibernate support | ✓ | ✓ | ✗ | ✗ | ✓ |
| Entity Framework v1 - v6 support | ✓ | ✗ | ✓ | ✗ | ✓ |
| Entity Framework Core support | ✓ | ✗ | ✓ | ✗ | ✓ |
| Telerik Data Access support | ✓ | ✗ | ✗ | ✗ | ✓ |
| LINQ to SQL support | ✓ | ✗ | ✗ | ✓ | ✓ |
| Visual schema modelling | ✓ | ✓ | ✓ | ✓ | ✓ |
| Reverse engineering | ✓ | ✓ | ✓ | ✓ | ✓ |
| Predefined templates | ✓ | ✓ | ✓ | ✓ | ✓ |
| Custom templates | ✓ | ✓ | ✓ | ✓ | ✗ |
| Unlimited number of entities in model | ✓ | ✓ | ✓ | ✓ | ✗ |
| Single License Price | $299.95 | $199.95 | $199.95 | $99.95 | Free |

***Figure 19-15.*** *Variants of Entity Developer*

When installing Entity Developer, the installer already provides a VSIX integration with Visual Studio 2015 and Visual Studio 2017 in addition to the stand-alone Entity Developer application. The application is quite slim, requiring around 60MB of disk space.

# Selecting the ORM Technique

Depending on the installed variant, Entity Developer offers different ORM techniques at startup. For Entity Framework Core, an `.efml` file is created, for the classic Entity Framework an `.edml` file is created, for Telerik Data Access a `.daml` file is created, and for NHibernate an `.hbml` file is created. After selecting EF Core Model (see Figure 19-16), the second wizard step is followed by the decision between reverse engineering (here called Database First) and forward engineering (here called Model First), as shown in Figure 19-17. In addition to Microsoft SQL Server, Entity Developer also supports Oracle, MySQL, PostgreSQL, SQLite, and IBMs DB2 as databases, each in conjunction with DevArt's own Entity Framework Core drivers (see `https://www.devart.com/dotconnect/#database`).

---

**Note**   When using Entity Developer within Visual Studio, there is no selection wizard for the ORM technique; instead, there are specific element templates such as DevArt EF Core Model and DevArt NHibernate Model.

---

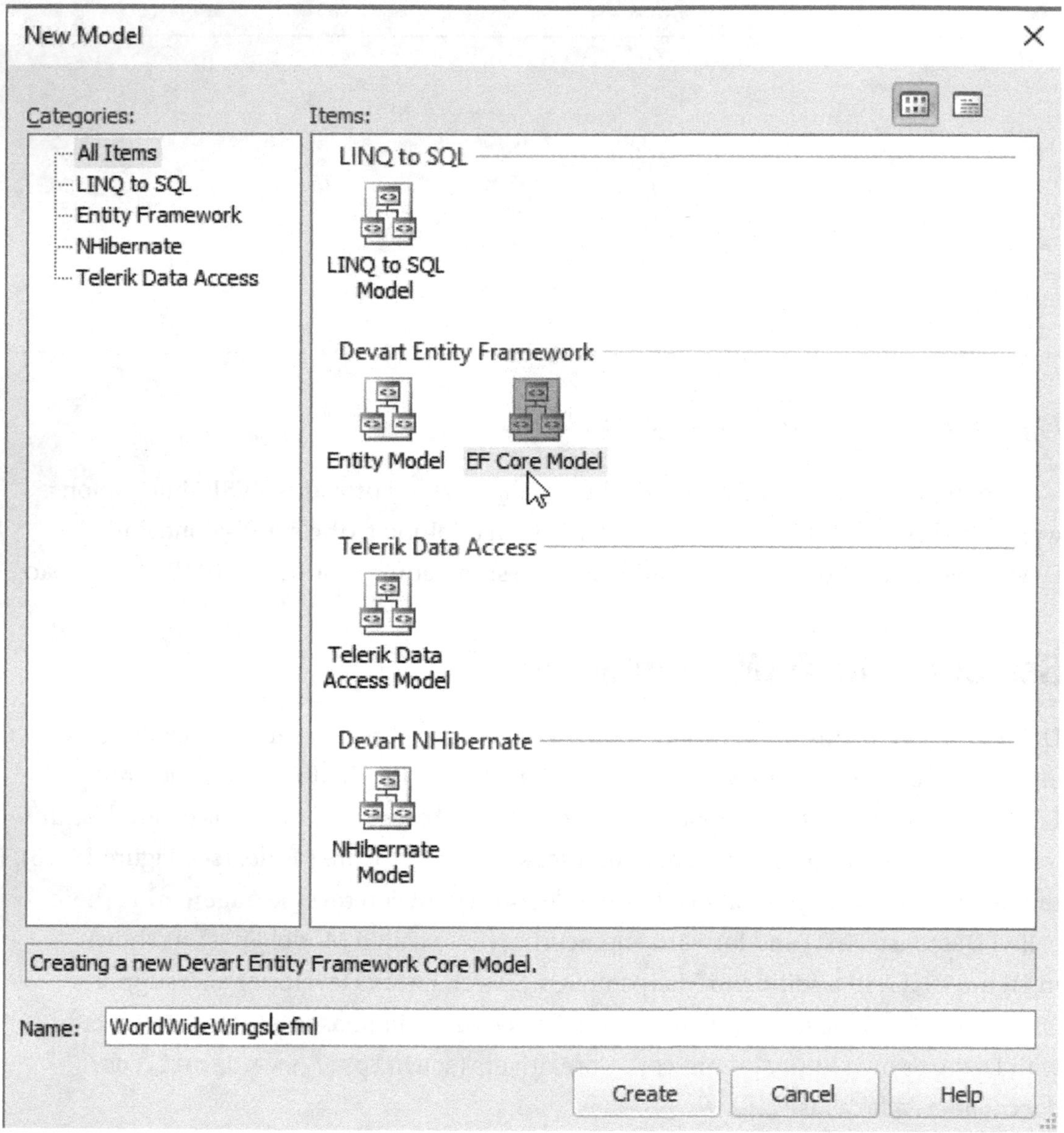

**Figure 19-16.** *Selecting the ORM technique in Entity Developer*

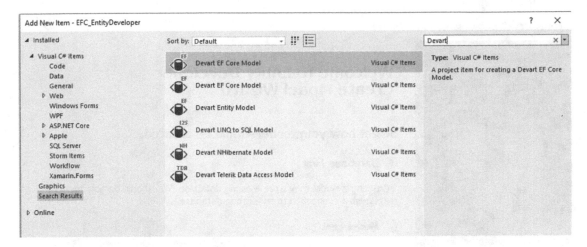

**Figure 19-17.**  *Entity Developer templates within Visual Studio*

# Reverse Engineering with Entity Developer

Database First then selects an existing database and selects the artifacts (tables, views, stored procedures, and table-valued functions), as in the classic Entity Framework wizard in Visual Studio, but with the benefit that the developer may choose down to the column level (Figure 19-18).

This is followed by a page in the code generation naming convention wizard that goes far beyond what Visual Studio has offered so far (Figure 19-19). In the following options page, some options such as the N:M relationships and the table per type inheritance are grayed out because Entity Framework Core does not yet have these mapping capabilities (Figure 19-20). In the penultimate step, you choose whether to have all the artifacts on a chart surface or just selected ones. It is also possible to create one diagram per schema name (Figure 19-21). For each diagram, a `.view` file is created.

In the last step, you select the code generation template. Entity Developer offers to directly apply multiple code generation templates (Figure 19-22).

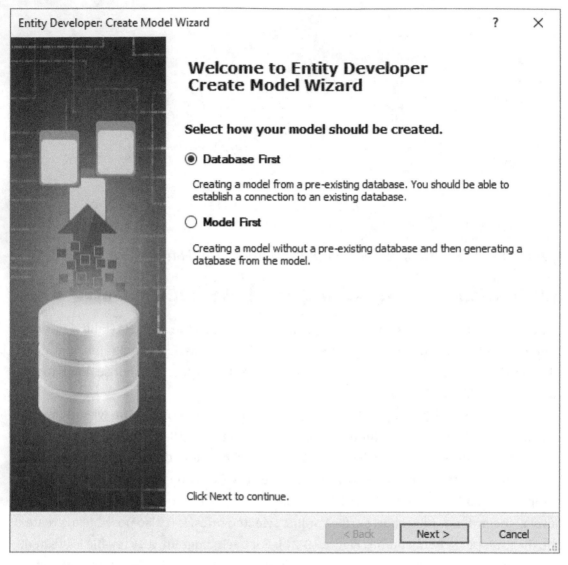

***Figure 19-18.***  *Selection of the process model*

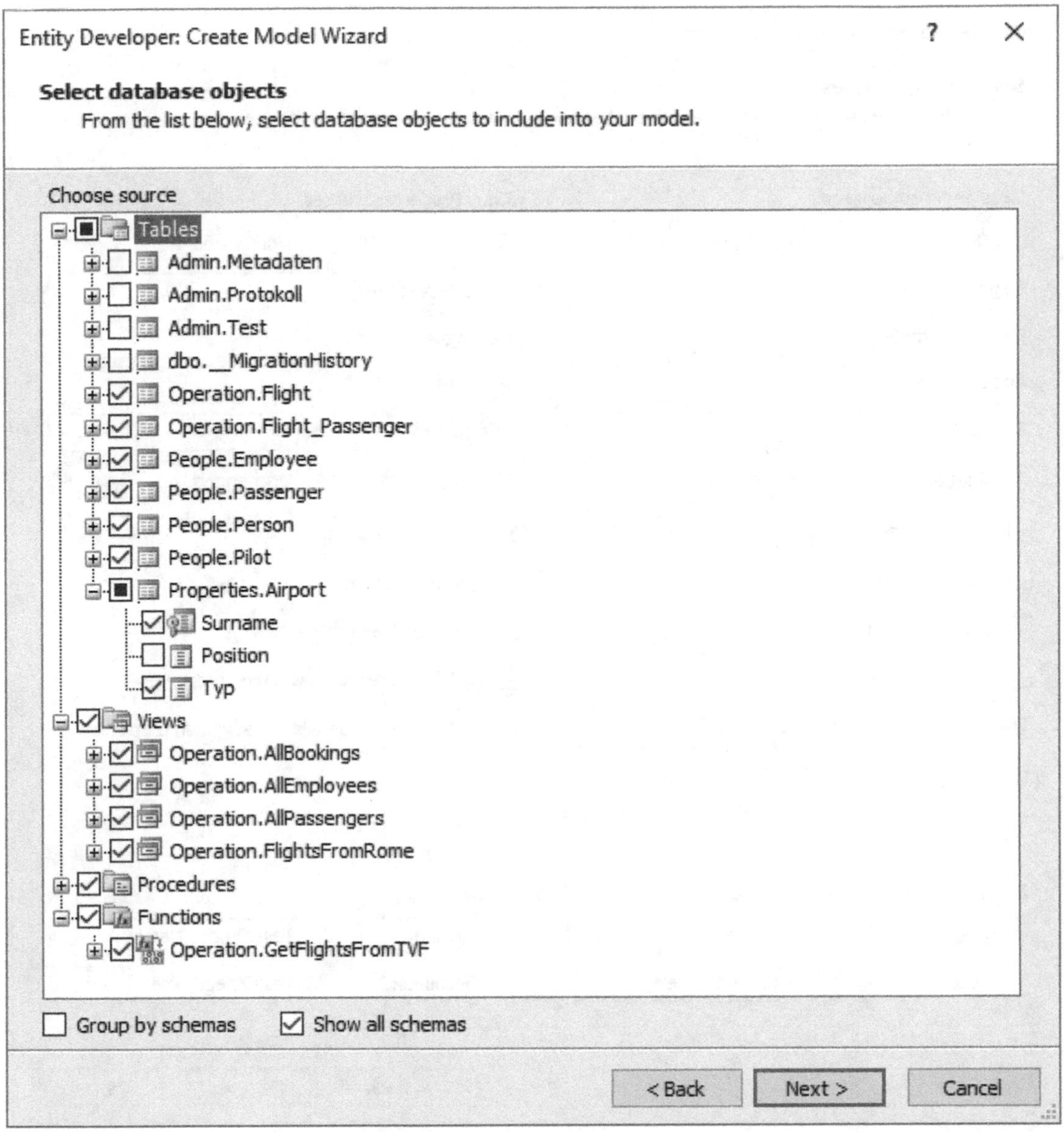

*Figure 19-19.*  *Selection of artifacts down to the column level*

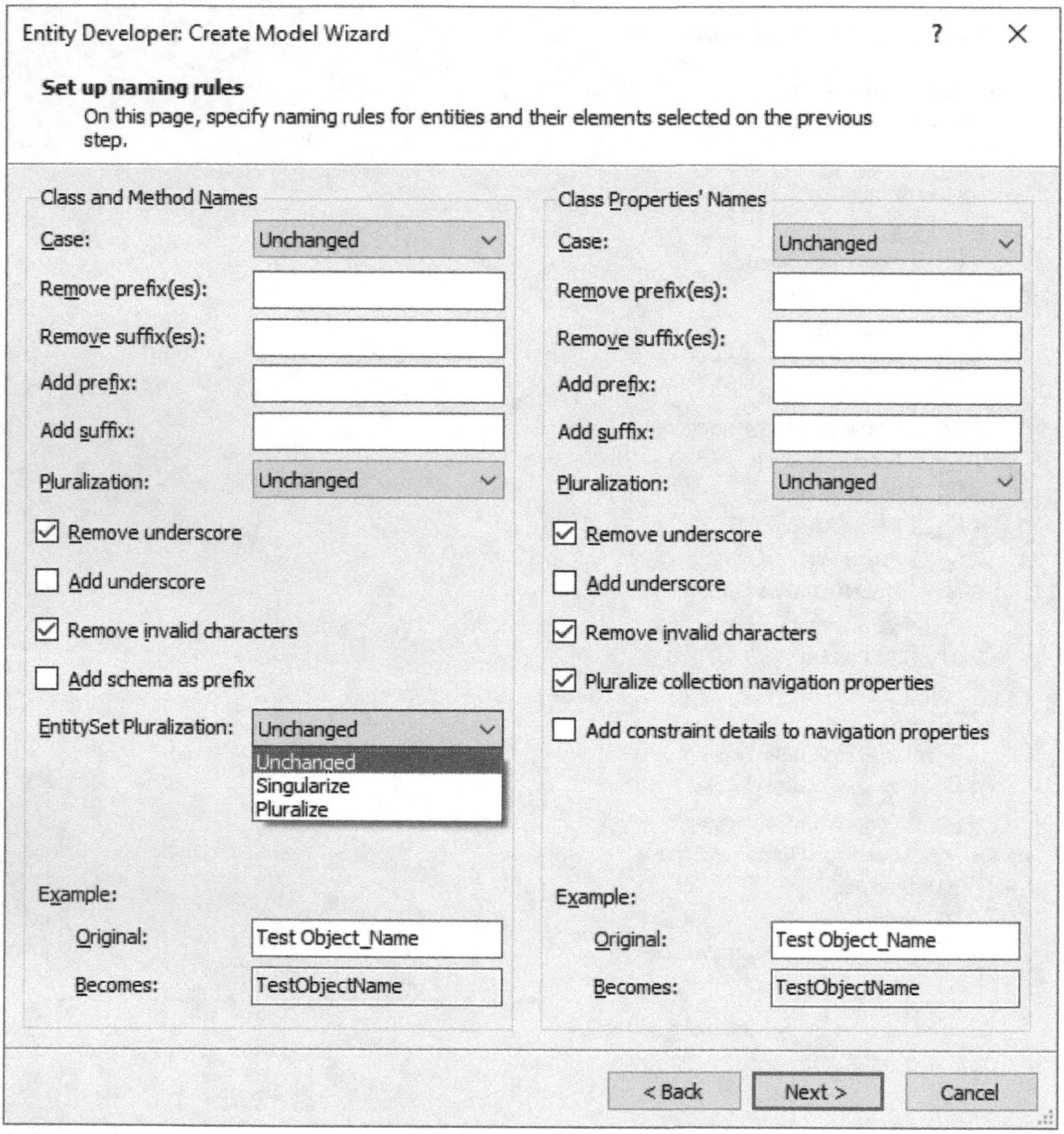

**Figure 19-20.**  *Many settings for the naming conventions of classes and class members in the code to be generated in Entity Developer*

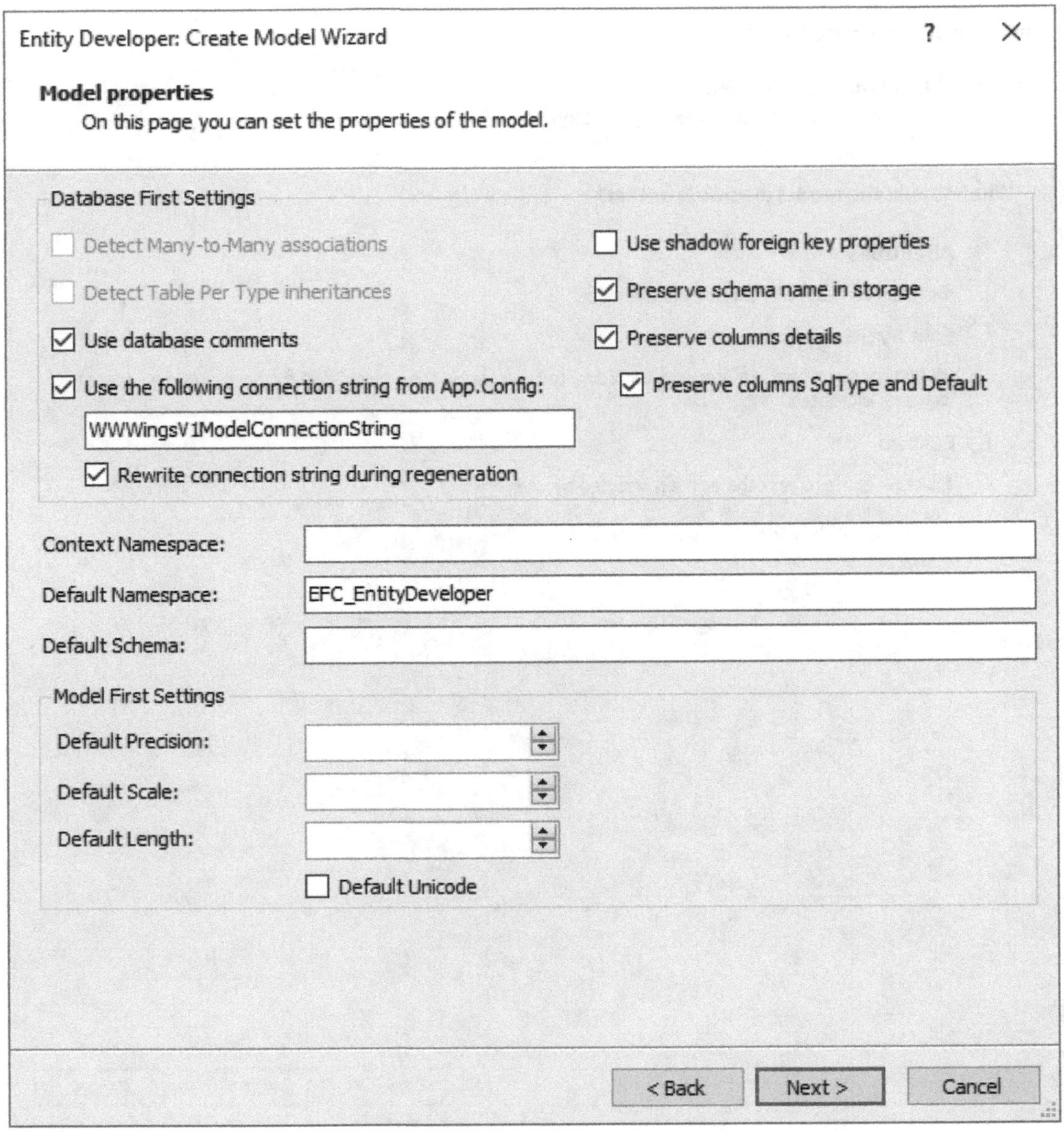

***Figure 19-21.*** *Selecting model properties*

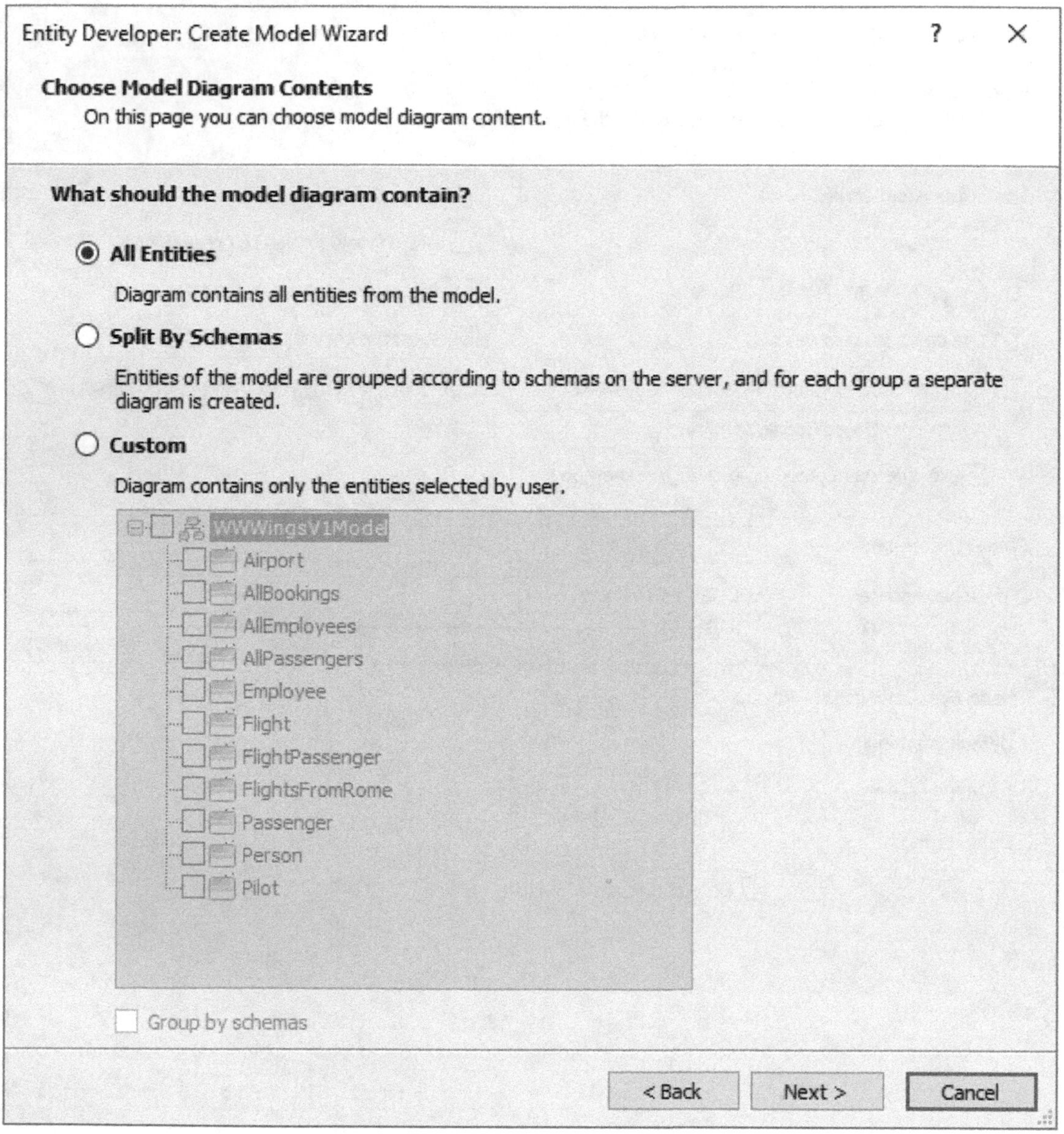

**Figure 19-22.**  *Selecting diagram content*

Figure 19-23 shows the templates supplied with Entity Developer, which are barely documented at `https://www.devart.com/entitydeveloper/templates.html`. You have to figure out whether the generated code fits your needs. The predefined templates can be copied with the function Copy to Model Folder into a template file in your own application folder and then modified there. The templates are similar to the Text Template Transformation Toolkit (T4) templates used in Visual Studio, but they are not compatible. Unlike the T4 templates, DevArt templates allow code generation to be influenced by the parameters set in a property grid. For example, you can specify the following for the selected template called EF Core:

- You can specify the landing of the entity classes and the context class in different folders (here you can capture a relative or absolute path).

- Partial classes may be generated.

- The interfaces `INotifyPropertyChanging` and `INotifyPropertyChanged` may be implemented in the entity classes.

- You can set, that the entity classes receive the annotations `[DataContract]` and `[DataMember]` for the Windows Communication Foundation (WCF).

- You can set, that the entity classes receive the annotation `[Serializable]`.

- You can override the entity classes `Equals()`.

- You can implement the entity classes of `IClonable`.

The names of the diagrams, the templates and their parameters, and the list of generated files are stored by Entity Developer in an `.edps` file.

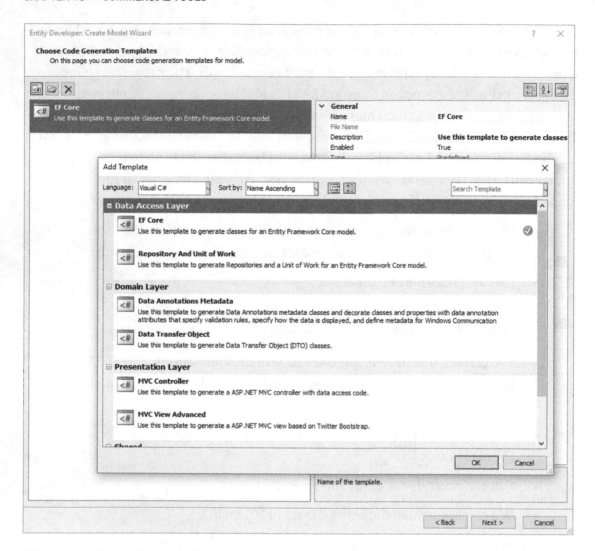

**Figure 19-23.** *Selecting the code generation template*

After completing the very flexible assistant, when looking at the model in the designer, you might be a little disillusioned, at least if you have selected database views. Entity Developer complains that there is no primary key for this. This is because Entity Framework Core is not yet set to mapping views and handles views like tables, which always need to have a primary key. You have to set one manually in the property window for each database view.

The relationships between the tables are also modeled as associations in Entity Developer in conjunction with Entity Framework Core, even if inheritance were possible. With the classic Entity Framework, Entity Developer has the option of recognizing table per type inheritance, but Entity Framework Core does not yet support table by type inheritance.

You can now adjust the graphs or create new graphs in the Model Explorer (see the left of Figure 19-24). You can drag and drop additional tables, views, procedures, and functions directly into the model from the Database Explorer (see the right of Figure 19-24) instead of having to run the wizard over and over again in Visual Studio. Entity Developer, like the classic Entity Framework tools in Visual Studio, can manage multiple charts per model of overlapping entities. You can change the order of the properties in an entity class by dragging and dropping, while the change of order with Microsoft's tool is, curiously enough, possible only with a cumbersome context menu or keyboard shortcut. Dragging and dropping properties is also possible between different entities. Entity Developer allows colors to be assigned to the entities in the model for better visual separation. The coloring then applies to all diagrams in which the entity occurs. On the chart surface, you can also make comments at any point.

The program code generation is triggered by the menu item Model ➤ Generate Code (key F7). The standard code generation template EF Core creates the following:

- A context class

- One entity class per table and per view

- A class for each return type of a stored procedure

The program code for using the stored procedures and table-valued functions is in the context class, which can be very long. It is interesting that Entity Developer does not rely on Entity Framework Core for the implementation but instead picks up the data records via `DataReader` and realizes the complete mapping itself (see Listing 19-4). After all, Entity Developer recognizes that the stored procedure `GetFlight` shown in the listing returns the same structure as the table `Flight` and therefore uses the entity class `Flight` in the return type. Entity Developer or the template could also have used Entity Framework Core with the extension method `FromSql()`. The advantage of DevArt's own implementation is that it also works for stored procedures that do not return entity types. Entity Framework Core cannot do that yet. In these cases, Entity Developer creates its own classes for the return type.

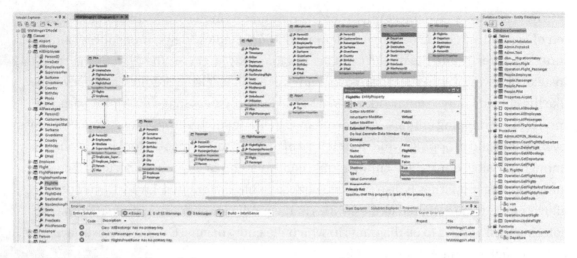

**Figure 19-24.** *Graphical designer in Entity Developer*

**Listing 19-4.** Mapping for the Stored Procedure GetFlight( )

```
public List<Flight> GetFlight(System.Nullable<int> FlightNo)
  {

   List<Flight> result = new List<Flight>();

   DbConnection connection = this.Database.GetDbConnection();
   bool needClose = false;
   if (connection.State != ConnectionState.Open)
   {
    connection.Open();
    needClose = true;
   }

   try
   {
    using (DbCommand cmd = connection.CreateCommand())
    {
     if (this.Database.GetCommandTimeout().HasValue)
      cmd.CommandTimeout = this.Database.GetCommandTimeout().Value;
     cmd.CommandType = CommandType.StoredProcedure;
     cmd.CommandText = @"Operation.GetFlight";
```

```
DbParameter FlightNoParameter = cmd.CreateParameter();
FlightNoParameter.ParameterName = "FlightNo";
FlightNoParameter.Direction = ParameterDirection.Input;
if (FlightNo.HasValue)
{
 FlightNoParameter.Value = FlightNo.Value;
}
else
{
 FlightNoParameter.DbType = DbType.Int32;
 FlightNoParameter.Size = -1;
 FlightNoParameter.Value = DBNull.Value;
}
cmd.Parameters.Add(FlightNoParameter);

using (IDataReader reader = cmd.ExecuteReader())
{
 while (reader.Read())
 {
  Flight row = new Flight();
  if (!reader.IsDBNull(reader.GetOrdinal("FlightNo")))
   row.FlightNo = (int)Convert.ChangeType(reader.GetValue(reader.
   GetOrdinal(@"FlightNo")), typeof(int));

  if (!reader.IsDBNull(reader.GetOrdinal("Timestamp")))
   row.Timestamp = (byte[])Convert.ChangeType(reader.GetValue(reader.
   GetOrdinal(@"Timestamp")), typeof(byte[]));
  else
   row.Timestamp = null;

  if (!reader.IsDBNull(reader.GetOrdinal("Airline")))
   row.Airline = (string)Convert.ChangeType(reader.GetValue(reader.
   GetOrdinal(@"Airline")), typeof(string));
  else
   row.Airline = null;

  if (!reader.IsDBNull(reader.GetOrdinal("Departure")))
```

```
row.Departure = (string)Convert.ChangeType(reader.GetValue(reader.
GetOrdinal(@"Departure")), typeof(string));

if (!reader.IsDBNull(reader.GetOrdinal("Destination")))
 row.Destination = (string)Convert.ChangeType(reader.
 GetValue(reader.GetOrdinal(@"Destination")), typeof(string));

if (!reader.IsDBNull(reader.GetOrdinal("FlightDate")))
 row.FlightDate = (System.DateTime)Convert.ChangeType(reader.
 GetValue(reader.GetOrdinal(@"FlightDate")), typeof(System.
 DateTime));

if (!reader.IsDBNull(reader.GetOrdinal("NonSmokingFlight")))
 row.NonSmokingFlight = (bool)Convert.ChangeType(reader.
 GetValue(reader.GetOrdinal(@"NonSmokingFlight")), typeof(bool));

if (!reader.IsDBNull(reader.GetOrdinal("Seats")))
 row.Seats = (short)Convert.ChangeType(reader.GetValue(reader.
 GetOrdinal(@"Seats")), typeof(short));

if (!reader.IsDBNull(reader.GetOrdinal("FreeSeats")))
 row.FreeSeats = (short)Convert.ChangeType(reader.GetValue(reader.
 GetOrdinal(@"FreeSeats")), typeof(short));
else
 row.FreeSeats = null;

if (!reader.IsDBNull(reader.GetOrdinal("Pilot_PersonID")))
 row.PilotPersonID = (int)Convert.ChangeType(reader.GetValue(reader.
 GetOrdinal(@"Pilot_PersonID")), typeof(int));
else
 row.PilotPersonID = null;

if (!reader.IsDBNull(reader.GetOrdinal("Memo")))
 row.Memo = (string)Convert.ChangeType(reader.GetValue(reader.
 GetOrdinal(@"Memo")), typeof(string));
else
 row.Memo = null;

if (!reader.IsDBNull(reader.GetOrdinal("Strikebound")))
```

```
    row.Strikebound = (bool)Convert.ChangeType(reader.GetValue(reader.
    GetOrdinal(@"Strikebound")), typeof(bool));
  else
    row.Strikebound = null;

  if (!reader.IsDBNull(reader.GetOrdinal("`Utilization `")))
    row.Utilization = (int)Convert.ChangeType(reader.GetValue(reader.
    GetOrdinal(@"`Utilization `")), typeof(int));
  else
    row.Utilization = null;

  result.Add(row);
   }
  }
 }
}
finally
{
 if (needClose)
  connection.Close();
}
 return result;
}
```

You can now include the generated program code in a Visual Studio project, if you have not already performed the steps there. The used template files can be adjusted in Entity Developer at any time (see the branch Templates in Model Explorer). Templates can also be edited in Entity Developer, including IntelliSense input support.

Alternatively, you can also use the installed Visual Studio extension. In Visual Studio you will find new entries like DevArt EF Core Model in the element templates under the category Data. Selecting one of these opens the same assistant and, at the end, the same designer (including Model Explorer and template editor) as the stand-alone application. The advantage is that the generated program code automatically belongs to the Visual Studio project, which also contains the .efml file.

If the database schema has changed, you can update the model with the menu item Model ➤ Update Model from Database. You define the general mapping rules for database types with .NET types under Tools ➤ Options ➤ Servers' Options.

The data preview is also helpful (select Retrieve Data in the context menu of an entity class); it includes navigation to linked data records and hierarchical unfolding. You can also access the data preview directly from the diagram in the context menu of each entity or from a table or view in the Database Explorer.

# Forward Engineering with Entity Developer

I'll now cover forward engineering with Entity Developer. After selecting Model First, the dialog Model Properties opens because there is no existing database nor are there its artifacts or any naming conventions to choose from. With the settings in Model First Settings, you can set the standards for database schema generation.

- *Default Precision*: For decimal numbers, the number of digits before the comma

- *Default Scale*: For decimal numbers, the number of digits after the decimal point

- *Default Length*: For strings, the maximum number of characters (empty means that the strings are unlimited)

The third and final step of the wizard in the case of Model First is the dialog for selecting the code generation templates.

The empty designer interface then appears, which you can fill with classes, enumerations, associations, and inheritance relationships using the symbols in the Model Explorer. Then you configure them through the Properties window (see Figure 19-25). You can, for example, set the primary key, enable the [ConcurrencyCheck] annotation, and specify that a property be a shadow property that exists in the database but not in the generated entity class. There are some unavailable options in Entity Framework Core, such as an N:M mapping, that Entity Developer does not even offer in a Entity Framework core model.

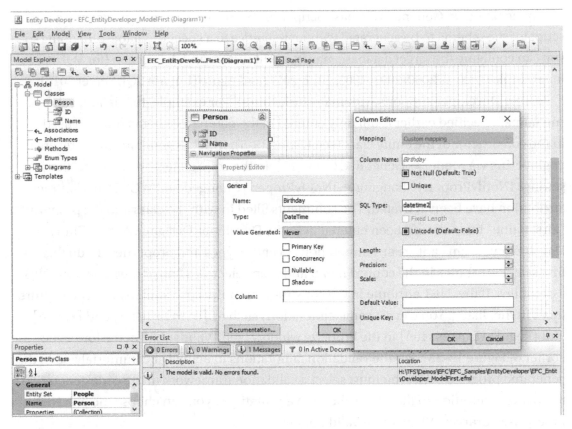

***Figure 19-25.*** *Creation and configuration of new properties in Entity Developer*

With the function Model ➤ Update Database from Model, you can create a database schema from it. This wizard asks for a target database, which must already exist. The wizard then displays which schema changes are to be transferred to the database and offers the choice not to transfer certain changes. In the last step, you can view the SQL script to be executed. Entity Developer does not use the command-line schema migration tools of Entity Framework Core (`dotnet ef` or the PowerShell cmdlets); instead, it uses its own method of comparing the existing schema with the target schema. However, Entity Developer also tries to obtain the data. With the Recreate Database Tables option, you can make existing tables, including their data, disappear. An additional table called `__EFMigrationsHistory` does not exist in the Entity Developer migrations. You define the conventions for the naming in the database schema to be generated under Model ➤ Settings ➤ Synchronization ➤ Database Naming.

In the Model ➤ Generate Database Script from Model menu, you can generate a SQL script for the schema to be generated without reference to a specific database. In doing so, you can configure the target database management system and the version (for SQL Server, for example) such as 2000, 2005, 2008, 2012, 2014, and Azure (but not SQL Server 2016).

Entity Developer supports with many little things. For example, the default value of "no name" specified earlier is not only entered in the database schema but also used in the constructor of the entity class (see Listing 19-5), which is generated as in reverse engineering via Model ➤ Generated Code (F7). In the code generation settings, the settings INotifyPropertyChanging ➤ INotifyPropertyChanged and WCF Data Contract attribution have been activated. The comments filled in with the class and the property "First name" have already been entered in Entity Developer Designer. Also in Entity Developer you can capture any annotations to entity types and properties. To do this, first select Attributes in the context menu and then select .NET attributes from any .NET assemblies. If the .NET attribute in the constructor requires parameters, you can capture them in the dialog. You can set some annotations such as [DisplayName] and [Range] and [RegularExpression] in the Property window (see "Validation" in the lower-left corner of Figure 19-25). For the validation annotations to actually be immortalized in the generated program code, you must select a validation framework in the code generation template. In addition to the .NET validation annotations, you can choose the old .NET Enterprise library or NHibernate Validator here.

It is also interesting that you can extend the property grid to any settings. These settings can then be considered during code generation. Additional settings are defined under Model ➤ Settings ➤ Model in the tree displayed there under Model ➤ Extended Properties for artifacts such as classes, properties, and associations. You then have to consider the meaning of these additional settings in a separate .tmpl code generation template.

***Listing 19-5.*** Example of an Entity Class Generated by Entity Developer

```
//------------------------------------------------------------------------
// This is auto-generated code.
//------------------------------------------------------------------------
// This code was generated by Entity Developer tool using EF Core template.
// Code is generated on: 31/12/2017 00:04:31
//
// Changes to this file may cause incorrect behavior and will be lost if
// the code is regenerated.
```

```csharp
//-----------------------------------------------------------------------------

using System;
using System.Data;
using System.ComponentModel;
using System.Linq;
using System.Linq.Expressions;
using System.Data.Common;
using System.Collections.Generic;

namespace Model
{
    public partial class Person {

        public Person()
        {
            OnCreated();
        }

        public virtual string ID
        {
            get;
            set;
        }

        public virtual string Name
        {
            get;
            set;
        }

        public virtual System.DateTime Birthday
        {
            get;
            set;
        }

        #region Extensibility Method Definitions

        partial void OnCreated();
```

```
        #endregion
    }

}
//---------------------------------------------------------------------
// This is auto-generated code.
//---------------------------------------------------------------------
// This code was generated by Entity Developer tool using EF Core template.
// Code is generated on: 31/12/2017 00:04:31
//
// Changes to this file may cause incorrect behavior and will be lost if
// the code is regenerated.
//---------------------------------------------------------------------

using System;
using System.Data;
using System.Linq;
using System.Linq.Expressions;
using System.ComponentModel;
using System.Reflection;
using System.Data.Common;
using System.Collections.Generic;
using Microsoft.EntityFrameworkCore;
using Microsoft.EntityFrameworkCore.Infrastructure;
using Microsoft.EntityFrameworkCore.Internal;
using Microsoft.EntityFrameworkCore.Metadata;

namespace Model
{

    public partial class Model : DbContext
    {

        public Model() :
            base()
        {
            OnCreated();
        }
```

```csharp
public Model(DbContextOptions<Model> options) :
    base(options)
{
    OnCreated();
}

protected override void OnConfiguring(DbContextOptionsBuilder
optionsBuilder)
{
    if (!optionsBuilder.Options.Extensions.OfType<RelationalO
    ptionsExtension>().Any(ext => !string.IsNullOrEmpty(ext.
    ConnectionString) || ext.Connection != null))
        CustomizeConfiguration(ref optionsBuilder);
    base.OnConfiguring(optionsBuilder);
}

partial void CustomizeConfiguration(ref DbContextOptionsBuilder
optionsBuilder);

public virtual DbSet<Person> People
{
    get;
    set;
}

protected override void OnModelCreating(ModelBuilder modelBuilder)
{
    this.PersonMapping(modelBuilder);
    this.CustomizePersonMapping(modelBuilder);

    RelationshipsMapping(modelBuilder);
    CustomizeMapping(ref modelBuilder);
}

#region Person Mapping

private void PersonMapping(ModelBuilder modelBuilder)
{
    modelBuilder.Entity<Person>().ToTable(@"People");
```

```csharp
        modelBuilder.Entity<Person>().Property<string>(x => x.ID).
        HasColumnName(@"ID").IsRequired().ValueGeneratedNever();
        modelBuilder.Entity<Person>().Property<string>(x => x.Name).
        HasColumnName(@"Name").IsRequired().ValueGeneratedNever();
        modelBuilder.Entity<Person>().Property<System.
        DateTime>(x => x.Birthday).HasColumnName(@"Birthday").
        HasColumnType(@"datetime2").IsRequired().ValueGeneratedNever();
        modelBuilder.Entity<Person>().HasKey(@"ID");
    }

    partial void CustomizePersonMapping(ModelBuilder modelBuilder);

    #endregion

    private void RelationshipsMapping(ModelBuilder modelBuilder)
    {
    }

    partial void CustomizeMapping(ref ModelBuilder modelBuilder);

    public bool HasChanges()
    {
        return ChangeTracker.Entries().Any(e => e.State == Microsoft.
        EntityFrameworkCore.EntityState.Added || e.State == Microsoft.
        EntityFrameworkCore.EntityState.Modified || e.State ==
        Microsoft.EntityFrameworkCore.EntityState.Deleted);
    }

    partial void OnCreated();
    }
}
```

# Entity Framework Profiler

Object-relational mapping means abstraction from SQL, and naturally the question arises as to which and how many commands are actually sent to the database management system. You can monitor the communication with a DBMS's own profiler, such as the Microsoft SQL Server Profiler, or with an ORM-specific tool such as the Entity Framework Profiler.

Almost all OR mappers use their own query languages, such as HQL on NHibernate and LINQ on Entity Framework and Entity Framework Core. These languages work on the database-neutral object model, and the OR mapper translates into the SQL dialect of each database management system. This automated generation of SQL commands is always a starting point for a fundamental criticism of ORM, especially regarding the warehouse of the SQL Optimizer. In fact, not all OR statements generated by the OR mapper are always optimal.

Tracking down nonoptimal SQL and unfavorable loading strategies is one of the responsibilities of a software developer using an OR mapper. This is where the Entity Framework Profiler from the company Hibernating Rhinos comes in. It works with both Entity Framework Core and the classic Entity Framework.

---

## Entity Framework Profiler

| | |
|---|---|
| Tool name | Entity Framework Profiler |
| Web site | www.efprof.com |
| Manufacturer | Hibernating Rhinos, Israel |
| Free version | No |
| Commercial version | From $45 each month |

---

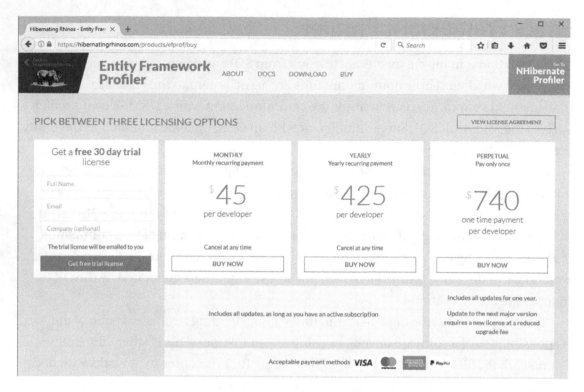

*Figure 19-26.* *Entity Framework Profiler licensing options*

# Integrating Entity Framework Profiler

For the Entity Framework Profiler to be able to record the activities between the OR mapper and the database management system, the application to be monitored must be "instrumented." This requires two changes to the program code.

- The developer must have a reference to the `HibernatingRhinos` assembly `.Profiler.CreateAppender.dll`.

---

**Tip**    This assembly ships with the Entity Framework Profiler (folder `/Appender`) in three variations: for .NET 3.5, for .NET 4.*x*, and as a .NET standard assembly (including .NET Core). While you can refer directly to this assembly in a project using the classic .NET Framework, in a .NET Core project you should not reference the assembly directly from the `/Appender/netstandard/` folder. Instead, you should install it via NuGet (`Install-Package EntityFrameworkProfiler.Appender`). Otherwise, you may be missing dependencies.

---

- At the beginning of your program (or where you want to start profiling in the program), the program line `HibernatingRhinos.Profiler.Appender.EntityFramework.EntityFrameworkProfiler.Initialize()` appears in the program code.

---

**Tip**    The Entity Framework Profiler does not require the application to run in the Visual Studio debugger. Recording also works if the application is started directly, even if it was compiled in Release mode. It's a one-line instrumentation code. So, you can make the application so that the instrumentation code is called when needed; for example, it can be controlled by a configuration file.

---

## Monitoring Commands with the Entity Framework Profiler

Start the WPF-based Entity Framework Profiler user interface (`EFProf.exe`) before the application to be monitored. After starting the application to be monitored, the Entity Framework Profiler (in the list on the left) shows all the created instances of the Entity Framework class `ObjectContext` (or all the derived classes). Unfortunately, the individual context instances are not named; you must designate them in the Entity Framework Profiler user interface yourself.

Each context contains the number of SQL commands executed via the context with the respective execution times, both those in the DBMS and the total time, including the materialization of the objects in RAM. In Figure 19-27, for example, the problem arises that many object contexts are created without any command at all being executed.

---

**Note**    The Entity Framework Profiler speaks of *object context*, which in the classic Entity Framework was the original base class for the Entity Framework context. In Entity Framework Core, there is only the more modern `DbContext`. The name in the Entity Framework Profiler has not been changed. However, the Entity Framework Profiler works with the `DbContext` base class in both Entity Framework and Entity Framework Core.

---

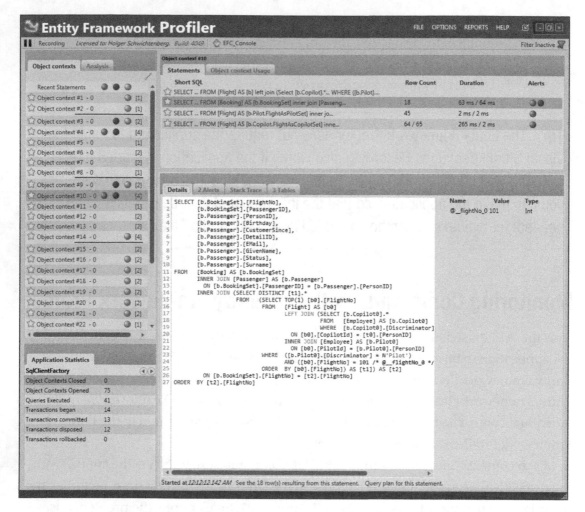

***Figure 19-27.***   *Entity Framework Profiler in action*

In the right part of the screen you will find a list of executed commands for the currently selected context. Under Details you can see the complete SQL command with parameters and the associated execution plan (see Figure 19-28) as well as the result set. To do this, however, you must enter the connection string in the Entity Framework Profiler (see Figure 19-29).

The Stack Trace tab reveals which method triggered a single SQL command. It's nice that double-clicking an entry in the Stack Trace tab leads directly to the matching code in the open Visual Studio window. This will help you quickly find the LINQ or SQL command that triggered the SQL command.

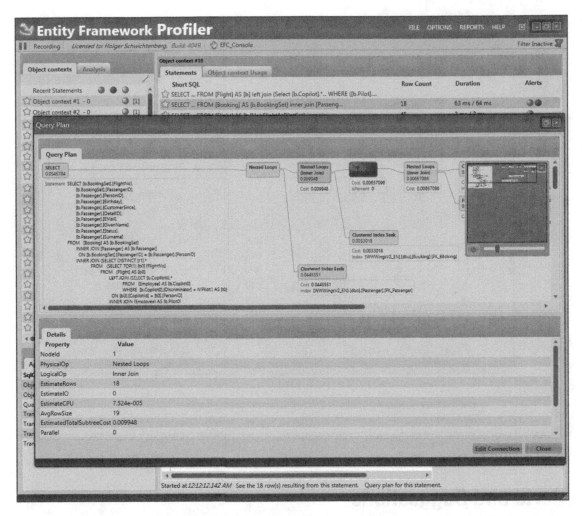

***Figure 19-28.*** *Execution plan of the database management system in the Entity Framework Profiler*

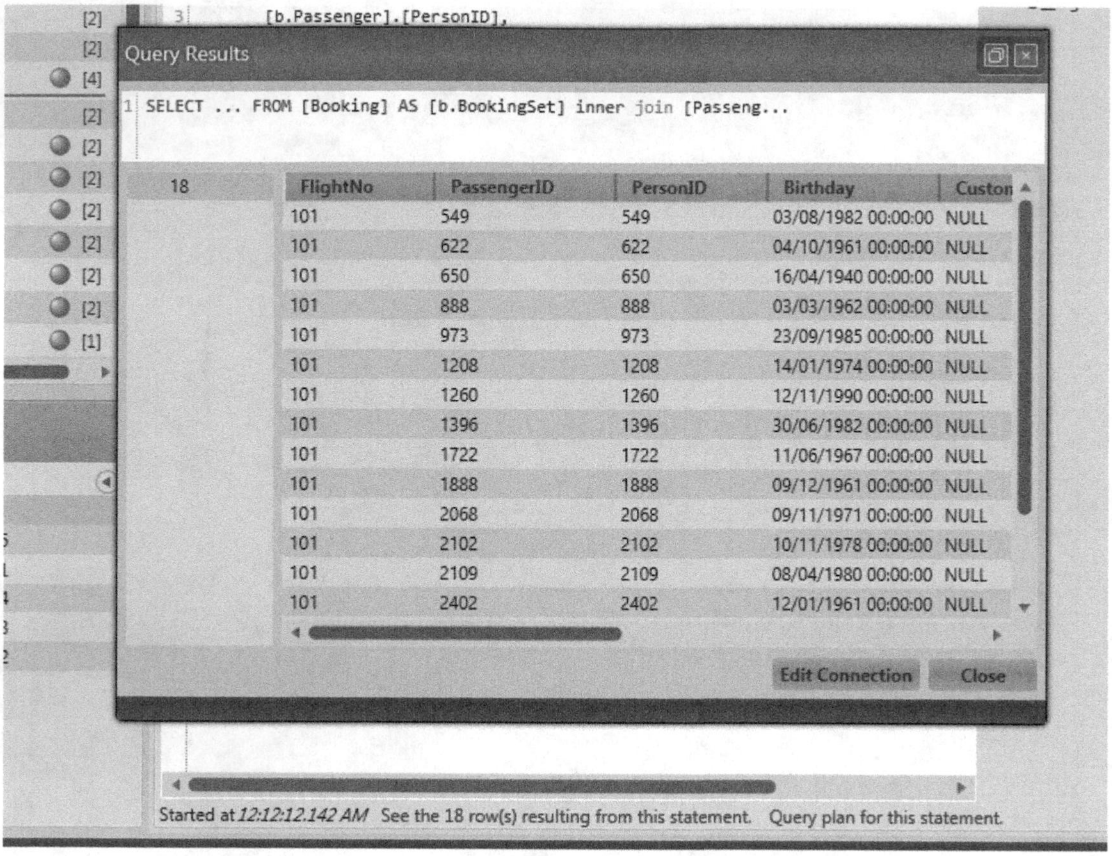

**Figure 19-29.** *Displaying the result in the Entity Framework Profiler*

## Alerts and Suggestions

Pay special attention to the gray circles (suggestions) and red circles (warnings) (see Figure 19-30). Here the Entity Framework Profiler is helping you discover potential problems. In Figure 19-30, this is what the Entity Framework Profiler calls a SELECT N + 1 problem. A large number of similar SQL commands that are executed one after the other indicates that lazy loading is erroneously used here. You should consider eager loading.

Another problem that the Entity Framework Profiler shows very well is the not recommended use of a context object in different threads. Other hints (see Figure 19-30) exist when a query uses many joins, starts with a wildcard (like % xy), returns many records, and does not contain a TOP statement (unbounded result sets). The last point is arguable. The intent of this proposal is that you should not run the risk of asking for more records than you really anticipate needing. But in practice (except for applications that explicitly display records using the scrolling feature), you often cannot set an upper limit that will apply permanently. There is also a warning when many INSERT, UPDATE, and DELETE commands are executed, and you should check that this is not mapped by a mass operation.

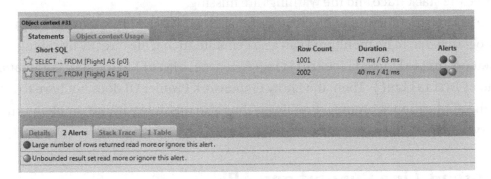

***Figure 19-30.*** *Alerts and suggestions in the Entity Framework Profiler*

## Analysis

The analysis functions in the Analysis tab are useful. You'll find evaluations that show the following:

- Which methods in the program code triggered which SQL commands (Queries By Method; see Figure 19-31)

- How many different commands exist (despite the name Unique Queries, INSERT, UPDATE, and DELETE commands appear here as well!)

- Which commands lasted the longest (expensive queries)

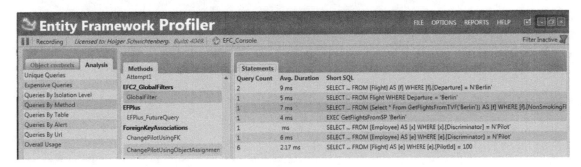

***Figure 19-31.*** *Analysis of Queries By Method*

An interesting feature is hidden in the File ➤ Export to File menu. This creates a JavaScript-enriched HTML page that looks like the Entity Framework Profiler. You can view all the context instances and the SQL commands and call the analysis results. However, the stack trace and the warnings are missing.

The normal memory function creates a binary file with the file extension `.efprof`. The program code to be monitored can also generate such a file directly by calling the method `InitializeOfflineProfiling(filename.efprof)` in the start command instead of `Initialize()`. Then, the Entity Framework Profiler UI does not have to run while the application is running. Profiling is thus also possible without problems on target systems.

## Command-Line Support and API

In terms of continuous integration, you can also run the Entity Framework Profiler from the command line. But you need one license per computer. The profiler itself also has a programming interface in `HibernatingRhinos.Profiler.Integration.dll`.

## Conclusion to Entity Framework Profiler

The Entity Framework Profiler is a useful tool to understand which SQL commands an Entity Framework Core–based application really executes. However, the price is high.

# CHAPTER 20

# Additional Components

This chapter introduces Entity Framework Core add-on components that extend the capabilities of Entity Framework Core. I am in no way involved in the development or distribution of these tools.

## Oracle Database Driver by DevArt

Oracle does not currently support Entity Framework Core for its databases. Basically, Oracle has said it is working on support (www.oracle.com/technetwork/topics/dotnet/tech-info/odpnet-dotnet-core-sod-3628981.pdf), but so far no solution is available. It took Oracle a few years to offer a solution for the classic Entity Framework.

A commercial Oracle driver for Entity Framework Core is available from DevArt as part of the dotConnect for Oracle product.

| Component name | dotConnect for Oracle |
|---|---|
| Web site | https://www.devart.com/dotconnect/oracle/ |
| Source code | No |
| NuGet | Setup: dcoracleXYpro.exe<br>Install-Package Devart.Data.Oracle.EFCore |
| Free version | No |
| Commercial version | $149.95 |

© Holger Schwichtenberg 2018
H. Schwichtenberg, *Modern Data Access with Entity Framework Core*,
https://doi.org/10.1007/978-1-4842-3552-2_20

# Installation

First, the DevArt installation package (dcoracleXYpro.exe, where XY stands for the version number) should be executed on the system. In addition, the NuGet package Devart.Data.Oracle.EFCore should be installed in the project in which the context class is located.

# Tools

The Oracle driver works in reverse engineering and forward engineering with the standard Entity Framework Core tools.

```
Scaffold-DbContext "User ID=WWWings; Password=secret; Direct=true;
Host=localhost; SID=ITVisions; Port=1521;" Devart.Data.Oracle.Entity.EFCore
-Tables DEPT,EMP
```

---

**Tip** Alternatively, you can use DevArt's Entity Developer if you want a graphical user interface for reverse engineering or forward engineering of Oracle databases.

---

# Context Class

In the context class in OnConfiguring(), you can call the method UseOracle() with a connection string.

```
protected override void OnConfiguring(DbContextOptionsBuilder builder)
  {
    builder.UseOracle(@"User ID=WWWings; Password=secret; Direct=true;
    Host=localhost; SID=ITVisions; Port=1521;");
}
```

> **Note**   The previous connection string uses the so-called Oracle Direct mode.
> This eliminates the need for Oracle client setup! If you have a connection string
> like `UserId = WWWings; Password = secret; Data Source = Name;`,
> you will need to install the Oracle client software for it. Otherwise, you will get the
> following error: "Cannot obtain Oracle Client information from registry. Make sure
> that Oracle Client Software is installed and that the bitness of your application (x86)
> matches the bitness of your Oracle Client, or use the Direct mode of connecting to
> a server." For information about the data source, see `https://docs.oracle.`
> `com/cd/B28359_01/win.111/b28375/featConnecting.htm`.

## Entity Classes

Note that the schema, table, and column names in Oracle can be only 30 characters each
(`https://docs.oracle.com/database/121/SQLRF/sql_elements008.htm#SQLRF51129`).
If a name is too long, the following runtime error will appear: "TableName
'EntityClassWithAllSupportedDataTypes' is too long. An identifier with more than 30
characters was specified."

## Data Types

Figure 20-1 and Figure 20-2 show the mapping between Oracle column types and .NET
data types in the DevArt Oracle driver.

| Oracle data types | SSDL[1] | CSDL[1] | .NET |
|---|---|---|---|
| NUMBER(1) | bool | Boolean | System.Boolean |
| NUMBER(2)..NUMBER(9)[2] | int | Int32 | System.Int32 |
| NUMBER(10)..NUMBER(18)[2] | int64 | Int64 | System.Int64 |
| NUMBER (p, s), where 0 < s < p < 16[2] | double | Double | System.Double |
| other NUMBERs | decimal | Decimal | System.Decimal |
| FLOAT, REAL, BINARY_FLOAT[3], BINARY_DOUBLE[3] | the same[3] | Decimal | System.Decimal |
| DATE, TIMESTAMP, TIMESTAMP WITH TIME ZONE, TIMESTAMP WITH LOCAL TIME ZONE | the same[4] | DateTime | System.DateTime |
| INTERVAL DAY TO SECOND | the same[4] | Time | System.TimeSpan |
| CHAR, NCHAR, VARCHAR2, NVARCHAR2, CLOB, NCLOB, ROWID, UROWID, XMLTYPE, INTERVAL YEAR TO MONTH, LONG[5] | the same[4] | String | System.String |
| BLOB, RAW[6], LONG RAW[5] | the same[4] | Binary | System.Byte[] |
| PL/SQL BOOLEAN | the same[4] | Boolean | System.Boolean |
| RAW(16) | guid | Guid | System.Guid |
| PL/SQL BOOLEAN | the same[4] | Boolean | System.Boolean |
| SDO_GEOMETRY[7] | sdo_geometry[4] | Geometry, GeometryCollection, GeometryLineString, GeometryMultiLineString, GeometryMultiPoint, GeometryMultiPolygon, GeometryPoint, GeometryPolygon | DbGeometry |
| SDO_GEOMETRY[7] | sdo_geography[4] | Geography, GeographyCollection, GeographyLineString, GeographyMultiLineString, GeographyMultiPoint, GeographyMultiPolygon, GeographyPoint, GeographyPolygon | DbGeography |

[1] Applicable only to Entity Framework v1 - v6. Not applicable to Entity Framework Core, because Entity Framework Core does not support XML mapping.

[2] The negative scale cases are taken into account.

[3] BINARY_DOUBLE and BINARY_FLOAT data types appeared in Oracle 10g.

[4] These SSDL types completely identical to the corresponding Oracle data types.

[5] According to official Oracle recomendations, using of LONG and LONG RAW data types is not recommended.

[6] All RAW types, except RAW(16).

[7] Supported in Entity Framework v5 and v6.

*Figure 20-1.* *Data type mapping during reverse engineering (source: https://www.devart.com/dotconnect/oracle/docs/)*

| .NET | CSDL[1] | SSDL[1] | Oracle data types |
|---|---|---|---|
| System.Boolean | Boolean | bool | NUMBER(1) |
| System.Byte | Byte | byte | NUMBER(3)[2] |
| System.Byte[] | Binary | BLOB | BLOB |
| DbGeometry[3] | Geometry | sdo_geometry | SDO_GEOMETRY |
| DbGeography[3] | Geography | sdo_geography | SDO_GEOMETRY |
| System.DateTime | DateTime | TIMESTAMP | TIMESTAMP |
| System.DateTimeOffset | DateTimeOffset | datetimeoffset | TIMESTAMP WITH TIME ZONE |
| System.Decimal | Decimal | decimal | NUMBER[4] |
| System.Double | Double | double | NUMBER |
| System.Guid | Guid | guid | RAW(16) |
| System.Int16 | Int16 | int16 | NUMBER(5)[2] |
| System.Int32 | Int32 | int | NUMBER(10)[2] |
| System.Int64 | Int64 | int64 | NUMBER(19)[2] |
| System.SByte | SByte | sbyte | NUMBER(3)[2] |
| System.Single | Single | single | NUMBER(15,5)[2] |
| System.String | String | VARCHAR2 | VARCHAR2 |
| System.TimeSpan | Time | INTERVAL DAY TO SECOND | INTERVAL DAY TO SECOND |

[1] Applicable only to Entity Framework v1 - v6. Not applicable to Entity Framework Core, because Entity Framework Core does not support XML mapping.

[2] Note that when mapping corresponding database data type, you will need the .NET type with larger precision. That is because, for example, any Int32 value can be stored in the NUMBER(10) column, but largest NUMBER(10) column value cannot be stored in the Int32 field, it requires Int64 field.

[3] Supported in Entity Framework v5 and v6.

[4] Mapping of this type depends on the DecimalPropertyConvention. If this convention is enabled (it is enabled by default), System.Decimal is mapped to NUMBER(18,2). Otherwise, it is mapped to NUMBER.

*Figure 20-2.* *Data type mapping during forward engineering (source: https:// www.devart.com/dotconnect/oracle/docs/)*

# Entity Framework Plus

Entity Framework Plus (EFPlus) is an additional component that exists for the classic Entity Framework. Even though the EFPlus web site talks only about Entity Framework, there is also a variant for Entity Framework Core available.

Entity Framework Plus provides several additional features to Entity Framework Core:

- Formulation of UPDATE and DELETE commands as lambda expressions (see Chapter 17)

- Auditing (all changes to records are automatically noted in a change table)

- Global query filter (Entity Framework Core has been able to do this since version 2.0; EFPlus also offers this for Entity Framework Core 1.*x*)

- A second-level cache as an alternative to EFSecondLevelCache.Core (see Chapter 17)

- Query batching (combining multiple SELECT queries in one go through the database management system)

## Entity Framework Plus
### EF Must-Have Features

| | |
|---|---|
| Component name | Entity Framework Plus for EFCore |
| Web site | http://entityframework-plus.net |
| Source code | https://github.com/zzzprojects/EntityFramework-Plus |
| NuGet | Install-Package Z.EntityFramework.Plus.EFCore |
| Free version | Yes |
| Commercial version | No |

**Note**   EFPlus 1.6.11 (and newer) supports Entity Framework Core version 2.0.

# Second-Level Caching with EFSecondLevelCache.Core

The component EFSecondLevelCache.Core provides an alternate second-level cache to the second-level cache included in Entity Framework Plus. EFSecondLevelCache.Core is much more complex in configuration but also more flexible because in addition to the main memory cache (MemoryCache), you can use Redis as a cache.

## EFSecondLevelCache.Core

Entity Framework Core Second Level Caching Library.

| | |
|---|---|
| Component name | EFSecondLevelCache.Core |
| Web site | https://github.com/VahidN/EFSecondLevelCache.Core |
| Source code | Yes |
| NuGet | Install-Package EFSecondLevelCache.Core |
| Necessary related packages | CacheManager.Core<br>CacheManager.Microsoft.Extensions.Caching.Memory<br>CacheManager.Serialization.Json<br>Optional: CacheManager.StackExchange.Redis |
| Free version | Yes |
| Commercial version | No |

# Object-Object Mapping with AutoMapper

In modern software architectures, typical tasks are the object-relational mapping of relational database structures to objects and the mapping of different object structures to each other. The open source tool AutoMapper facilitates object-to-object mapping (OOM).

The requirement to transform an object type into another type of object is common, for example, in data transfer objects (DTOs) between layers or in ViewModels that contain rendered data for display or expression (see Figure 20-3 and Figure 20-4). The object types to be imaged are often similar but not completely identical. And they usually do not have a common base class or interface that would permit type conversion at the programming language level, that is, by type conversion expression.

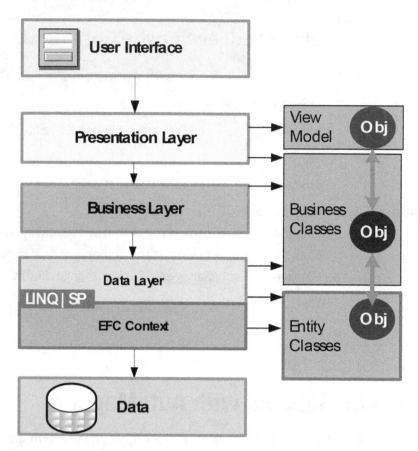

**Figure 20-3.**  *Object-to-object mapping is used between entity classes, business objects, and ViewModel classes*

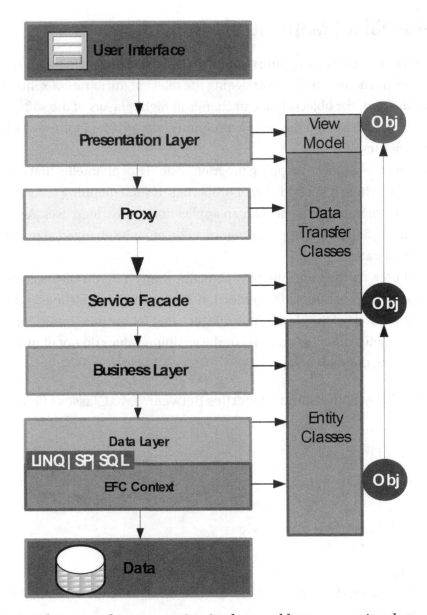

**Figure 20-4.** *Object-to-object mapping is also used between entity classes and data transfer classes in modern software architectures*

The .NET Framework and .NET Core class libraries do not include functions that support the mapping of different types of objects (object-to-object mapping). The Type Converter in the .NET Framework Class Library (http://msdn.microsoft.com/en-us/library/system.componentmodel.typeconverter.asp) merely defines a common interface for object type mappings. But it does not help with the actual imaging work.

# Object-to-Object Mapping via Reflection

Writing the object-to-object mapping manually means creating an instance of y for each instance of x within an iteration and assigning the relevant attributes of x individually to the properties of y. For objects that can change in higher layers of the software architecture, you will also find the reverse program code, which maps the attributes of y back to the attributes of x.

Writing this object-object-mapping program code is not an intellectual challenge but an annoying task, where it's easy to forget properties. If such mapping tasks are manually programmed, the maintenance effort for an application always increases. After all, with each new database field, the mapping program code must be changed at many different points in the application.

If the attributes are the same and have the same data type, you can easily do the object-to-object mapping yourself via reflection. Listing 20-1 and Listing 20-2 show the two extension methods of the class System.Object that follow this simple convention. However, if the names differ (irregularly) or the attribute values do not map 1:1, the primitive approach does not help.

***Listing 20-1.*** Copy of the Same Properties Between Two Classes via Reflection

```
using System;
using System.Reflection;

namespace EFC_Console.OOM
{
 public static class ObjectExtensions
 {
  /// <summary>
  /// Copy the properties and fields of the same name to another, new object
  /// </summary>
  public static T CopyTo<T>(this object from)
   where T : new()
  {
   T to = new T();
   return CopyTo<T>(from, to);
  }
```

```csharp
/// <summary>
/// Copy the properties and fields with the same name to another,
existing object
/// </summary>
public static T CopyTo<T>(this object from, T to)
 where T : new()
{
 Type fromType = from.GetType();
 Type toType = to.GetType();

 // Copy fields
 foreach (FieldInfo f in fromType.GetFields())
 {
  FieldInfo t = toType.GetField(f.Name);
  if (t != null)
  {
   t.SetValue(to, f.GetValue(from));
  }
 }

 // Copy properties
 foreach (PropertyInfo f in fromType.GetProperties())
 {
  object[] Empty = new object[0];
  PropertyInfo t = toType.GetProperty(f.Name);
  if (t != null)
  {
   t.SetValue(to, f.GetValue(from, Empty), Empty);
  }
 }
 return to;
 }
 }
}
```

***Listing 20-2.***  Using the Extension Methods from Listing 20-1

```
using System;
using System.Linq;
using DA;

namespace EFC_Console.OOM
{
 public class FlightDTO
 {
  public int FlightNo { get; set; }
  public string Departure { get; set; }
  public string Destination { get; set; }
  public DateTime Date { get; set; }
 }

 public static class ReflectionMapping
 {
  public static void Run()
  {
   using (var ctx = new WWWingsContext())
   {
    var flightSet = ctx.FlightSet.Where(x => x.Departure == "Berlin").
    ToList();
    foreach (var flight in flightSet)
    {
     var dto = flight.CopyTo<FlightDTO>();
     Console.WriteLine(dto.FlightNo + ": " + dto.Departure +"->" + dto.
     Destination + ": " + dto.Date.ToShortDateString());
    }
   }
  }
 }
}
```

# AutoMapper

The open source library AutoMapper by Jimmy Bogard has established itself in the
.NET developer world for object-to-object mapping. The NuGet package AutoMapper
(`https://www.nuget.org/packages/AutoMapper`) consists of `AutoMapper.dll`, whose
central class is `AutoMapper.Mapper`.

| Component name | AutoMapper |
| --- | --- |
| Web site | `https://github.com/AutoMapper` |
| NuGet | `Install-Package Automapper` |
| Free version | Yes |
| Commercial version | No |

AutoMapper runs on the following .NET variants:

- .NET

- .NET Core

- Silverlight

- .NET for Windows Store apps/Windows Runtime

- Universal Windows Platform (UWP) apps

- Xamarin.iOS

- Xamarin.Android

You can find the source code (`https://github.com/AutoMapper/AutoMapper`)
and a wiki (`https://github.com/AutoMapper/AutoMapper/wiki`) on GitHub. You
can find other resources (for example, a video) on the AutoMapper web site (`http://
automapper.org`). Overall, however, the available documentation is (as with so many
open source projects) scarce, incomplete, and sometimes outdated. Many functions of
AutoMapper are not described in the wiki, so look at blog entries and forums to fully
understand AutoMapper. So far, there is no documentation on the differences between
the stable version 3.3.1 and the current prerelease version 4.0. Even I, having been
working with AutoMapper on my projects for a long time, had to spend many hours
researching to discover further, undocumented features in the software component.

507

**Note**   This chapter describes version 6.1.1 of AutoMapper. Unfortunately, AutoMapper has had major changes in the past, so the commands shown here only partially work in older versions.

## Looking at an Example

In this example, parts of the World Wide Wings version 2 object model (shown in Figure 20-5) will be mapped onto the simplified object model in Figure 20-6. The classes Pilot, Employee, and Person are dissolved. The information about the Pilot is shown on a string directly in the FlightView class and on a detail object called PilotDetailView. The new PassengerView class also includes personal data from the Person class. A lot of information (e.g., from the entity Employee ) is deliberately no longer used here.

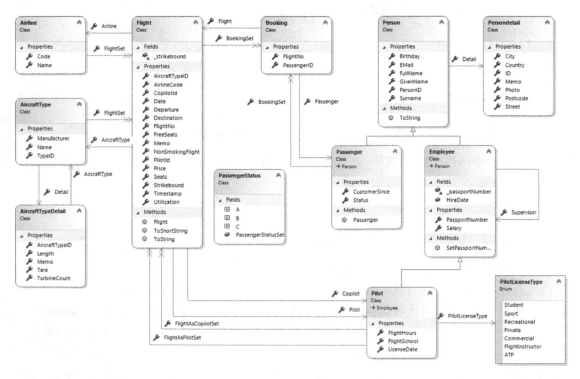

***Figure 20-5.***   *The World Wide Wings version 2 object model that uses Entity Framework Core*

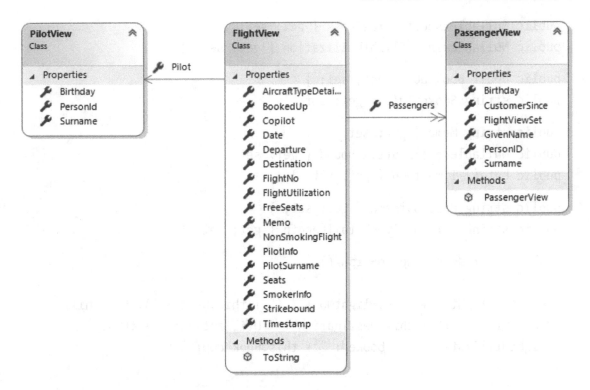

***Figure 20-6.*** *The simplified target model to be created from the model from Figure 20-5 using object-to-object mapping*

Listing 20-3 shows these three classes.

***Listing 20-3.*** ViewModel Classes

```
using System;
using System.Collections.Generic;

namespace EFC_Console.ViewModels
{
 public class FlightView
 {
  public int FlightNo { get; set; }
  public string Departure { get; set; }
  public string Destination { get; set; }
  public string Date { get; set; }
  public bool NonSmokingFlight { get; set; }
  public short Seats { get; set; }
```

```csharp
   public Nullable<short> FreeSeats { get; set; }
   public Nullable<int> FlightUtilization { get; set; }

   public bool? BookedUp { get; set; }
   public string SmokerInfo { get; set; }

   public string Memo { get; set; }
   public Nullable<bool> Strikebound { get; set; }
   public byte[] Timestamp { get; set; }

   public string PilotSurname { get; set; }
   public string AircraftTypeDetailLength { get; set; }
   public override string ToString()
   {
    return "Flight " + this.FlightNo + " (" + this.Date + "): " + this.
    Departure + "->" + this.Destination + " Utilization: " + this.
    FlightUtilization + "% booked: " + this.BookedUp;
   }

   public string PilotInfo { get; set; }


   /// <summary>
   /// Pilot 1:1
   /// </summary>
   public PilotView Pilot { get; set; }

   /// <summary>
   /// Passengers 1:n
   /// </summary>
   public List<PassengerView> Passengers{ get; set; }
  }

 public class PilotView
 {
  public int PersonId { get; set; }
  public string Surname { get; set; }
  public DateTime Birthday { get; set; }
 }
```

```
public class PassengerView
{
 public PassengerView()
 {

  this.FlightViewSet = new HashSet<FlightView>();
 }

 public int PersonID { get; set; }
 public Nullable<System.DateTime> CustomerSince { get; set; }
 public int Birthday { get; set; }
 public string GivenName { get; set; }
 public string Surname { get; set; }
 public virtual ICollection<FlightView> FlightViewSet { get; set; }
 }
}
```

## Configuring Mappings

Before you can do a mapping with AutoMapper, you have to register the mapping with the classes involved in AutoMapper once per application domain. You do this with the method Initialize().

---

**Attention**   If Initialize() is called several times in a method, only the last executed configuration is valid!

---

Within Initialize(), the CreateMap() method is used for the concrete mapping definition between two classes. CreateMap() expects two type parameters.

- The first parameter is always the source type.

- The second is the target type.

If you need a conversion in both directions, you have to create it explicitly.

```
Mapper.Initialize(cfg =>
{
 cfg.CreateMap<Flight, FlightView>();
 cfg.CreateMap<FlightView, Flight>();
```

```
cfg.CreateMap<Passenger, PassengerView>();
cfg.CreateMap<PassengerView, Passenger>();
cfg.CreateMap<Pilot, PilotDetailView>();
cfg.CreateMap<PilotDetailView, Pilot>();
});
```

Alternatively, you can use the ReverseMap() method to create the mapping in both directions in a row.

```
Mapper.Initialize(cfg =>
{
cfg.CreateMap<Flight, FlightView>().ReverseMap();
cfg.CreateMap<Passenger, PassengerView>().ReverseMap();
cfg.CreateMap<Pilot, PilotDetailView>().ReverseMap();
});
```

The order of multiple calls to CreateMap() within Initialize() is not relevant.

For a single class, there can be a mapping to a class as well as several other classes. Here's an example:

```
Mapper.Initialize(cfg =>
{
cfg.CreateMap<Flight, FlightView>();
cfg.CreateMap<Flight, FlightDTOShort>();
cfg.CreateMap<Flight, FlightDTO>();
});
```

Which of these mappings (that is to say, on which target type) is then actually used is decided by the parameters of the Map() method, which performs the actual mapping.

## Running the Mapping with Map()

AutoMapper's Map() method has three options.

For option 1, you can map to a new object and then specify the target type as the generic type parameter and the source object as the parameter of the method.

```
FlightView flightView1 = Mapper.Map<FlightView>(flight);
```

For option 2, if you use the nongeneric variant of Map(), the program code becomes more extensive. Now, enter the source object type after the source object as the second parameter and the target type as the third parameter. In addition, Map() returns only the type System.Object, making a type cast with FlightView necessary.

```
FlightView FlightView2 = (FlightView) AutoMapper.Mapper .Map
(Flight,    Flight.GetType(),   typeof ( FlightView ));
```

For option 3, you map an object to another existing object. This is the third variant of Map(). In this case, no type parameter is necessary for the method, but the source and target objects are to be transferred as parameters.

```
var flightView3 = new FlightView();
flightView3.Memo = "test";
Mapper.Map(flight, flightView3);
```

## Using the Nonstatic API

In addition to the static API, AutoMapper has a nonstatic API (the Instance API) for configuring the mapping (Listing 20-4). You configure an instance of the MapperConfiguration class and use it to create an object with an IMapper interface using CreateMapper(). This object then has the Map() method.

*Listing 20-4.*  Nonstatic API

```
var config = new MapperConfiguration(cfg => {
 cfg.CreateMap<Flight, FlightView>();
 cfg.CreateMap<Pilot, PilotView>();
 cfg.CreateMap<Passenger, PassengerView>();
 cfg.AddProfile<AutoMapperProfile2>();
});
config.AssertConfigurationIsValid();

IMapper mapper = config.CreateMapper();
var flightView4 = mapper.Map<Flight, FlightView>(flight);
```

# Mapping Conventions

AutoMapper not only maps attributes of the same name but also contains other standard conventions, listed here:

- If an attribute x is not found in the source object, a GetX() function is searched for and, if necessary, called.

- If the name of the attribute contains multiple capital letters and there are dependent objects, then each word is understood as a level and separated by a period. For example, the name obj. AircraftTypeDetailLength is the mapped to obj.AircraftType. Detail.Length. AutoMapper calls the feature *flattening*.

- AutoMapper ignores any null reference runtime errors.

- AutoMapper accesses private getters and setters, but only if the individual getter or setter is private. If the whole property is declared private, it will be ignored.

- In AutoMapper, the treatment of uppercase and lowercase letters as well as underscores is very exciting. AutoMapper looks for suitable properties also in changed cases and with or without an underscore. Even the order of the property declaration in the classes is relevant!

Table 20-1 shows several cases of a property Free Spaces with four different spellings: FreeSeats, freeSeats, Free_Seats, and free_Seats. It is always assumed that there is only the variant of the FreePoints property in the source object, which is also set in column 1 of the table. Also, the target object always contains all four variants of the property, and the properties are in the same order in which they are in the table.

***Table 20-1.*** *AutoMappers Convention-Based Mapping Behavior Regarding Underscore and Case*

| Values in the Source Object | Values in the Target Object | Comment |
|---|---|---|
| f.FreeSeats = 1 | f.FreeSeats = 1<br>f.freeSeats = 1<br>f.Free_Seats = 1<br>f.free_Seats = 1 | The value from one property in the source object is copied by AutoMapper into all four variants in the target object. |
| f.FreeSeats = 1<br>f.freeSeats = 2 | f.FreeSeats = 1<br>f.freeSeats = 1<br>f.Free_Seats = 1<br>f.free_Seats = 1 | The value of a property FreeSeats is ignored because FreeSeats has already mapped its value to all target properties. |
| f.FreeSeats = 1<br>f.freeSeats = 2<br>f.Free_Seats = 3 | f.FreeSeats = 1<br>f.freeSeats = 1<br>f.Free_Seats = 3<br>f.free_Seats = 3 | The first property without an underscore maps to all properties without an underscore, and the first one with an underscore maps to all properties with an underscore. |
| f.FreeSeats = 1<br>f.freeSeats = 2<br>f.free_Seats = 4 | f.FreeSeats = 1<br>f.freeSeats = 1<br>f.Free_Seats = 4<br>f.free_Seats = 4 | The first property without an underscore maps to properties without an underscore, and the first one with an underscore maps to all properties with an underscore. |
| f.FreeSeats = 1<br>f.freeSeats = 2<br>f.Free_Seats = 3<br>f.free_Seats = 4 | f.FreeSeats = 1<br>f.freeSeats = 1<br>f.Free_Seats = 3<br>f.free_Seats = 3 | The first property without an underscore maps to all properties without an underscore, and the first one with an underscore maps to all properties with an underscore. |

Listing 20-5 shows how to map a `Flight` object to `FlightView`. Figure 20-7 shows the output.

***Listing 20-5.*** Mapping from Flight to FlightView

```
public class AutoMapperBasics
 {
  public static void Demo_SingleObject()
  {
   CUI.Headline(nameof(Demo_SingleObject));

   // take the first flight as an example
   var ctx = new WWWingsContext();
   var flight = ctx.FlightSet.Include(x=>x.Pilot).Include(x =>
   x.AircraftType).ThenInclude(y=>y.Detail).FirstOrDefault();
   Console.WriteLine(flight);

   //#################################################

   Mapper.Initialize(cfg =>
   {
    cfg.CreateMap<Flight, FlightView>();
    cfg.CreateMap<FlightView, Flight>();
    cfg.CreateMap<Passenger, PassengerView>();
    cfg.CreateMap<PassengerView, Passenger>();
    cfg.CreateMap<Pilot, PilotDetailView>();
    cfg.CreateMap<PilotDetailView, Pilot>();

    cfg.CreateMap<Flight, FlightDTOShort>();
    cfg.CreateMap<Flight, FlightDTO>();
   });

   Mapper.Initialize(cfg =>
   {
    cfg.CreateMap<Flight, FlightView>().ReverseMap();
    cfg.CreateMap<Passenger, PassengerView>().ReverseMap();
    cfg.CreateMap<Pilot, PilotDetailView>().ReverseMap();
   });
```

```
Mapper.Initialize(cfg =>
{
 cfg.SourceMemberNamingConvention = new NoNamingConvention();
 cfg.DestinationMemberNamingConvention = new NoNamingConvention();

 cfg.CreateMap<Flight, FlightView>().ReverseMap();
 cfg.CreateMap<Passenger, PassengerView>().ReverseMap();
 cfg.CreateMap<Pilot, PilotDetailView>().ReverseMap();

 cfg.CreateMap<Flight, FlightDTOShort>();
 cfg.CreateMap<Flight, FlightDTO>();
});

Mapper.Initialize(cfg =>
{
 cfg.AddProfile<AutoMapperProfileEinfach>();
});

Mapper.Initialize(cfg =>
{
 cfg.AddProfile<AutoMapperProfileKomplex>();
});

// ----------------------
CUI.Headline("Mapping to new object");
FlightView flightView1 = Mapper.Map<FlightView>(flight);

Console.WriteLine(flightView1);
Console.WriteLine(flightView1.PilotSurname);
Console.WriteLine(flightView1.SmokerInfo);
Console.WriteLine(flightView1.PilotInfo);
if (flightView1.Pilot == null) CUI.PrintError("No pilot!");
else
{
 Console.WriteLine(flightView1.Pilot?.Surname + " born " + flightView1.
 Pilot?.Birthday);
}
Console.WriteLine(flightView1.Memo);
Console.WriteLine(flightView1.AircraftTypeDetailLength);
```

```csharp
FlightView flightView2 = (FlightView)Mapper.Map(flight, flight.
GetType(), typeof(FlightView));

Console.WriteLine(flightView2);
Console.WriteLine(flightView2.PilotSurname);
Console.WriteLine(flightView2.SmokerInfo);
Console.WriteLine(flightView2.PilotInfo);
if (flightView2.Pilot == null) CUI.PrintError("No pilot!");
else
{
 Console.WriteLine(flightView2.Pilot?.Surname + " born " + flightView2.
 Pilot?.Birthday);
}
Console.WriteLine(flightView2.AircraftTypeDetailLength);
Console.WriteLine(flightView2.Memo);

// ----------------------
CUI.Headline("Mapping to existing object");
var flightView3 = new FlightView();
Mapper.Map(flight, flightView3);

Console.WriteLine(flightView3);
Console.WriteLine(flightView3.PilotSurname);
Console.WriteLine(flightView3.SmokerInfo);
Console.WriteLine(flightView3.PilotInfo);
if (flightView3.Pilot == null) CUI.PrintError("No pilot!");
else
{
 Console.WriteLine(flightView3.Pilot?.Surname + " born " + flightView3.
 Pilot?.Birthday);
}
Console.WriteLine(flightView3.Memo);
}
}
```

```
Select H:\TFS\Demos\EFC\EFC_Samples\EFC_Console\bin\Debug\EFC_Console.exe

Demo_SingleObject
Flight #100: from Berlin to Paris on 23.08.18 03:31: 199 free Seats.
Mapping to new object
Flight 100 (23/08/2018 03:31:43): Berlin->Paris Utilization: 79% booked: False
Özdemir
Rauchen ist erlaubt.
#81: Christian Özdemir
Christian Özdemir
Loaded from Database: 01/01/2018 18:06:51
72,3
Flight 100 (23/08/2018 03:31:43): Berlin->Paris Utilization: 79% booked: False
Özdemir
Rauchen ist erlaubt.
#81: Christian Özdemir
Christian Özdemir
72,3
Mapping to existing object
Flight 100 (23/08/2018 03:31:43): Berlin->Paris Utilization: 79% booked: False
Özdemir
Rauchen ist erlaubt.
#81: Christian Özdemir
Christian Özdemir
Loaded from Database: 01/01/2018 18:06:51
```

*Figure 20-7.* *Output of Listing 20-5*

# Changing Mapping Conventions

You can override the convention of accepting underscores as delimiters. To do this, you
write your own convention class (see Listing 20-6), which overrides the rendering of
underscores by setting `SeparatorCharacter` to an empty string and using no regular
expression in `SplittingExpression`.

*Listing 20-6.* A Separate Convention Class for AutoMapper That Overrides the
Rendering of Underscores

```
using AutoMapper;
using System.Text.RegularExpressions;

namespace EFC_Console.AutoMapper
{
 /// <summary>
 /// No use of underscores when mapping
```

```
/// </summary>
class NoNamingConvention : INamingConvention
{
 #region INamingConvention Members
 public string ReplaceValue(Match match)
 {
  return "";
 }

 public string SeparatorCharacter
 {
  get { return ""; }
 }
 public Regex SplittingExpression
 {
  get { return new Regex(""); }
 }
 #endregion
 }
}
```

Your own convention has to be included in the configuration, as shown here:

```
Mapper.Initialize(cfg =>
{
 cfg.SourceMemberNamingConvention = new NoNamingConvention();
 cfg.DestinationMemberNamingConvention = new NoNamingConvention();

 cfg.CreateMap<Flight, FlightView>().ReverseMap();
 cfg.CreateMap<Passenger, PassengerView>().ReverseMap();
 cfg.CreateMap<Pilot, PilotDetailView>().ReverseMap();
});
```

## Profile Classes

You can outsource the AutoMapper configuration to so-called profile classes, which inherit from the base class Profile.

```
using AutoMapper;
using BO;

namespace EFC_Console.AutoMapper
{
 /// <summary>
 /// Simple profile class for AutoMapper
 /// </summary

 public class AutoMapperProfile1 : Profile
  {
   public AutoMapperProfile1()
   {
    this.SourceMemberNamingConvention = new NoNamingConvention();
    this.DestinationMemberNamingConvention = new NoNamingConvention();
    this.CreateMap<Flight, FlightView>().ReverseMap();
    this.CreateMap<Passenger, PassengerView>().ReverseMap();
    this.CreateMap<Pilot, PilotDetailView>().ReverseMap();
    this.CreateMap<Flight, FlightDTOShort>();
    this.CreateMap<Flight, FlightDTO>();
   }
  }
}
```

This profile class is then called in `Initialize()` via `AddProfile()`.

```
Mapper.Initialize(cfg =>
{
 cfg.AddProfile<AutoMapperProfile1>();
});
```

## Ignoring Subobjects

If the source and target objects all have a subobject and the property name maps according to one of the AutoMapper conventions but there is no mapping for that object type, AutoMapper complains with the following error: "Missing type map configuration or unsupported mapping." Either you have to create a mapping with `CreateMap()` for the subobject type or you have to tell AutoMapper explicitly that you do not want to map the subobjects.

Ignoring is done through the Fluent API of the `CreateMap()` method using the `Ignore()` method after calling `ForMember()`.

```
AutoMapper.Mapper.CreateMap<Passenger, PassengerView>().ForMember(z =>
z.PilotView, m => m.Ignore());
```

---

**Note**   Flattening still works after an `Ignore()` call for a subobject. That is, AutoMapper will continue to populate the property `PilotSurname` in class `FlightView` with the values from `Pilot.Surname`, even if `flightView.PilotView` is set to null by the `Ignore()` statement.

---

## Custom Mappings

The developer of AutoMapper offers numerous ways to manipulate a mapping. Using the Fluent API of the `CreateMap()` method, you can define the mappings of the properties of the source object to mappings of the properties of the target object, called *projections*. A manual mapping uses the `ForMember()` method. The first parameter to be specified is a lambda expression for the target property (variable name z for the target), and the second parameter is a lambda expression for the value (variable name q for the source), whereby one or more properties of the source object can be referenced.

Listing 20-7 shows the following ten possibilities:

- Mapping a property to a static value using `UseValue()`.

- Mapping a property to the result of an expression to a source property using `MapFrom()`, where the result is a Boolean value.

- Mapping a property to a calculation of multiple source properties using `MapFrom()`, where the value is a number.

- Mapping a property with `MapFrom()` to the result of the `ToString()` method in a subobject of the source object.

- Mapping a property with `MapFrom()`to an object containing values from multiple source properties.

- Mapping a property with a `ValueResolver` class using `ResolveUsing()` and the `IValueResolver` interface.

- Mapping zero values to another value with the NullSubstitute() method.

- Specify that the properties of the target object must not be overwritten with a value of the source object. This is done with the UseDestinationValue() method.

- The penultimate case shows how Condition() maps only if the source value (SourceValue) meets a specific condition.

- The last case shows the conversion of an N:M mapping into a 1:N mapping. Here the intermediate entity booking, which connects Flight and Passenger, is eliminated. The destination class FlightView has a property of the type List<Passenger>.

Since such mapping definitions can often become very extensive, it is usually advisable to outsource them to a profile class (see Listing 20-7) instead of spreading them somewhere in the program code. Listing 20-8 shows the resolver class.

*Listing 20-7.* Manual AutoMapper Mappings with ForMember( )

```
public AutoMapperProfile2()
  {
  #region Mappings for class Flight
  CreateMap<Flight, FlightView>()

  // 1. Set Memo to static value
  .ForMember(z => z.Memo,
          q => q.UseValue("Loaded from Database: " + DateTime.Now))

  // 2. Mapping for a bool property
  .ForMember(z => z.BookedUp, q => q.MapFrom(f => f.FreeSeats <= 0))

  // 3. Mapping with calculation
  .ForMember(z => z.FlightUtilization,
          q => q.MapFrom(f => (int)Math.Abs(((decimal)f.FreeSeats /
          (decimal)f.Seats) * 100)))

  // 4. Mapping to a method result
  .ForMember(z => z.PilotInfo, m => m.MapFrom(
          q => q.Pilot.ToString())))
```

```csharp
        // 5. Mapping to a method result with object construction
        .ForMember(z => z.Pilot,
          m => m.MapFrom(
            q => new Pilot { PersonID = q.Pilot.PersonID, Surname = q.Pilot.
            FullName, Birthday = q.Pilot.Birthday.GetValueOrDefault() }))

        // 6. Mapping with a value resolver
        .ForMember(z => z.SmokerInfo,
                      m => m.ResolveUsing<SmokerInfoResolver>())

        // 7. Mapping if source value is null
        .ForMember(z => z.Destination, q => q.NullSubstitute("unknown"))

        // 8. No Mapping for existing values
        .ForMember(z => z.Timestamp, q => q.UseDestinationValue())

        // 9. Conditional Mapping
        .ForMember(z => z.Seats, x => x.Condition(q => q.FreeSeats < 250))

        // 10. Map n:m to zu 1:n (for Flight->Booking->Passenger)
        .ForMember(dto => dto.PassengerViewSet, opt => opt.MapFrom(x =>
        x.BookingSet.Select(y => y.Passenger).ToList()))

        // 11. Include reverse Mapping
        .ReverseMap();
      #endregion

      #region Other class mappings
      CreateMap<Pilot, string>().ConvertUsing<PilotStringConverter>();
      // Map n:m to zu 1:n (for Passenger->Booking->Flight)
      CreateMap<Passenger, PassengerView>()
        .ForMember(z => z.FlightViewSet, m => m.MapFrom(q => q.BookingSet.
        Select(y => y.Flight)));
      #endregion

      #region Typkonvertierungen
      CreateMap<byte, long>().ConvertUsing(Convert.ToInt64);
      CreateMap<byte, long>().ConvertUsing(ConvertByteToLong);
      #endregion
    }
```

*Listing 20-8.* A Value Resolver Class for AutoMapper

```
namespace EFC_Console.AutoMapper
{
 /// <summary>
 /// Value Resolver for Automapper, converts true/false to
 /// string property "SmokerInfo"
 /// </summary>
 public class SmokerInfoResolver : IValueResolver<Flight, FlightView,
 string>
 {
  public string Resolve(Flight source, FlightView destination, string
  member, ResolutionContext context)
  {
   if (source.NonSmokingFlight.GetValueOrDefault()) destination.SmokerInfo
   = "This is a non-smoking flight!";
   else destination.SmokerInfo = "Smoking is allowed.";
   return destination.SmokerInfo;
  }
 }
}
```

# Type Conversions

When mapping elementary data types (string, int, decimal, bool, and so on),
AutoMapper maps easily if the types are the same or if the target type is string. In
the case of the target type string, AutoMapper can always get to a string by calling
ToString(). Number types are automatically converted both up and down. This allows
AutoMapper to map from byte to long, but also from long to byte. However, some of
this flexibility is new since version 4.0. Version 3.3.1 responded to the attempt to map
long to byte with the following error: "Missing type map configuration or unsupported
mapping. Mapping types: System.Byte -> System.Int64." Also in AutoMapper 4.0, if the
value to be mapped does not fit into the target number type, the following runtime error
occurs: "AutoMapper.AutoMapperMappingException: Value was either too large or too
small for an unsigned byte."

Of course, AutoMapper cannot map if the types are completely different. For example, if the property `Birthday` has the type `DateTime` in the source object but `Integer` is used in the target object, the runtime error will always occur (`AutoMapper.AutoMapperMappingException`). In the error message, you will find detailed information about the problem, as shown here:

*System.DateTime* ➤ *System.Int32*

*Destination path:*

*PassengerView.Birthday*

*Source value:*

*01.10.1980 00:00:00*

For type images that AutoMapper does not carry out automatically or that should be done differently than AutoMapper does it in the standard version, you must provide AutoMapper with a type converter (Listing 20-9 and Listing 20-10). Such a type converter can be implemented in a simple method that takes type x and returns y. This converter method is then registered with AutoMapper.

```
CreateMap<byte, long>().ConvertUsing(ConvertByteToLong);
CreateMap<DateTime, Int32>().ConvertUsing(ConvertDateTimeToInt);
```

If necessary, you can call a standard converter method of the .NET Framework.

```
CreateMap<byte, long>().ConvertUsing(Convert.ToInt64);
```

***Listing 20-9.*** Method-Based Type Converter for AutoMapper

```
/// <summary>
/// Converts bytes to long with special case 0
/// </summary>
/// <param name="b">Byte value</param>
/// <returns></returns>
public static long ConvertByteToLong(byte b)
{
 if (b == 0) return -1;
 else return (long) b;
}
```

**Listing 20-10.** Another Method-Based Type Converter for AutoMapper

```
/// <summary>
/// Converts bytes to long with special case 0
/// </summary>
/// <param name="d">DateTime value</param>
/// <returns></returns>
public static Int32 ConvertDateTimeToInt(DateTime d)
{
 return d.Year;
}
```

You can also implement the type converter as a class that implements the ITypeConverter interface using the Convert() method (see Listing 20-11). This custom converter class is then registered with a generic variant of ConvertUsing().

```
CreateMap<Pilot, string>().ConvertUsing<PilotStringConverter>();
```

**Listing 20-11.** Class-Based Type Converter for AutoMapper

```
/// <summary>
 /// Converts a Pilot to a string
 /// </summary>
 public class PilotStringConverter : ITypeConverter<Pilot, string>
 {
  public string Convert(Pilot pilot, string s, ResolutionContext context)
  {
   if (pilot == null) return "(Not assigned)";
   return "Pilot # " + pilot.PersonID;
  }
 }
}
```

The conversions shown so far are global for all images in all classes. This, of course, is a powerful feature because it avoids having to repeat some mappings. But you should also be careful here because you may create images that you do not want so that data can be lost.

It may also be that you do not want such a conversion at all globally, but only for a single property image in individual classes. In this case, you can write a ValueResolver (see subchapter before 'Custom Mappings').

# Collections

Even if you always configure AutoMapper to map individual classes, AutoMapper can map not only individual instances but also arbitrarily large amounts of these classes to each other using Map().

Here's an example:

```
List<FlightView> FlightviewList = AutoMapper.Mapper.Map <List<Flight View
>> (Flight list);
```

AutoMapper supports the following types of volumes (Listing 20-12):

- IEnumerable

- IEnumerable <T>

- ICollection

- ICollection <T>

- IList

- IList<T>

- List<T>

- Arrays

***Listing 20-12.*** Mapping of an Entire List

```
public static void Demo_ListMapping()
{
 CUI.Headline(nameof(Demo_ListMapping));

 Mapper.Initialize(cfg =>
 {
  cfg.AddProfile<AutoMapperProfile2>();
 });

 using (var ctx2 = new WWWingsContext())
 {
  var flightSet = ctx2.FlightSet.Include(f => f.Pilot).Include(f =>
  f.BookingSet).ThenInclude(x => x.Passenger).Where(f => f.Departure ==
  "Berlin").OrderBy(f => f.FlightNo).Take(5).ToList();
```

```
// map all objects in this list
List<FlightView> flightviewListe = Mapper.Map<List<FlightView>>
(flightSet);
foreach (var f in flightviewListe)
{
 Console.WriteLine(f.ToString());
 if (f.Passengers!= null)
 {
  foreach (var pas in f.PassengerViewSet)
  {
   Console.WriteLine("   - " + pas.GivenName + " " + pas.Surname + " has
   " + pas.FlightViewSet.Count + " Flights!");
  }
 }
}
}
```

# Inheritance

To illustrate the behavior of AutoMapper in inheritance relationships, the example from Listing 20-13 uses the Person, Woman, and Man classes, as well as the associated data transfer object (DTO) classes PersonDTO, MsDTO, and MannDTO. The differences between Man and Woman are, according to an old cliché, based on the possession of a significant number of cars (in Man) or shoes (in Woman). The DTO classes are distinguished by the data type of the number values (bytes instead of integers) as well as the combination of the first name and last name as property names. In addition, the birthday in the DTO classes saves only the year and not the complete date.

***Listing 20-13.*** Class Hierarchy for the Inheritance Example

```
class Person
 {
 public string GivenName { get; set; }
 public string Surname { get; set; }
 public DateTime Birthday { get; set; }
 }
```

529

```
class Man : Person
{
 public int NumberOfCars { get; set; }
}

class Woman : Person
{
 public int NumberOfShoes { get; set; }
}

class PersonDTO
{
 public string Name { get; set; }
 public int YearOfBirth { get; set; }
}

class ManDTO : PersonDTO
{
 public byte NumberOfCars{ get; set; }
}

class WomanDTO : PersonDTO
{
 public byte NumberOfShoes{ get; set; }
}
```

Basically, you have to define a mapping for each individual class in the inheritance hierarchy in an inheritance relationship.

```
Mapper.Initialize(cfg =>
   {
    cfg.CreateMap<Person, PersonDTO>();
    cfg.CreateMap<Woman, WomanDTO>();
    cfg.CreateMap<Man, ManDTO>();
   });
```

While AutoMapper automatically handles type casting integers in bytes, the name and birth date mapping is missing. The type conflict in the case of the date of birth leads to a runtime error during mapping (`AutoMapper.AutoMapperMappingException`).

It is not enough to set the manual mappings with `ForMember()` and `MapFrom()` only on the base class.

```
cfg.CreateMap<Person, PersonDTO>()
.ForMember(z => z.Name, map => map.MapFrom(q => q.GivenName + " " +
q.Surname))
.ForMember(z => z.YearOfBirth, map => map.MapFrom(q => q.Birthday.Year));
cfg.CreateMap<Woman, WomanDTO>();
cfg.CreateMap<Man, ManDTO>();
```

After that, only the mapping for the class `Person` is done correctly. The classes `Man` and `Woman` continue to produce a runtime error. AutoMapper expects that there will be manual mapping configuration for each class in the inheritance hierarchy, as shown here:

```
Mapper.Initialize(cfg =>
 {
   cfg.CreateMap<Person, PersonDTO>()
      .ForMember(z => z.Name, map => map.MapFrom(q => q.GivenName + " " +
      q.Surname))
      .ForMember(z => z.YearOfBirth, map => map.MapFrom(q => q.Birthday.
      Year));
   cfg.CreateMap<Man, ManDTO>()
      .ForMember(z => z.Name, map => map.MapFrom(q => q.GivenName + " " +
      q.Surname))
      .ForMember(z => z.YearOfBirth, map => map.MapFrom(q => q.Birthday.
      Year));
   cfg.CreateMap<Woman, WomanDTO>()
      .ForMember(z => z.Name, map => map.MapFrom(q => q.GivenName + " " +
      q.Surname))
      .ForMember(z => z.YearOfBirth, map => map.MapFrom(q => q.Birthday.
      Year));
 });
```

But you can avoid this program code repetition by using AutoMapper's `Include()` methods (not to be confused with the Entity Framework's `Include()` method!).

```
Mapper.Initialize(cfg =>
{
 cfg.CreateMap<Person, PersonDTO>()
        .Include<Man, ManDTO>()
        .Include<Woman, WomanDTO>()
        .ForMember(z => z.Name, map => map.MapFrom(q => q.GivenName + " " +
        q.Surname))
        .ForMember(z => z.YearOfBirth, map => map.MapFrom(q => q.Birthday.
        Year));
 cfg.CreateMap<Man, ManDTO>();
 cfg.CreateMap<Woman, WomanDTO>();
});
```

Listing 20-14 shows an example of the Person, Man, and Woman mapping, including a gender transformation of a man into a woman, with the number of cars being converted into ten times the number of shoes. After defining this mapping, the actual transformation from Man to Woman with AutoMapper is comparatively painless.

***Listing 20-14.*** Mapping with Person, Man, and Woman

```
public static void Inheritance()
{

  CUI.Headline(nameof(Inheritance));

  Mapper.Initialize(cfg =>
  {
    cfg.CreateMap<Person, PersonDTO>()
        .ForMember(z => z.Name, map => map.MapFrom(q => q.GivenName + " " +
        q.Surname))
        .ForMember(z => z.YearOfBirth, map => map.MapFrom(q => q.Birthday.
        Year));
    cfg.CreateMap<Man, ManDTO>()
        .ForMember(z => z.Name, map => map.MapFrom(q => q.GivenName + " " +
        q.Surname))
        .ForMember(z => z.YearOfBirth, map => map.MapFrom(q => q.Birthday.
        Year));
```

```
  cfg.CreateMap<Woman, WomanDTO>()
      .ForMember(z => z.Name, map => map.MapFrom(q => q.GivenName + " " +
      q.Surname))
      .ForMember(z => z.YearOfBirth, map => map.MapFrom(q => q.Birthday.
      Year));
});

// or shorter using include()
Mapper.Initialize(cfg =>
{
 cfg.CreateMap<Person, PersonDTO>()
       .Include<Man, ManDTO>()
       .Include<Woman, WomanDTO>()
       .ForMember(z => z.Name, map => map.MapFrom(q => q.GivenName + " "
       + q.Surname))
       .ForMember(z => z.YearOfBirth, map => map.MapFrom(q =>
       q.Birthday.Year));
 cfg.CreateMap<Man, ManDTO>();
 cfg.CreateMap<Woman, WomanDTO>();
});

var m = new Man()
{
 GivenName = "John",
 Surname = "Doe",
 Birthday = new DateTime(1980, 10, 1),
 NumberOfCars = 40
};

PersonDTO mDTO1 = Mapper.Map<PersonDTO>(m);
Console.WriteLine(mDTO1.Name + " *" + mDTO1.YearOfBirth);

ManDTO mDTO1b = Mapper.Map<ManDTO>(m);
Console.WriteLine(mDTO1b.Name + " *" + mDTO1b.YearOfBirth);

ManDTO mDTO2 = (ManDTO)Mapper.Map(m, m.GetType(), typeof(ManDTO));
Console.WriteLine(mDTO2.Name + " *" + mDTO2.YearOfBirth + " owns " +
mDTO2.NumberOfCars + " cars.");
```

```
ManDTO mDTO3 = Mapper.Map<ManDTO>(m);
Console.WriteLine(mDTO3.Name + " *" + mDTO3.YearOfBirth + " owns " +
mDTO3.NumberOfCars + " cars.");

// gender transformation: man -> woman
Mapper.Initialize(cfg =>
{
 cfg.CreateMap<Man, Woman>()
        .ForMember(z => z.NumberOfShoes, map => map.MapFrom(q =>
        q.NumberOfCars * 10));
});

Woman f = Mapper.Map<Woman>(m);
Console.WriteLine(f.GivenName + " " + f.Surname + " *" + f.Birthday + "
owns " + f.NumberOfShoes + " shoes.");
}
```

What happens if a derived class has a manual mapping that contradicts a base class mapping? According to the documentation (`https://github.com/AutoMapper/AutoMapper/wiki/Mapping-inheritance`), the evaluation is done by AutoMapper with the following priorities:

- Explicit mapping in the derived class

- Inherited explicit mapping

- Mappings with `Ignore()`

- AutoMapper conventions that work only in the last step

# Generic Classes

AutoMapper also helps with generic classes. It's pretty basic for AutoMapper to map generic lists (see Listing 20-15).

*Listing 20-15.* Mapping Generic Lists, Here the Example List<T>

```
public static void GenericHomogeneousList()
  {
  CUI.Headline(nameof(GenericHomogeneousList));
```

```
   var PersonSet = new List<Person>();
   for (int i = 0; i < 100; i++)
   {
    PersonSet.Add(new Person() { GivenName="John", Surname="Doe"});
   }

   // define Mapping
   Mapper.Initialize(cfg =>
   {
    cfg.CreateMap<Person, PersonDTO>()
.ForMember(z => z.Name, map => map.MapFrom(q => q.GivenName + " " +
q.Surname))
.ForMember(z => z.YearOfBirth, map => map.MapFrom(q -> q.Birthday.Year));
   });

   // Convert list
   var PersonDTOSet = Mapper.Map<List<PersonDTO>>(PersonSet);

   Console.WriteLine(PersonDTOSet.Count());
   foreach (var p in PersonDTOSet.Take(5))
   {
    Console.WriteLine(p.Name + ": "+ p.YearOfBirth);
   }
  }
```

AutoMapper can also map entire generic types to other generic types. Sticking to the example with Person, Woman, and Man, Listing 20-16 defines two generic types for a partnership: a registered partnership and a marriage.

***Listing 20-16.*** Generic Types for Partnership and Marriage

```
// see https://europa.eu/youreurope/citizens/family/couple/registered-
partners/index_en.htm
 class RegisteredPartnership<T1, T2>
  where T1 : Person
  where T2 : Person
 {
  public T1 Partner1 { get; set; }
```

```
  public T2 Partner2 { get; set; }
  public DateTime Date { get; set; }
}

class Marriage<T1, T2>
 where T1 : Person
 where T2 : Person
{
 public T1 Partner1 { get; set; }
 public T2 Partner2 { get; set; }
 public DateTime Date { get; set; }
}
```

This allows same-sex registered partnerships as well as same-sex marriages! You can even allow a registered partnership to be automatically converted into a marriage. For example, converting the type of registered partnership <husband, husband> into marriage <husband, husband> is no problem for AutoMapper. All you have to do is define a general mapping between the generic classes. By no means do you have to write a mapping for all possible variants of type parameters of these generic classes.

```
Mapper.Initialize (cfg =>
{
 cfg.CreateMap (typeof (RegisteredPartnership <,>), typeof (Marriage <,>));
30.4
```

Listing 20-17 shows the application of this mapping.

***Listing 20-17.*** Mapping Your Own Generic Types from Listing 20-16

```
// A registered partnership between two men
var m1 = new   Man() {first name = "Heinz" , last name = "Müller" };
var m2 = new   Man() {first name = "Gerd" , last name = "Meier" };
var ep = new   RegisteredPartnership < Man , Man >() {Partner1 = m1,
Partner2 = m2, Date = new   DateTime (2015,5,28)};

// The general mapping between the generic classes
Mapper.Initialize (cfg =>
   {
```

```
      cfg.CreateMap (typeof (RegisteredPartnership <,>), typeof (Marriage <,>));
      30.4
```

```
// Then every figure with concrete type parameters is allowed!
Marriage < husband , husband > marriage = AutoMapper.Mapper .Map < marriage
< man , man >> (ep);
Console .WriteLine (before.Partner1.Name + "+" + marriage.Partner2.Name +
":" + marriage.- DateToShortDateString());
```

An additional mapping of the generic parameters is possible, such as
RegisteredPartnership<man, man> to Marriage <ManDTO, ManDTO>. Of course this
requires the following:

- That the generic class marriage <T1, T2> as a type parameter also
  allows ManDTO or PersonDTO (which it does not yet do). So, you have
  to change it like this:

```
class Marriage<T1, T2>
  where T1 : PersonDTO
  where T2 : PersonDTO
 {
  public T1 Partner1 { get; set; }
  public T2 Partner2 { get; set; }
  public DateTime Date { get; set; }
 }
```

- It is also necessary to define a mapping between Man and ManDTO, as
  shown here:

```
    Mapper.Initialize(cfg =>
    {
     cfg.CreateMap(typeof(RegisteredPartnership<,>),
     typeof(Marriage<,>));
     cfg.CreateMap<Man, ManDTO>()
      .ForMember(z => z.NumberOfCars, map => map.MapFrom(q =>
      q.GivenName + " " + q.Surname))
      .ForMember(z => z.YearOfBirth, map => map.MapFrom(q =>
      q.Birthday.Year));
    });
```

After that, the mapping between `RegisteredPartnership<Man, Man>` and `Marriage<ManDTO, ManDTO>` can be done.

```
Marriage<ManDTO, ManDTO> marriageDTO = Mapper.Map<Marriage<ManDTO,
ManDTO>>(ep);
Console.WriteLine(marriageDTO.Partner1.Name + " + " + marriageDTO.
Partner2.Name + ": " + marriage.Date.ToShortDateString());
```

# Additional Actions Before and After the Mapping

AutoMapper allows mapping actions to be performed before mapping or after mapping. The `BeforeMap()` method in the Fluent API of `CreateMap()` sets the upstream actions; `AfterMap()` sets the downstream. Both methods can be called multiple times, as Listing 20-18 shows with the example of `BeforeMap()`. Both methods each expect an expression that gets both the source object (briefly named `q` in the listing) and the target object (briefly named `z` in the listing). It is possible to call a method as part of the expression. This is also shown in Listing 20-18 using the example of `AfterMap()`.

The example in Listing 20-18 does the following when mapping `Person` to `PersonDTO`:

- If the first name or last name is empty, the entry in the source object is replaced with a question mark.

- If the name `??` appears in the target object because the first and last names are empty, the value `"error"` or `"no information"` will be delivered depending on the year of birth. The business process rule behind it is as follows: All people born before 1980 may remain anonymous. All people born after that are required to give a name. If the name is still missing, there must be an error.

Of course, you can set such business logic outside of AutoMapper. These are the advantages of integrating with AutoMapper:

- You have all the mapping actions in one place.

- You do not have to explicitly preprogram an iteration over the objects before or after the mapping.

***Listing 20-18.*** BeforeMap( )and AfterMap( ) in Action

```
public static void BeforeAfterDemo()
  {
   CUI.Headline(nameof(BeforeAfterDemo));

   var PersonSet = new List<Person>();
   for (int I = 0; i < 10; i++)
   {
    PersonSet.Add(new Person()
    {
     GivenName =""Joh"",
     Surname =""Do"",
     Birthday = new DateTime(1980, 10, 1),
    });
   }

   // Define mapping
   Mapper.Initialize(cfg =>
   {
    cfg.CreateMap<Person, PersonDTO>()
       .ForMember(z => z.Name, map => map.MapFrom(q => q.GivenName +"""" +
       q.Surname))
       .ForMember(z => z.YearOfBirth, map => map.MapFrom(q => q.Birthday.
       Year))
       .BeforeMap((q, z) => q.GivenName = (String.IsNullOrEmpty(q.
       GivenName) ? q.GivenName ="""" : q.GivenName))
       .BeforeMap((q, z) => q.Surname = (String.IsNullOrEmpty(q.Surname) ?
       q.Surname ="""" : q.Surname))
       .AfterMap((q, z) => z.Name = GetName(z.Name, z.YearOfBirth));
    cfg.CreateMap<DateTime, Int32>().ConvertUsing(ConvertDateTimeToInt);
   });

   // Map list
   var PersonDTOSet = Mapper.Map<List<PersonDTO>>(PersonSet);

   foreach (var p in PersonDTOSet)
   {
```

```
  Console.WriteLine(p.Name +"" in born in year"" + p.YearOfBirth);
 }
}

/// <summary>
/// Converges DateTime into integer (only extracts year)
/// </summary>
/// <returns></returns>
public static Int32 ConvertDateTimeToInt(DateTime d)
{
 return d.Year;
}

/// <summary>
/// Method called as part of AfterMap()
/// </summary>
/// <param name"""">Surname</param>
/// <param name""yearOfBirt"">YearOfBirth</param>
/// <returns></returns>
public static string GetName(string name, int yearOfBirth)
{
 if (yearOfBirth == 0) return name;
 if (yearOfBirth <= 1980) return name +"" (too young"";
 return name +"" "" + yearOfBirth """";
}
```

# Performance

AutoMapper does not use reflection for each fetch and set of values for the mapping. Rather, `CreateMap()` generates the program code at runtime using `Reflection.Emit()`. This raises the question of how long it takes to map large amounts of data.

In Table 20-2, three mapping paths are compared.

- Explicit, hard-coded object-object mapping (i.e., xa = ya, for each property)

- Reflection-based object-object mapping with the program code at the beginning of this chapter

- Object-object mapping with AutoMapper

So that you do not compare apples with pears, in Table 20-2 a mapping takes place between two exactly identically constructed types. In other words, all properties are called the same way in both classes and also have the same data type. The performance test measures the values for 1, 10, 100, 1,000, 10,000 and 100,000 objects that are in a generic list.

When looking at the results table, it is clear that AutoMapper is significantly slower. Generating the mapping code with CreateMap() always takes about 208 milliseconds. If a mapping of a type occurs repeatedly in a process, this happens only once. With repeated calls, it takes approximately 7 milliseconds. However, AutoMapper is slower than explicit mapping in all cases and even slower than reflection-based mapping for data sets up to 1,000 objects.

***Table 20-2.*** *Speed Comparison of Three Methods for Object-to-Object Mapping (in Milliseconds)*

| Number of Objects | Explicit (Hard-Coded) Mapping | | Reflection Mapping | | AutoMapper | |
|---|---|---|---|---|---|---|
| | One-time initialization effort per application domain | Effort for the mapping | One-time initialization effort per application domain | Effort for the mapping | One-time initialization effort per application domain | Effort for the mapping |
| 1 | 0 | 0 | 0 | 0 | 208 | 18 |
| 10 | 0 | 0 | 0 | 0 | 208 | 18 |
| 100 | 0 | 0 | 0 | 1 | 208 | 18 |
| 1,000 | 0 | 0 | 0 | 10 | 208 | 19 |
| 10,000 | 0 | 1 | 0 | 104 | 208 | 30 |
| 100,000 | 0 | 29 | 0 | 1010 | 208 | 63 |

# Conclusion to AutoMapper

AutoMapper offers flexible imaging options between different object structures. The performance of AutoMapper is very disappointing at first glance. However, you must not forget that AutoMapper saves a lot of programming work compared to explicit mapping and can do much more than reflection mapping.

However, developers such as Andrew Harcourt (`http://www.uglybugger.org/software/post/friends_dont_let_friends_use_automapper`) criticize not only the performance of AutoMapper but also dislike the conventions. Mapping properties of the same name becomes a problem when you rename a property that will be mapped, unless you also think about writing a custom mapping with `ForMember()`. Harcourt advocates explicitly programming all mappings, which makes automatic name refactoring possible. To reduce the programming work for the explicit mapping, he has written a code generator for the matching mapping code. Unfortunately, he does not provide the code generator to the public.

Tools that generate an explicit mapping include OTIS-LIB (`http://code.google.com/p/otis-lib`) and Wayne Hartmann's Object To Object Mapping Utility (`http://waynehartman.com/download?file=d2333998-c0cc-4bd4-8f02-82bef57d463c`). However, not everyone likes generator-generated program code. Therefore, this chapter ends without a clear recommendation for or against AutoMapper. It depends on the application (the size and number of objects) and your own preference.

# Case Studies

This chapter describes some practical applications for Entity Framework Core that are outside the World Wide Wings example.

## Using Entity Framework Core in an ASP.NET Core Application

In 2015, Microsoft paid more than $100 million to acquire the Berlin-based app publisher Wunderlist (`https://www.theverge.com/2015/6/2/8707883/microsoft-wunderlist-acquisition-announced`). MiracleList is a reprogramming of the Wunderlist task management app as a web application and cross-platform application for Windows, Linux, macOS, Android, and iOS with a cross-platform back end in the cloud. See Figure A-1, Figure A-2, and Figure A-3.

The logged-in user can create a list of task categories and then create a list of tasks in each category. A task consists of a title, a note, an entry date, and a due date, and it can be marked as done. Beyond the functions of Wunderlist, in MiracleList a task can have three (A, B, or C) instead of just two levels of importance (yes/no) and an effort (number). The effort has no unit of measure; the user can decide whether to set the effort to hours, days, or a relative value such as 1 (for low) to 10 (for high).

Like with Wunderlist, a task can have subtasks, with a subtask having only one title and one status. Some details from the original are missing in MiracleList, such as the ability to upload files to tasks, move tasks between categories, search for hashtags, duplicate and print lists, and exchange tasks among users. Some features, such as clickable hyperlinks in the text of a task, are not implemented to prevent misuse.

© Holger Schwichtenberg 2018
H. Schwichtenberg, *Modern Data Access with Entity Framework Core*,
https://doi.org/10.1007/978-1-4842-3552-2_21

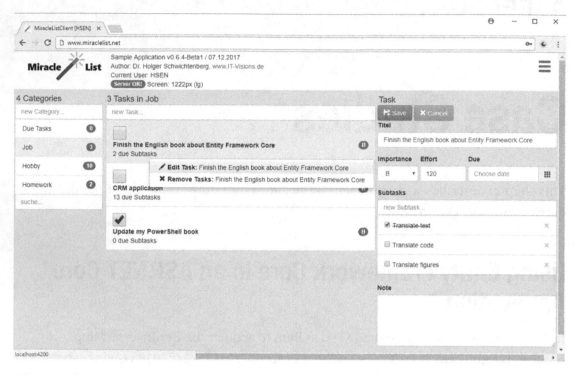

***Figure A-1.*** *MiracleList web application*

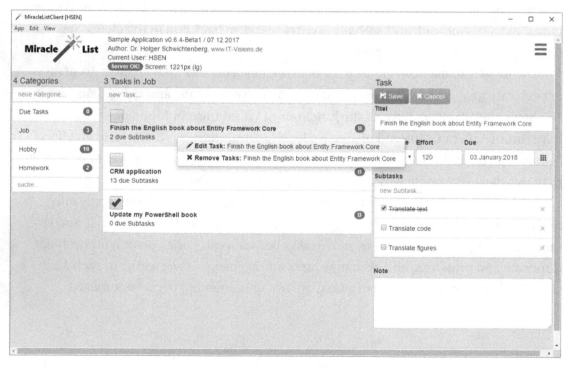

***Figure A-2.*** *MiracleList desktop client for Windows*

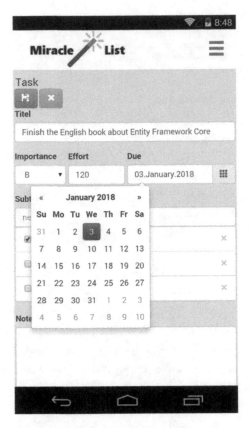

*Figure A-3.* *MiracleList client for Android*

In addition to the web API, the back end also has a web interface that offers the following (Figure A-4):

- Version information for the web API

- OpenAPI specification for the web API

- Help page for the web API

- The ability to request a client ID to create your own client

- The ability to download the desktop clients

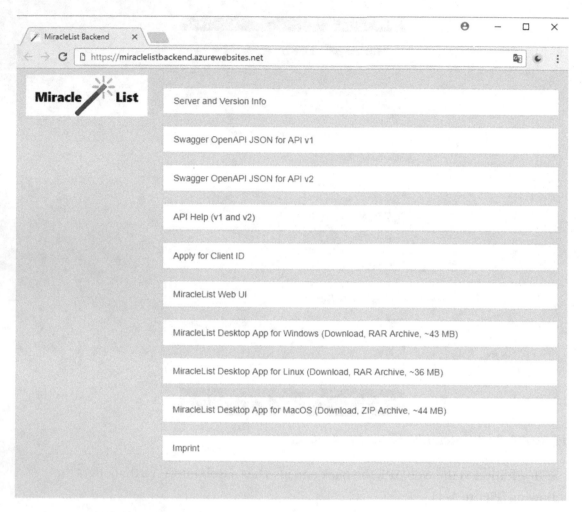

*Figure A-4.* *Web interface of the back end*

In this book, only excerpts from the back end are discussed because it is there that Entity Framework Core is used.

You can find more information at the following sites:

- *Web application and client download*: http://www.miraclelist.net

- *Back end*: https://miraclelistbackend.azurewebsites.net

- *Source code for the back end*: https://github.com/ HSchwichtenberg/MiracleListBackend

- *Source code for the front end*: https://github.com/ HSchwichtenberg/MiracleListClient

# Architecture

MiracleList uses the following techniques:

- *Back end*: .NET Core, C# , ASP.NET Core Web API, Entity Framework Core, SQL Azure, Azure Web App, Swagger/Swashbuckle. AspNetCore, Application Insights

- *Front end*: SPA with HTML, CSS, TypeScript, Angular, Bootstrap, MomentJS, ng2-datetime, angular2-moment, angular2-contextmenu, angular2-modal, Electron, Cordova

The back end for MiracleList is available for anyone to use at `https://miraclelistbackend.azurewebsites.net`. It runs on C# 6.0 and .NET Core 2.0 and was built using SQL Azure as the database, Entity Framework Core 2.0 as the OR mapper, and ASP.NET Core 2.0 as the web server framework. It is hosted in Microsoft's Azure cloud as a web app.

The MiracleList back end provides a clear separation of layers. The solution (see Figure A-5) consists of the following:

- *Business objects (BOs)*: This contains the entity classes used for Entity Framework Core. These classes are deliberately implemented in such a way that they can also be used as input and output types in the Web API, meaning that additional object-to-object mapping with AutoMapper or other tools is unnecessary.

- *Data access layer (DAL)*: This layer implements the Entity Framework core context (`Context.cs`).

- *Business logic (BL)*: Here, "manager" classes are implemented that implement the back-end functionality using the Entity Framework Core context.

- *MiracleList_WebAPI*: Here the controllers for the web API are realized.

- *EFTools*: This contains Entity Framework Core tools for forward engineering.

- *UnitTests*: This contains unit tests with XUnit.

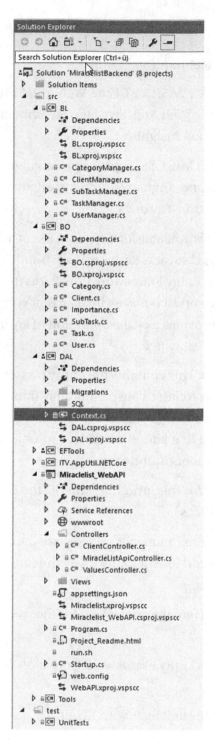

***Figure A-5.*** *Projects in the MiracleList back-end solution*

# Entity

Figure A-6 shows the MiracleList object model, consisting of five classes and an enumeration type. Listing A-1, Listing A-2, Listing A-3, Listing A-4, and Listing A-5 show the entities.

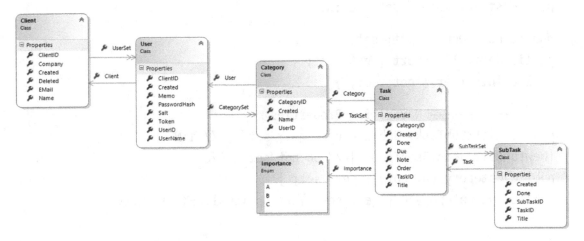

***Figure A-6.*** *MiracleList object model*

***Listing A-1.*** Task.cs

```
using System;
using System.Collections.Generic;
using System.ComponentModel.DataAnnotations;
using System.ComponentModel.DataAnnotations.Schema;

namespace BO
{

  /// <summary>
  /// Entity class representing a task
  /// Used on the server up to the WebAPI
  /// Corresponding proxy class in TypeScript is used on client
  /// </summary>
  public class Task
  {
    public int TaskID { get; set; } // PK per Konvention
```

```
    [MaxLength(250)] // alias: StringLength
    public string Title { get; set; }
    public DateTime Created { get; set; } = DateTime.Now;
    public DateTime? Due { get; set; }
    public Importance? Importance { get; set; }
    public string Note { get; set; }

    public bool Done { get; set; }
    public decimal? Effort { get; set; }
    public int Order { get; set; }

    // ------------- Navigation Properties
    public List<SubTask> SubTaskSet { get; set; } // 1:N
    [Newtonsoft.Json.JsonIgnore] // Do not serialize
    public Category Category { get; set; }
    public int CategoryID { get; set; } // optional: FK Property
  }
}
```

***Listing A-2.*** Importance.cs

```
namespace BO

{
 public enum Importance
 {
  A, B, C
 }
}
```

***Listing A-3.*** SubTask.cs

```
using System;
using System.ComponentModel.DataAnnotations;

namespace BO
{
 /// <summary>
 /// Entity class representing a subtask
```

```
/// Used on the server up to the WebAPI
/// Corresponding proxy class in TypeScript is used on client
/// </summary>
public class SubTask
{
 public int SubTaskID { get; set; } // PK
 [MaxLength(250)]
 public string Title { get; set; }
 public bool Done { get; set; }
 public DateTime Created { get; set; } = DateTime.Now;
 // -------------- Navigation Properties
 public Task Task { get; set; }
 public int TaskID { get; set; }
 }
}
```

**_Listing A-4._** Category.cs

```
using System;
using System.Collections.Generic;
using System.ComponentModel.DataAnnotations;

namespace BO
{
 /// <summary>
 /// Entity class representing a category of tasks
 /// Used on the server up to the WebAPI
 /// Corresponding proxy class in TypeScript is used on client
 /// </summary>
 public class Category
 {
  public int CategoryID { get; set; } // PK

  [MaxLength(50)]
  public string Name { get; set; }

  public DateTime Created { get; set; } = DateTime.Now;
```

```
// -------------- Navigation Properties
public List<Task> TaskSet { get; set; }
[Newtonsoft.Json.JsonIgnore] // Do not serialize
public User User { get; set; }
[Newtonsoft.Json.JsonIgnore] // Do not serialize
public int UserID { get; set; }

}
}
```

***Listing A-5.*** Client.cs

```
using System;
using System.Collections.Generic;
using System.ComponentModel.DataAnnotations;
namespace BO
{
 /// <summary>
 /// Entity class representing a category of tasks
 /// Used on the server up to the WebAPI
 /// Not used in the client!
 /// </summary>
 public class Client
 {
  public Guid ClientID { get; set; }
  [StringLength(50)]
  public string Name { get; set; }
  [StringLength(50)]
  public string Company { get; set; }
  [StringLength(50)]
  public string EMail { get; set; }
  public DateTime Created { get; set; } = DateTime.Now;
  public DateTime? Deleted { get; set; }
  public string Memo { get; set; }
  [StringLength(10)]
  public string Type { get; set; }
```

```
    // -------------- Navigation Properties
    public List<User> UserSet { get; set; }
 }
}
```

# Entity Framework Core Context Class

The context class derived from DbContext always has a property of type DbSet<T> for the four entity classes (Listing A-6). In the OnConfiguring() method, UseSqlServer() sets the Entity Framework Core database provider, passing in the connection string. The connection string is realized as a public static class member, so it can be set externally.

In the OnModelCreating() method additional indexes on columns are set, which allow searching. In addition, it is specified globally for all entity classes that the table names in the database should not be named like the dbSet<T> properties but like the classes (in other words, table Task instead of table TaskSet). The exceptions are only the classes that have a [Table] annotation, so you have the opportunity to set individual deviations from the rule.

***Listing A-6.*** Context.cs

```
using BO;
using Microsoft.EntityFrameworkCore;
using System;
using System.ComponentModel.DataAnnotations.Schema;
using System.Reflection;
using Microsoft.EntityFrameworkCore.Metadata;
using Microsoft.EntityFrameworkCore.Metadata.Internal;

namespace DAL
{
 /// <summary>
 /// Context class for Entity Framework Core
 /// Forms the DAL that is used the BL manager classes
 /// </summary>
 public class Context : DbContext
 {
  // Register the entity classes in the context
  public DbSet<Client> ClientSet { get; set; }
```

```csharp
public DbSet<User> UserSet { get; set; }
public DbSet<Task> TaskSet { get; set; }
public DbSet<Category> CategorySet { get; set; }
public DbSet<Log> LogSet { get; set; }

// This connection string is just for testing. Is filled at runtime from
configuration file
public static string ConnectionString { get; set; } = "Data
Source=.;Initial Catalog = MiracleList_TEST; Integrated Security = True;
Connect Timeout = 15; Encrypt=False;TrustServerCertificate=True;
ApplicationIntent=ReadWrite;MultiSubnetFailover=False;Application
Name=EntityFramework";

protected override void OnConfiguring(DbContextOptionsBuilder builder)
{
 builder.UseSqlServer(Context.ConnectionString);
}

protected override void OnModelCreating(ModelBuilder builder)
{
 // In this case, EFCore can derive the database schema from the entity
 classes by convention and annotation.
 // The following Fluent API configurations only change the default
 behavior!

 #region Mass configuration via model class
 foreach (IMutableEntityType entity in builder.Model.GetEntityTypes())
 {
  // all table names = class names (as with EF 6.x),
  // except the classes that have [Table] annotation
  var annotation = entity.ClrType.GetCustomAttribute<TableAttribute>();
  if (annotation == null)
  {
   entity.Relational().TableName = entity.DisplayName();
  }
 }
 #endregion
```

```
#region Custom Indices
builder.Entity<Category>().HasIndex(x => x.Name);
builder.Entity<Task>().HasIndex(x => x.Title);
builder.Entity<Task>().HasIndex(x => x.Done);
builder.Entity<Task>().HasIndex(x => x.Due);
builder.Entity<Task>().HasIndex(x => new { x.Title, x.Due });
#endregion
  }
 }
}
```

# Lifetime of Context Class in ASP.NET Core Applications

When using Entity Framework Core in ASP.NET and ASP.NET Core–based web applications and web services/web APIs, the lifetime of an instance of the context class must not exceed the lifetime of processing a single HTTP request.

For each incoming HTTP request, ASP.NET Core creates a new instance of the controller class using different threads. If you were to use an instance of the context class in more than one HTTP request, you would use the context instance in different threads, which does not support it. The context class is not thread-safe, so it does not support multithreading. Multithreading with a context instance causes the following error at runtime: "System.InvalidOperationException : 'A second operation started on this context before a previous operation completed. Any instance members are not guaranteed to be thread safe.'"

---

**Note**    Also, the context classes `ObjectContext` and `DbContext` in the classic Entity Framework are not thread-safe. In the classic Entity Framework, however, there was not a clear error message when using the multithreading context. There were just curious crashes somewhere in the depths of Entity Framework.

---

In addition, you would encounter a problem with the Entity Framework core cache because a second request would find in the context instance the data of a previous request. However, a second HTTP request may affect another user. Setting an Entity

Framework core context in a user-based variable is not a good solution because it greatly reduces the scalability of the web application.

It is therefore appropriate to limit the lifetime of the context instance to the lifetime of processing a single HTTP request. As part of an HTTP request, you create one or possibly more instances of the context class, which are then destroyed when the request is completed.

The architecture shown in this practice solution elegantly accommodates this without directly using the context classes in the ASP.NET/ASP.NET core controller. The usage happens indirectly through manager classes that provide the business logic.

Each instance of a manager class creates a new context instance. The lifetime of instances of the manager classes is tied to the lifetime of the instance of the WebAPI controller. As a result, a context instance does not live beyond processing the HTTP request.

# Business Logic

Listing A-7 shows an example of the class TaskManager. This implementation is based on the generic base class EntityManagerBase <contextType, entity type>. In turn, EntityManagerBase is based on DataManagerBase <contextType>.

These two helper classes provide basic functions that do not always have to be realized in every manager class. This includes the following (Listing A-7, Listing A-8, and Listing A-9):

- Creating a context instance when creating the Manager class

- Destroying the context instance when calling Dispose()

- Update(): Adding an object to the context in the "modified" state and saving the change

- New(): Adding a new object and saving the new object directly

- Remove(): Deleting an object and executing the deletion directly

- IsLoaded(): Checking whether an object exists in the local cache

*Listing A-7.* TaskManager.cs

```
using System;
using System.Collections.Generic;
using System.Linq;
```

```csharp
using BO;
using DAL;
using ITVisions.EFC;

using Microsoft.EntityFrameworkCore;
using System.Diagnostics;
using ITVisions.EFCore;

namespace BL
{
 /// <summary>
 /// Business Logic manager for Tasks entities
 /// </summary>
 public class TaskManager : EntityManagerBase<Context, Task>
 {
  // To manage the subtasks
  private SubTaskManager stm = new SubTaskManager();
  // Current user
  private int userID;

  /// <summary>
  /// Instantiation specifying the user ID to which all operations in this
  instance refer
  /// </summary>
  /// <param name="userID"></param>
  public TaskManager(int userID)
  {
   this.userID = userID;
  }

  /// <summary>
  /// Get a task list of one category for the current user
  /// </summary>
  public List<Task> GetTaskSet(int categoryID)
  {
   return ctx.TaskSet.Include(x => x.SubTaskSet).Where(x => x.Category.
   UserID == this.userID && x.CategoryID == categoryID).ToList();
  }
```

```csharp
/// <summary>
/// Get a task including its subtasks
/// </summary>
public Task GetTask(int taskID)
{
 var t = ctx.TaskSet.Include(x => x.SubTaskSet).Where(x => x.Category.
 UserID == this.userID && x.TaskID == taskID).SingleOrDefault();
 return t;
}

/// <summary>
/// Create a new task from Task object
/// </summary>
public Task CreateTask(Task t)
{
 ValidateTask(t);
 return this.New(t);
}

/// <summary>
/// Create a new task from details
/// </summary>
public Task CreateTask(int categoryID, string title, string note,
DateTime due, Importance importance, decimal? effort, List<SubTask>
subtasks = null)
{
 this.StartTracking();
 var t = new Task();
 t.CategoryID = categoryID;
 t.Created = DateTime.Now;
 SetTaskDetails(t, title, note, due, importance, false, effort,
 subtasks);
 this.New(t);
 this.SetTracking();
 return t;
}
```

```
private static void SetTaskDetails(Task t, string title, string note,
DateTime? due, Importance? importance, bool done, decimal? effort,
List<SubTask> subtasks)
{
 t.Title = title;
 t.Note = note;
 t.Due = due;
 t.Importance = importance;
 t.SubTaskSet = subtasks;
 t.Effort = effort;
 t.Done = done;
}

/// <summary>
/// Change a task
/// </summary>
public Task ChangeTask(int taskID, string title, string note, DateTime
due, Importance? importance, bool done, decimal? effort, List<SubTask>
subtasks)
{
 ctx = new Context();
 ctx.Log();
 // Delete subtasks and then create new ones instead of change detection!
 stm.DeleteSubTasks(taskID);

 var t = ctx.TaskSet.SingleOrDefault(x => x.TaskID == taskID);
 SetTaskDetails(t, title, note, due, importance, done, effort, null);
 ctx.SaveChanges();

 t.SubTaskSet = subtasks;
 ctx.SaveChanges();
 return t;
}

public void Log(string s)
{
 Debug.WriteLine(s);
}
```

```csharp
/// <summary>
/// Change a task including subtasks
/// </summary>
public Task ChangeTask(Task tnew)
{
 if (tnew == null) return null;

 // Validate of the sent data!
 if (tnew.Category != null) tnew.Category = null; // user cannot change
 the category this way!
 ValidateTask(tnew);

 var ctx1 = new Context();
 ctx1.Log(Log);
 stm.DeleteSubTasks(tnew.TaskID);

 if (tnew.SubTaskSet != null) tnew.SubTaskSet.ForEach(x => x.SubTaskID = 0);
 // delete ID, so that EFCore regards this as a new object

 tnew.CategoryID = this.GetByID(tnew.TaskID).CategoryID; // Use existing
 category

 ctx1.TaskSet.Update(tnew);

 var count = ctx1.SaveChanges();
 return tnew;
}

/// <summary>
/// Checks if the TaskID exists and belongs to the current user
/// </summary>
private void ValidateTask(int taskID)
{
 var taskAusDB = ctx.TaskSet.Include(t => t.Category).SingleOrDefault
 (x => x.TaskID == taskID);
 if (taskAusDB == null) throw new UnauthorizedAccessException("Task nicht
 vorhanden!");
```

```csharp
  if (taskAusDB.Category.UserID != this.userID) throw new Unauthorized
  AccessException("Task gehört nicht zu diesem User!");
}

/// <summary>
/// Checks if transferred task object is valid
/// </summary>
private void ValidateTask(Task tnew = null)
{
 ValidateTask(tnew.TaskID);
 if (tnew.CategoryID > 0)
 {
  var catAusDB = new CategoryManager(this.userID).GetByID(tnew.CategoryID);
  if (catAusDB.UserID != this.userID) throw new UnauthorizedAccessException
  ("Task gehört nicht zu diesem User!");
 }
}

/// <summary>
/// Full-text search in tasks and subtasks, return tasks grouped by
category
/// </summary>
public List<Category> Search(string text)
{
 var r = new List<Category>();
 text = text.ToLower();
 var taskSet = ctx.TaskSet.Include(x => x.SubTaskSet).Include(x =>
 x.Category).
  Where(x => x.Category.UserID == this.userID && // nur von diesem User
  !!!
  (x.Title.ToLower().Contains(text) || x.Note.ToLower().Contains(text) ||
  x.SubTaskSet.Any(y => y.Title.Contains(text))))).ToList();

 foreach (var t in taskSet)
 {
  if (!r.Any(x => x.CategoryID == t.CategoryID)) r.Add(t.Category);
 }
```

```csharp
  return r;
}

/// <summary>
/// Returns all tasks due, including tomorrow, grouped by category,
sorted by date
/// </summary>
public List<Category> GetDueTaskSet()
{
 var tomorrow = DateTime.Now.Date.AddDays(1);
 var r = new List<Category>();
 var taskSet = ctx.TaskSet.Include(x => x.SubTaskSet).Include(x =>
 x.Category).
  Where(x => x.Category.UserID == this.userID && // nur von diesem User
  !!!
  (x.Done == false && x.Due != null && x.Due.Value.Date <= tomorrow)).
  OrderByDescending(x => x.Due).ToList();

 foreach (var t in taskSet)
 {
  if (!r.Any(x => x.CategoryID == t.CategoryID)) r.Add(t.Category);
 }
 return r;
}

/// <summary>
/// Remove Task with its subtasks
/// </summary>
public void RemoveTask(int id)
{
 ValidateTask(id);
 this.Remove(id);
 }
 }
}
```

**Listing A-8.** EntityBaseManager.cs

```
using Microsoft.EntityFrameworkCore;
using System.Linq;

namespace ITVisions.EFC
{
 /// <summary>
 /// Base class for all data managers to manage a specific entity type,
 even if they are detached
 /// V1.3
 /// Assumption: There is always only one primary key column!
 /// </summary>
 public abstract class EntityManagerBase<TDbContext, TEntity> :
 DataManagerBase<TDbContext>
  where TDbContext : DbContext, new()
  where TEntity : class
 {
  public EntityManagerBase() : base(false)
  {
  }
  public EntityManagerBase(bool tracking) : base(tracking)
  {
  }
  protected EntityManagerBase(TDbContext kontext = null, bool tracking =
  false) : base(kontext, tracking)
  {
  }

  /// <summary>
  /// Get object based on the primary key
  /// </summary>
  /// <returns></returns>
  public virtual TEntity GetByID(object id)
  {
   return ctx.Set<TEntity>().Find(id);
  }
```

```
/// <summary>
/// Saves changed object
/// </summary>
public TEntity Update(TEntity obj)
{
 if (!this.tracking) this.StartTracking(); // Start change tracking if
 no-tracking is on
 ctx.Set<TEntity>().Attach(obj);
 ctx.Entry(obj).State = EntityState.Modified;
 ctx.SaveChanges();
 this.SetTracking();
 return obj;
}

/// <summary>
/// Adds a new object
/// </summary>
public TEntity New(TEntity obj)
{
 if (!this.tracking) this.StartTracking(); // Start change tracking if
 no-tracking is on
 ctx.Set<TEntity>().Add(obj);
 ctx.SaveChanges();
 this.SetTracking();
 return obj;
}

/// <summary>
/// Deletes an object based on the primary key
/// </summary>
public virtual void Remove(object id)
{
 if (!this.tracking) this.StartTracking(); // Start change tracking if
 no-tracking is on
 TEntity obj = ctx.Set<TEntity>().Find(id);
 Remove(obj);
```

```
    this.SetTracking();
   }

   /// <summary>
   /// Deletes an object
   /// </summary>
   public bool Remove(TEntity obj)
   {
    if (!this.tracking) this.StartTracking(); // Switch on tracking for a
    short time
    if (!this.IsLoaded(obj)) ctx.Set<TEntity>().Attach(obj);
    ctx.Set<TEntity>().Remove(obj);
    ctx.SaveChanges();
    this.SetTracking();
    return true;
   }

   /// <summary>
   /// Checks if an object is already in the local cache
   /// </summary>
   public bool IsLoaded(TEntity obj)
   {
    return ctx.Set<TEntity>().Local.Any(e => e == obj);
   }
  }
}
```

***Listing A-9.*** DataManagerBase.cs

```
using System;
using System.Collections.Generic;
using Microsoft.EntityFrameworkCore;
using System.Linq;

namespace ITVisions.EFC
{
```

```csharp
/// <summary>
/// Base class for all data managers
/// </summary>
abstract public class DataManagerBase<TDbContext> : IDisposable
 where TDbContext : DbContext, new()
{
 // One instance of the framework context per manager instance
 protected TDbContext ctx;
 protected bool disposeContext = true;
 protected bool tracking = false;

 protected DataManagerBase(bool tracking) : this(null, tracking)
 {
 }

 public void StartTracking()
 {
  ctx.ChangeTracker.QueryTrackingBehavior = Microsoft.EntityFrameworkCore.
  QueryTrackingBehavior.TrackAll;
 }

 public void SetTracking()
 {
  if (tracking) ctx.ChangeTracker.QueryTrackingBehavior =
  QueryTrackingBehavior.TrackAll;
  else ctx.ChangeTracker.QueryTrackingBehavior = QueryTrackingBehavior.
  NoTracking;
 }

 public DataManagerBase()
 {
  this.ctx = new TDbContext();
 }
 protected DataManagerBase(TDbContext kontext = null, bool tracking =
 false)
 {
  this.tracking = tracking;
```

```
// If a context has been handed in, take this!
if (kontext != null) { this.ctx = kontext; disposeContext = false; }
else
{
 this.ctx = new TDbContext();
}

SetTracking();
}

/// <summary>
/// Destroy DataManager (also destroys the EF context)
/// </summary>
public void Dispose()
{
 // If the context was submitted from the outside, we should not call
 Dispose() on the context! That's up to the caller!
 if (disposeContext) ctx.Dispose();
}

/// <summary>
/// Save Changes in context
/// </summary>
/// <returns>a string that contains information about the number of new,
changed, and deleted records</returns>
public string Save()
{
 string ergebnis = GetChangeTrackerStatistics();
 var count = ctx.SaveChanges();
 return ergebnis;
}

/// <summary>
/// Save for detached entity objects with auto increment primary key
named ID
// The newly added objects must return the store routine because the IDs
for the
```

```
/// </summary>
protected List<TEntity> Save<TEntity>(IEnumerable<TEntity> menge, out
string Statistik)
where TEntity : class
{
 StartTracking();
 var newObjects = new List<TEntity>();

 foreach (dynamic o in menge)
 {
  // Attach to the context
  ctx.Set<TEntity>().Attach((TEntity)o);
  if (o.ID == 0) // No value -> new object
  {
   ctx.Entry(o).State = EntityState.Added;
   if (o.ID < 0) o.ID = 0; // Necessary hack, because EFCore writes a big
   negative number in ID after the added and considers that as key :-(
   // Remember new records because they have to be returned after saving
   (they will have their IDs!)
   newObjects.Add(o);
  }
  else // existing object --> UPDATE
  {
   ctx.Entry(o).State = EntityState.Modified;
  }
  SetTracking();
 }

 // Get statistics of changes
 Statistik = GetChangeTrackerStatistics<TEntity>();
 var e = ctx.SaveChanges();
 return newObjects;
}

/// <summary>
/// Save for detached entity objects with an EntityState property
/// </summary>
```

```csharp
  protected List<TEntity> SaveEx<TEntity>(IEnumerable<TEntity> menge, out
  string Statistik)
where TEntity : class
  {
   StartTracking();
   var newObjects = new List<TEntity>();

   foreach (dynamic o in menge)
   {
    if (o.EntityState == ITVEntityState.Added)
    {
     ctx.Entry(o).State = EntityState.Added;
     newObjects.Add(o);
    }
    if (o.EntityState == ITVEntityState.Deleted)
    {
     ctx.Set<TEntity>().Attach((TEntity)o);
     ctx.Set<TEntity>().Remove(o);
    }
    if (o.EntityState == ITVEntityState.Modified)
    {
     ctx.Set<TEntity>().Attach((TEntity)o);
     ctx.Entry(o).State = EntityState.Modified;
    }
   }

   Statistik = GetChangeTrackerStatistics<TEntity>();
   ctx.SaveChanges();
   SetTracking();
   return newObjects;
  }

  /// <summary>
  /// Provides statistics from the ChangeTracker as a string
  /// </summary>
  protected string GetChangeTrackerStatistics<TEntity>()
where TEntity : class
```

```csharp
{
 string Statistik = "";
 Statistik += "Changed: " + ctx.ChangeTracker.Entries<TEntity>().Where
 (x => x.State == EntityState.Modified).Count();
 Statistik += " New: " + ctx.ChangeTracker.Entries<TEntity>().Where
 (x => x.State == EntityState.Added).Count();
 Statistik += " Deleted: " + ctx.ChangeTracker.Entries<TEntity>().Where
 (x => x.State == EntityState.Deleted).Count();
 return Statistik;
}

/// <summary>
///  Provides statistics from the ChangeTracker as a string
/// </summary>
protected string GetChangeTrackerStatistics()
{
 string Statistik = "";
 Statistik += "Changed: " + ctx.ChangeTracker.Entries().Where(x =>
 x.State == EntityState.Modified).Count();
 Statistik += " New: " + ctx.ChangeTracker.Entries().Where(x =>
 x.State == EntityState.Added).Count();
 Statistik += " Deleted: " + ctx.ChangeTracker.Entries().Where(x =>
 x.State == EntityState.Deleted).Count();
 return Statistik;
 }
 }
}
```

# Web API

The MiracleList back end offers an HTTPS-based REST service in two versions.

- In version 1 of the REST service, the authentication token is passed in the URL.

- In version 2 of the REST service, the authentication token is passed in the HTTP header.

Version 1 of the REST service provides the following operations:

- `POST/Login`: Logging in with a client ID, username, and password. This operation/login sends back a GUID as a session token, to be given in all following operations.

- `GET/Logoff/{token}`: Logging out the user.

- `GET/CategorySet/{token}`: Listing categories.

- `GET/TaskSet/{token}/{id}`: Listing tasks in a category.

- `GET/Task/{token}/{id}`: Listing details about a subtask task.

- `GET/Search/{token}/{text}`: Full-text searching in tasks and subtasks.

- `GET/DueTaskSet/{token}`: Listing tasks due.

- `POST/CreateCategory/{token}/{name}`: Creating a category.

- `POST/CreateTask/{token}`: Creating a task to submit in the body in JSON format (including subtasks).

- `PUT/ChangeTask/{token}`: Changing a task to be submitted in the body in JSON format (including subtasks).

- `DELETE/DeleteTask/{token}/{id}`: Deleting a task with all subtasks.

- `DELETE/DeleteCategory/{token}/{id}`: Deleting a category with all tasks and subtasks.

Version 2 of the REST service also offers the following operations:

- `POST/Login`: Logging in with a client ID, username, and password. This operation returns a GUID as the session token to be included in all subsequent operations.

- `GET/CategorySet/`: Listing categories.

- `GET/TaskSet/{id}`: Listing tasks in a category.

- `GET/Task/{id}`: Listing details of a task with subtasks.

- `POST/CreateCategory/{name}`: Creating a category.

- `POST/CreateTask` and `PUT/ChangeTask`: Creating or changing a task to be submitted in the body in JSON format (including subtasks).

571

For all REST operations, metadata is available in the Swagger OpenAPI Specification for RESTful APIs (`http://swagger.io`). See `https://miraclelistbackend.azurewebsites.net/swagger/v1/swagger.json` for a formal description of REST services, and see `https://miraclelistbackend.azurewebsites.net/swagger` for an appropriate help page. The back end also supports cross-origin resource sharing (CORS) to allow access to any other hosted web site (`https://www.w3.org/TR/cors`).

The client ID specified at `/Login` must be requested once by each client developer at `https://miraclelistbackend.azurewebsites.net/Client`. The number of tasks is limited to 1,000 per client ID. On the other hand, it is not necessary to create user accounts in the back end. Since this is an example application, a user is automatically created if the one you have submitted does not already exist. Each new account automatically has three categories (Work, Home, and Leisure) with example tasks such as Set up a Team Meeting, Bringing out Garbage, and Training for MTB Marathon.

Listing A-10 shows the startup code of the ASP.NET Core application, which enables and configures the various components, listed here:

- ASP.NET Core MVC.

- CORS to allow any web client to access the web API.

- Application Insights for monitoring and telemetry data. Application Insights is a cloud service from Microsoft.

- Disabling serialization of circular references in the JSON serializer. (Circular references are not standardized in JSON. There are community solutions, but you should avoid this if you do not need to be dependent on these community solutions.)

- Swagger for creating an OpenAPI specification and help pages for the REST operations.

***Listing A-10.*** Startup.cs

```
using BL;
using Microsoft.AspNetCore.Builder;
using Microsoft.AspNetCore.Diagnostics;
using Microsoft.AspNetCore.Hosting;
using Microsoft.AspNetCore.Http;
using Microsoft.AspNetCore.Mvc.Filters;
```

```csharp
using Microsoft.Extensions.Configuration;
using Microsoft.Extensions.DependencyInjection;
using Microsoft.Extensions.Logging;
using Microsoft.Extensions.PlatformAbstractions;
using Swashbuckle.AspNetCore.Swagger;
using System;
using System.IO;
using System.Threading.Tasks;

namespace Miraclelist
{

 public class Startup
 {
  public Startup(IHostingEnvironment env)
  {
   var builder = new ConfigurationBuilder()
       .SetBasePath(env.ContentRootPath)
       .AddJsonFile("appsettings.json", optional: true, reloadOnChange:
       true)
       .AddJsonFile($"appsettings.{env.EnvironmentName}.json", optional:
       true);

   if (env.IsEnvironment("Development"))
   {
   // This will push telemetry data through Application Insights pipeline
   faster, allowing you to view results immediately.
   builder.AddApplicationInsightsSettings(developerMode: true);
   // Connect to EFCore Profiler
HibernatingRhinos.Profiler.Appender.EntityFramework.
EntityFrameworkProfiler.Initialize();
   }

   builder.AddEnvironmentVariables();
   Configuration = builder.Build();

   // inject connection string into DAL
   DAL.Context.ConnectionString = Configuration.GetConnectionString("Miracle
   ListDB");
```

```
 #region testuser
 if (env.IsEnvironment("Development"))
 {
  var um2 = new UserManager("unittest", "unittest");
  um2.InitDefaultTasks();
 }
 #endregion
}

public IConfigurationRoot Configuration { get; }

/// <summary>
/// Called by ASP.NET Core during startup
/// </summary>
public void ConfigureServices(IServiceCollection services)
{
 #region Enable Auth service for MLToken in the HTTP header
 services.AddAuthentication().AddMLToken();
 #endregion

 #region Enable App Insights
 services.AddApplicationInsightsTelemetry(Configuration);
 #endregion

 #region JSON configuration: no circular references and ISO date format
 services.AddMvc().AddJsonOptions(options =>
 {
  options.SerializerSettings.ReferenceLoopHandling = Newtonsoft.Json.
  ReferenceLoopHandling.Ignore;
  options.SerializerSettings.PreserveReferencesHandling = Newtonsoft.
  Json.PreserveReferencesHandling.None;
  options.SerializerSettings.DateFormatHandling = Newtonsoft.Json.
  DateFormatHandling.IsoDateFormat;
 });
 #endregion

 #region Enable MVC
 services.AddMvc(options =>
```

```
{
 // Exception Filter
 options.Filters.Add(typeof(GlobalExceptionFilter));
 //options.Filters.Add(typeof(GlobalExceptionAsyncFilter));
 options.Filters.Add(typeof(LoggingActionFilter));
});
#endregion

#region Enable CORS
services.AddCors();
#endregion

// Make configuration available everywhere
services.AddSingleton(Configuration);

#region Swagger
services.AddSwaggerGen(c =>
{
 c.DescribeAllEnumsAsStrings(); // Important for Enums!

 c.SwaggerDoc("v1", new Info
 {
  Version = "v1",
  Title = "MiracleList API",
  Description = "Backend for MiracleList.de with token in URL",
  TermsOfService = "None",
  Contact = new Contact { Name = "Holger Schwichtenberg", Email = "",
  Url = "http://it-visions.de/kontakt" }
 });

 c.SwaggerDoc("v2", new Info
 {
  Version = "v2",
  Title = "MiracleList API",
  Description = "Backend for MiracleList.de with token in HTTP header",
  TermsOfService = "None",
  Contact = new Contact { Name = "Holger Schwichtenberg", Email = "",
  Url = "http://it-visions.de/kontakt" }
 });
```

```
    // Adds tokens as header parameters
    c.OperationFilter<SwaggerTokenHeaderParameter>();

    // include XML comments in Swagger doc
    var basePath = PlatformServices.Default.Application.
    ApplicationBasePath;
    var xmlPath = Path.Combine(basePath, "Miraclelist_WebAPI.xml");
    c.IncludeXmlComments(xmlPath);
  });
  #endregion
}

/// <summary>
/// Called by ASP.NET Core during startup
/// </summary>
public void Configure(IApplicationBuilder app, IHostingEnvironment env,
ILoggerFactory loggerFactory)
{

  #region Error handling

  app.UseExceptionHandler(errorApp =>
  {
    errorApp.Run(async context =>
    {
      context.Response.StatusCode = 500;
      context.Response.ContentType = "text/plain";

      var error = context.Features.Get<IExceptionHandlerFeature>();
      if (error != null)
      {
        var ex = error.Error;
        await context.Response.WriteAsync("ASP.NET Core Exception
        Middleware:" + ex.ToString());
      }
    });
  });
```

```
// --------------------------- letzte Fehlerbehandlung: Fehlerseite für
HTTP-Statuscode
app.UseStatusCodePages();

#endregion

#region ASP.NET Core services
app.UseDefaultFiles();
app.UseStaticFiles();
app.UseDirectoryBrowser();
loggerFactory.AddConsole(Configuration.GetSection("Logging"));
loggerFactory.AddDebug();
#endregion

#region CORS
// NUGET: install-Package Microsoft.AspNet.Cors
// Namespace: using Microsoft.AspNet.Cors;
app.UseCors(builder =>
 builder.AllowAnyOrigin()
        .AllowAnyHeader()
        .AllowAnyMethod()
        .AllowCredentials()
 );
#endregion

#region Swagger
// NUGET: Install-Package Swashbuckle.AspNetCore
// Namespace: using Swashbuckle.AspNetCore.Swagger;
app.UseSwagger(c =>
{
});

// Enable middleware to serve swagger-ui (HTML, JS, CSS etc.),
specifying the Swagger JSON endpoint.
app.UseSwaggerUI(c =>
{
 c.SwaggerEndpoint("/swagger/v1/swagger.json", "MiracleList v1");
 c.SwaggerEndpoint("/swagger/v2/swagger.json", "MiracleList v2");
```

```csharp
        });
        #endregion

        #region  MVC with Routing
        app.UseMvc(routes =>
      {
      routes.MapRoute(
                name: "default",
                template: "{controller}/{action}/{id?}",
                defaults: new { controller = "Home", action = "Index" });
      });
        #endregion

      }
    }

    public class GlobalExceptionFilter : IExceptionFilter
    {
      public void OnException(ExceptionContext context)
      {
        if (context.Exception is UnauthorizedAccessException)
        {
          context.HttpContext.Response.StatusCode = 403;
        }
        else
        {
          context.HttpContext.Response.StatusCode = 500;
        }
        context.HttpContext.Response.ContentType = "text/plain";
        context.HttpContext.Response.WriteAsync("GlobalExceptionFilter:" +
        context.Exception.ToString());
      }
    }

    public class GlobalExceptionAsyncFilter : IAsyncExceptionFilter
    {
      public Task OnExceptionAsync(ExceptionContext context)
      {
```

```
context.HttpContext.Response.StatusCode = 500;
context.HttpContext.Response.ContentType = "text/plain";
return context.HttpContext.Response.WriteAsync("MVC
GlobalExceptionAsyncFilter:" + context.Exception.ToString());
 }
 }
}
```

Listing A-11 shows the implementation of the WebAPI controller in version 1 of the REST service, including the use of Application Insights to collect telemetry data. The WebAPI controller is completely free of data access code. So, Entity Framework Core is not being used here. All data operations are encapsulated in the business logic layer. The WebAPI controller only uses the manager classes implemented there.

***Listing A-11.*** MiracleListApiController.cs (Version 1 of the REST Service)

```
using BL;
using BO;
using ITVisions;
using Microsoft.ApplicationInsights;
using Microsoft.AspNetCore.Mvc;
using System;
using System.Collections.Generic;
using System.Linq;
using System.Reflection;
using System.Runtime.CompilerServices;

namespace Miraclelist.Controllers
{
 /// <summary>
 /// DTO
 /// </summary>
 public class LoginInfo
 {
  public string ClientID;
  public string Username;
  public string Password;
```

```csharp
 public string Token;
 public string Message;
}

/// <summary>
/// API v1
/// </summary>
[Route("")]
[ApiExplorerSettings(GroupName = "v1")]
public class MiracleListApiController : Controller
{
 private TelemetryClient telemetry = new TelemetryClient();
 TaskManager tm;
 UserManager um;
 CategoryManager cm;

 public MiracleListApiController()
 {
 }

 /// <summary>
 /// Helper for all actions to check the token and save telemetry data
 /// </summary>
 private bool CheckToken(string token, [CallerMemberName] string caller = "?")
 {
  if (token == null || token.Length < 2)
  {
   // save telemetry data
   var p2 = new Dictionary<string, string>();
   p2.Add("token", token);
   telemetry.TrackEvent("TOKENERROR_" + caller, p2);
   new LogManager().Log(Event.TokenCheckError, Severity.Warning,
   "Ungültiges Token", caller, token);
   throw new Exception("Ungültiges Token!");
  }
```

```csharp
// validate tokne
um = new UserManager(token);
var checkResult = um.IsValid();
if (checkResult != UserManager.TokenValidationResult.Ok)
{
 // save telemetry data
 var p2 = new Dictionary<string, string>();
 p2.Add("token", token);
 p2.Add("checkResult", checkResult.ToString());
 telemetry.TrackEvent("USERERROR_" + caller, p2);

 new LogManager().Log(Event.TokenCheckError, Severity.Warning,
 checkResult.ToString(), caller, token, um.CurrentUser?.UserID);

 throw new Exception(checkResult.ToString());
}
um.InitDefaultTasks();

// Create manager objects
cm = new CategoryManager(um.CurrentUser.UserID);
tm = new TaskManager(um.CurrentUser.UserID);

// save telemetry data
var p = new Dictionary<string, string>();
p.Add("token", token);
p.Add("user", um.CurrentUser.UserName);
telemetry.TrackEvent(caller, p);

new LogManager().Log(Event.TokenCheckOK, Severity.Information, null,
caller, token, um.CurrentUser?.UserID);
 return true;
}

/// <summary>
/// About this server
/// </summary>
/// <returns></returns>
[Route("/About")]
[HttpGet]
```

```
public IEnumerable<string> About()
{
 return new AppManager().GetAppInfo().Append("API-Version: v1");
}

/// <summary>
/// Get version of server
/// </summary>
/// <returns></returns>
[Route("/Version")]
[HttpGet]
public string Version()
{
 return
 Assembly.GetEntryAssembly()
.GetCustomAttribute<AssemblyInformationalVersionAttribute>()
.InformationalVersion.ToString();
}

/// <summary>
/// Nur für einen Test
/// </summary>
/// <returns></returns>
[Route("/About2")]
[ApiExplorerSettings(IgnoreApi = true)]
[HttpGet]
public JsonResult GetAbout2()
{
 var v = Assembly.GetEntryAssembly().GetCustomAttribute<Assembly
 InformationalVersionAttribute>().InformationalVersion;
 var e = new string[] { "MiracleListBackend", "(C) Dr. Holger
 Schwichtenberg, www.IT-Visions.de", "Version: " + v };
 var r = new JsonResult(e);
 this.Response.Headers.Add("X-Version", v);
 r.StatusCode = 202;
 return r;
}
```

```
/// <summary>
/// Login with a client ID, username and password. This operation sends
back a GUID as a session token, to be used in all following operations.
/// </summary>
[HttpPost("Login")] // neu
public async System.Threading.Tasks.Task<LoginInfo> Login([FromBody]
LoginInfo loginInfo)
{
 if (string.IsNullOrEmpty(loginInfo.Password))
 {
  new LogManager().Log(Event.LogginError, Severity.Warning, "", "password
  empty");
  throw new Exception("ERROR: password empty!");
 }

 var cm = new ClientManager();
 var e = cm.CheckClient(loginInfo.ClientID);
 if (e.CheckClientResultCode != ClientManager.CheckClientResultCode.Ok)
 {
  new LogManager().Log(Event.LogginError, Severity.Warning,
  Enum.GetName(typeof(ClientManager.CheckClientResultCode),
  e.CheckClientResultCode) + "\n" + e.client?.ToNameValueString(),
  "ClientIDCheck", "", um?.CurrentUser?.UserID);
  return new LoginInfo()
  {
   Message = "Client-ID-Check: " + Enum.GetName(typeof(ClientManager.
   CheckClientResultCode), e.CheckClientResultCode)
  };
 }

 var u = new UserManager(loginInfo.Username, loginInfo.Password).
 CurrentUser;
 if (u == null)
 {
  new LogManager().Log(Event.LogginError, Severity.Warning, loginInfo.
  ToNameValueString() + "\n" + e.client?.ToNameValueString(),
  "UserCheck", u?.Token, um?.CurrentUser?.UserID);
```

583

```csharp
    return new LoginInfo() { Message = "Access denied!" };
  }
  loginInfo.Token = u.Token;
  new LogManager().Log(Event.LoginOK, Severity.Information, null,
  "UserCheck", u.Token, u.UserID);
  loginInfo.Password = "";
  return loginInfo;
}

/// <summary>
/// Delete token
/// </summary>
[HttpGet("Logoff/{token}")]
public bool Logoff(string token)
{
  return UserManager.Logoff(token);
}

/// <summary>
/// Get a list of all categories
/// </summary>
[HttpGet("CategorySet/{token}")]
public IEnumerable<Category> GetCategorySet(string token)
{
  if (!CheckToken(token)) return null;
  return cm.GetCategorySet();
}

/// <summary>
/// Get a list of tasks in one category
/// </summary>
[HttpGet("TaskSet/{token}/{id}")]
public IEnumerable<Task> GetTaskSet(string token, int id)
{
  if (id <= 0) throw new Exception("Invalid ID!");
  if (!CheckToken(token)) return null;
  return tm.GetTaskSet(id);
}
```

```csharp
/// <summary>
/// Get details of one task
/// </summary>
[HttpGet("Task/{token}/{id}")]
public Task Task(string token, int id)
{
 if (id <= 0) throw new Exception("Invalid ID!");
 if (!CheckToken(token)) return null;
 return tm.GetTask(id);
}

/// <summary>
/// Search in tasks and subtasks
/// </summary>
[HttpGet("Search/{token}/{text}")]
public IEnumerable<Category> Search(string token, string text)
{
 if (!CheckToken(token)) return null;
 return tm.Search(text);
}

/// <summary>
/// Returns all tasks due, including tomorrow, grouped by category,
sorted by date
/// </summary>
[HttpGet("DueTaskSet/{token}")]
public IEnumerable<Category> GetDueTaskSet(string token)
{
 if (!CheckToken(token)) return null;
 return tm.GetDueTaskSet();
}

/// <summary>
/// Create a new category
/// </summary>
[HttpPost("CreateCategory/{token}/{name}")]
public Category CreateCategory(string token, string name)
```

```
  {
   if (!CheckToken(token)) return null;
   return cm.CreateCategory(name);
  }

  /// <summary>
  /// Create a task to be submitted in body in JSON format (including
  subtasks)
  /// </summary>
  /// <param name="token"></param>
  /// <param name="t"></param>
  /// <returns></returns>
  [HttpPost("CreateTask/{token}")] // neu
  public Task CreateTask(string token, [FromBody]Task t)
  {
   if (!CheckToken(token)) return null;
   return tm.New(t);
  }

  /// <summary>
  /// Create a task to be submitted in body in JSON format (including
  subtasks)
  /// </summary>
  [HttpPut("ChangeTask/{token}")] // geändert
  public Task ChangeTask(string token, [FromBody]Task t)
  {
   if (!CheckToken(token)) return null;
   return tm.ChangeTask(t);
  }

  /// <summary>
  /// Set a task to "done"
  /// </summary>
  [HttpPut("ChangeTaskDone/{token}")]
  public Task ChangeTaskDone(string token, int id, bool done)
  {
   throw new UnauthorizedAccessException("du kommst hier nicht rein!");
  }
```

```
/// <summary>
/// Change a subtask
/// </summary>
[HttpPut("ChangeSubTask/{token}")]
public SubTask ChangeSubTask(string token, [FromBody]SubTask st)
{
  throw new UnauthorizedAccessException("du kommst hier nicht rein!");
}

/// <summary>
/// Delete a task with all subtasks
/// </summary>
[HttpDelete("DeleteTask/{token}/{id}")]
public void DeleteTask(string token, int id)
{
  if (!CheckToken(token)) return;
  tm.RemoveTask(id);
}

/// <summary>
/// Delete a category with all tasks and subtasks
/// </summary>
[HttpDelete("[action]/{token}/{id}")]
public void DeleteCategory(string token, int id)
{
  if (!CheckToken(token)) return;
  cm.RemoveCategory(id);
 }
 }
}
```

# Using Entity Framework Core via Dependency Injection

When creating a new ASP.NET Core application in Visual Studio with the option Individual User Accounts or Store User Accounts In-App, Entity Framework Core also creates an Entity Framework Core context (`ApplicationDbContext`), which is created by the base class `Microsoft.AspNetCore.Identity.EntityFrameworkCore.`

`IdentityDbContext<T>` inherits and uses the class `ApplicationUser` as the type parameter, which in turn is provided by `Microsoft.AspNetCore.Identity`. `IdentityUser` inherits. The class `ApplicationUser` can be extended if required. In addition, a schema migration is created. The connection string is stored in `appsettings.json`. It points to a local database (Microsoft SQL Server `LocalDB`).

```
{
  "ConnectionStrings": {
    "DefaultConnection": "Server=(localdb)\\mssqllocaldb;Database=aspnet-
    ASPNETCore20-53bc9b9d-9d6a-45d4-8429-2a2761773502;Trusted_Connection=Tr
    ue;MultipleActiveResultSets=true"
  }
}
```

You can change a connection string if necessary. After an `Update-Database` command, the database is created to manage the local users (Figure A-7).

***Figure A-7.*** *Created database for managing users of the ASP.NET Core web application*

There is no instantiation of the context class `ApplicationDbContext` in the web application. Rather, this is used by dependency injection. In `Startup.cs`, you will find two lines (see Listing A-12).

- The `AddDbContext()` extension method registers the context class as a dependency injection service and passes a provider and connection string. The method `AddDbContext()` is provided by `Microsoft.EntityFrameworkCore.dll` in the Microsoft class `Microsoft.Extensions.DependencyInjection.EntityFrameworkServiceCollectionExtensions`.

- The `AddEntityFrameworkStores()` extension method tells the ASP.NET Identity component which context class to use. `AddEntityFrameworkStores()` is provided by `Microsoft.AspNetCore.Identity.EntityFrameworkCore.dll` in the Microsoft class `Extensions.Dependency injection.IdentityEntityFrameworkBuilderExtensions`.

---

**Note**    ASP.NET Identity ensures that the context class is instantiated when needed and, in addition, lives no longer than it takes to process an HTTP request.

---

*Listing A-12.*    An Extract from Startup.cs, an ASP.NET Core Web Application Created with the Project Template in Visual Studio, with Individual Local User Accounts

```
public void ConfigureServices(IServiceCollection services)
{
services.AddDbContext<ApplicationDbContext>(options =>
            options.UseSqlServer(Configuration.GetConnectionString("Default
            Connection")));

services.AddIdentity<ApplicationUser, IdentityRole>()
   .AddEntityFrameworkStores<ApplicationDbContext>()
   .AddDefaultTokenProviders();
}
```

> **Tip**   You can also use `AddDbContext()` for your own context classes. In this case, you register the context class in the startup class with `AddDbContext()`, as shown here:
>
> ```
> services.AddDbContext<EFCore_Kontext.WWWingsContext>(options =>
>
> options.UseSqlServer(Configuration.GetConnectionString("WWWings
> Connection")));
> ```

When using the standard dependency injection component of ASP.NET Core (`Microsoft.Extensions.DependencyInjection`), dependency injection can be done only via constructor injection (see Listing A-13).

***Listing A-13.*** The Class FlightManager Receives the Context Instance via Dependency Injection

```
using GO;
using EFCore_Context;
using System.Collections.Generic;
using System.Linq;

namespace ASPNETCore_NETCore.BL
{
public class FlightManager
{
  private WWWingsContext ctx;

  /// <summary>
  /// constructor
  /// </ summary>
  /// <param name="ctx">context instance comes via DI!</ Param>
  public FlightManager(WWWingsContext ctx)
  {
   this.ctx = ctx;
  }

  public List<Flight> GetFlightSet string departure, int from, int to)
  {
```

```
    var FlightSet = ctx.Flight
      .Where(x => x.Departure == departure)
      .Skip(from).Take(to - from).ToList();
    return FlightSet.ToList();
  }
}
}
```

Note, however, that ASP.NET Core injects the context instance only to the constructor of the FlightManager class if the instance of the FlightManager class itself is generated by the dependency injection (DI) container. To do this, the class FlightManager must be registered in the startup class for dependency injection with AddTransient().

```
ServicesAddTransient <Flight manager>();
```

AddTransient() causes a new instance of FlightManager to be generated each time an instance is requested. AddScoped() would ensure that the same instance is always returned as part of an HTTP request; this may be desired because then the cache of the Entity Framework core context is filled. AddSingleton() would always provide the same instance across multiple HTTP requests. This cannot work because the Entity Framework core context does not support multithreading.

An ASP.NET MVC controller then expects an instance of FlightManager via constructor injection (Listing A-14).

***Listing A-14.*** The WWWingsController Class Receives an Instance of FlightManager via Dependency Injection

```
public class WWWingsController: Controller
{
  private FlightManager fm;
  public WWWingsController(FlightManager fm)
  {
   this.fm = fm;
  }
  ...
}
```

# Practical Example: Context Instance Pooling (DbContext Pooling)

Ever since Entity Framework Core 2.0, you can use `AddDbContextPool()` instead of `AddDbContext()`. This method creates a set of context instances managed in a pool, similar to the ADO.NET connection pool. When dependency injection asks for a context instance, a free context instance is taken out of the pool. Entity Framework Core can also reset an already used context instance and release it for reuse. This somewhat improves the performance of web applications that use Entity Framework Core.

```
int poolSize = 40;
services.AddDbContextPool<ApplicationDbContext>(options => options.
UseSqlServer(Configuration.GetConnectionString("DefaultConnection")),
poolSize);
```

# Using Entity Framework Core in a Universal Windows Platform App

The MiracleList Light app used in this section is a sample application for simple local task management, implemented as a Universal Windows Platform (UWP) app for Windows 10 with a local SQLite database as the data store and no back end in the cloud (Figure A-8). This app allows you to save tasks that are due on a specific date. In the current version of the software, tasks always have exactly three subtasks (planning, execution, retrospective) that cannot be changed. Tasks can be removed from the list by clicking Done or Remove All.

---

**Note**   To use Entity Framework Core 2.0 in a UWP, you need UWP version 10.0.16299 in the Windows 10 Creators Fall 2017 update.

---

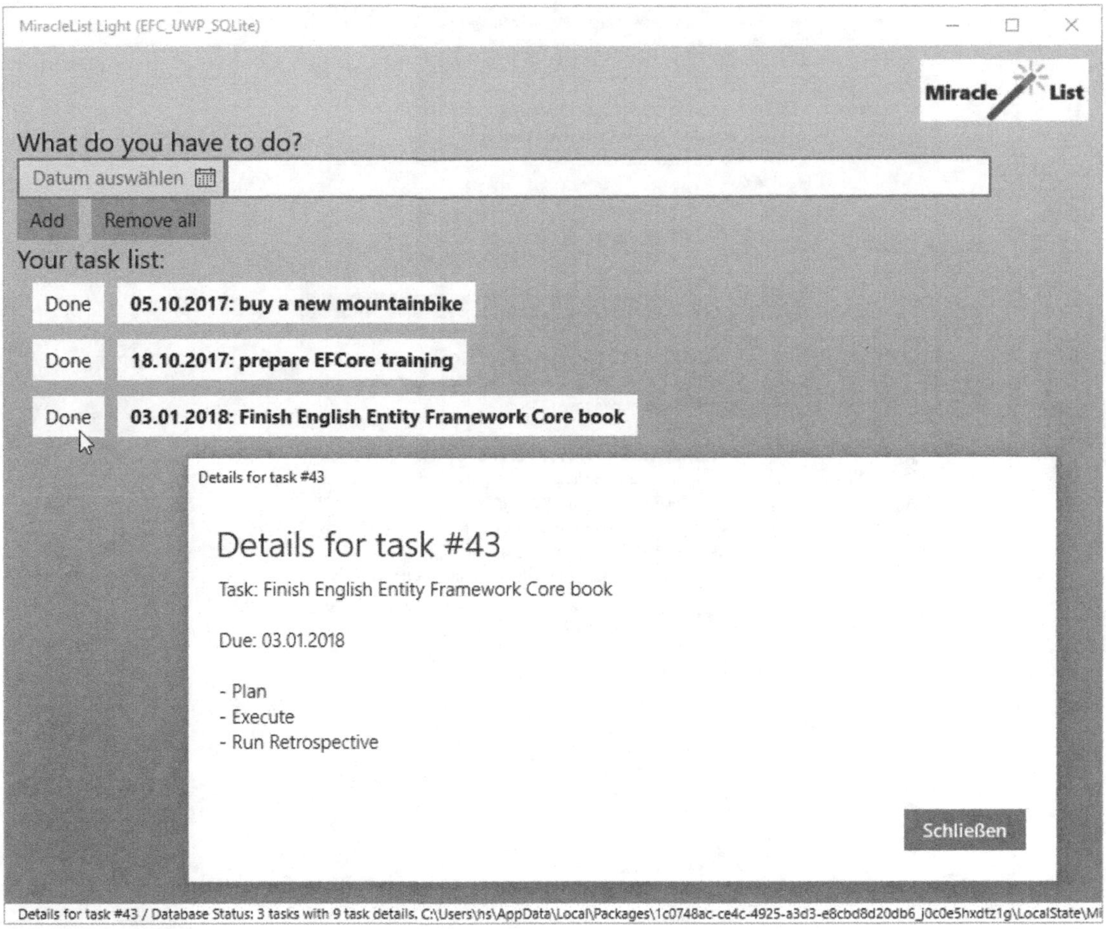

***Figure A-8.*** *The MiracleList Light app for Windows 10*

## Architecture

The app is implemented as a monolithic app in a Visual Studio project because of the limited code size. The project was created with the project template Windows Universal/ Blank App. For this template, the Windows 10 SDK must be installed in the version appropriate for your Windows installation. Using or compiling the program on an older operating systems is not supported by Microsoft.

The app references the NuGet package `Microsoft.EntityFrameworkCore.Sqlite`.

The app uses forward engineering for the database. If necessary, the database file with the appropriate database schema is generated at runtime if it does not exist when the app starts. Figure A-9 shows the structure of the project.

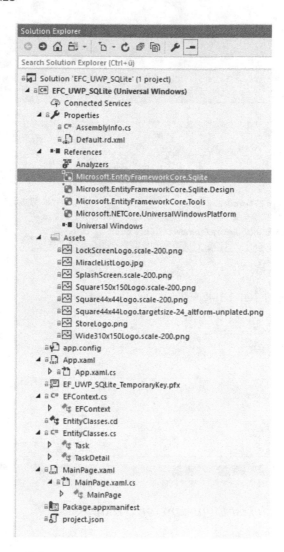

***Figure A-9.*** *Structure of the project*

# Entity

There are only two entity classes necessary.

- *A Task class for a task*: A `Task` object has a `List<TaskDetail>` in the `Details` property.

- *TaskDetail class for a subtask*: Each `TaskDetail` object uses the `Task` property to point to the task to which the subtask belongs. In addition, the `TaskDetail` class in the `TaskID` foreign key property knows the primary key of the parent task.

Data annotations of entity classes are not necessary in this case because Entity Framework Core can create the database schema based purely on the built-in conventions. Figure A-10 shows the object model of the application, and Listing A-15 shows the implementation.

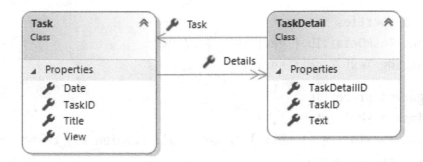

***Figure A-10.*** *Object model of the application*

***Listing A-15.*** Implementation of the Two Entity Classes in the EntityClasses.cs File

```csharp
using System;
using System.Collections.Generic;

namespace EFC_UWP_SQLite
{
 /// <summary>
 /// Entity class for tasks
 /// </summary>
 public class Task
 {
  // Basic properties
  public int TaskID { get; set; } // PK
  public string Title { get; set; } // TEXT
  public DateTime Date { get; set; } // DateTime

  // Navigation properties
  public List<TaskDetail> Details { get; set; } = new List<TaskDetail>();

  public string View { get { return Date.ToString("d") + ": " + Title; } }
 }
```

```
/// <summary>
/// Entity class for subtasks
/// </summary>
public class TaskDetail
{
 // Basic properties
 public int TaskDetailID { get; set; } // PK
 public string Text { get; set; }

 // Navigation properties
 public Task Task { get; set; }
 public int TaskID { get; set; } // optional: Foreign key column for
 navigation relationship
 }
}
```

# Entity Framework Core Context Class

The context class derived from DbContext always has a property of type DbSet<T> for the two entity classes. In the OnConfiguring() method, UseSqlite() sets the Entity Framework Core database provider, passing in as a parameter only the name you want for the SQLite database file. An implementation of OnModelCreating() is not necessary in this case because Entity Framework Core can create the database schema based purely on the built-in conventions (Listing A-16).

***Listing A-16.*** Implementation of the Context Class in the File EFContext.cs

```
using Microsoft.EntityFrameworkCore;

namespace EFC_UWP_SQLite
{
 /// <summary>
 /// Entity Framework core context
 /// </summary>
 public class EFContext : DbContext
 {
  public static string FileName = "MiracleList.db";
```

```
public DbSet<Task> TaskSet { get; set; }
public DbSet<TaskDetail> TaskDetailSet { get; set; }

protected override void OnConfiguring(DbContextOptionsBuilder
optionsBuilder)
{
  // Set provider and database filename
  optionsBuilder.UseSqlite($"Filename={FileName}");
}
}
}
```

# Start Code

When the application starts, the `App.xaml.cs` database file uses the `Database` method. `EnsureCreated()` is created if the database file does not yet exist. The English source code comments come from the project template of Microsoft. See Listing A-17.

***Listing A-17.*** Extract from the App.xaml.cs File

```
using System;
using Windows.ApplicationModel;
using Windows.ApplicationModel.Activation;
using Windows.Foundation;
using Windows.UI.Xaml;
using Windows.UI.Xaml.Controls;
using Windows.UI.Xaml.Navigation;

namespace EFC_UWP_SQLite
{
 /// <summary>
 /// Provides application-specific behavior to supplement the default
 Application class.
 /// </summary>
 sealed partial class App : Application
 {
  /// <summary>
```

```
/// Initializes the singleton application object.  This is the first line
of authored code
/// executed, and as such is the logical equivalent of main() or WinMain().
/// </summary>
public App()
{
 this.InitializeComponent();
 this.Suspending += OnSuspending;

 // Create DB, if not exists!
 using (var db = new EFContext())
 {
  db.Database.EnsureCreated();
 }
}
...
 }
}
```

## Generated Database

The generated database can be made visible and used interactively with the tool DB Browser for SQLite, as shown in Figure A-11, Figure A-12, and Figure A-13.

| Tool name | DB Browser for SQLite |
|---|---|
| Web site | www.sqlitebrowser.org |
| Free version | Yes |
| Commercial version | No |

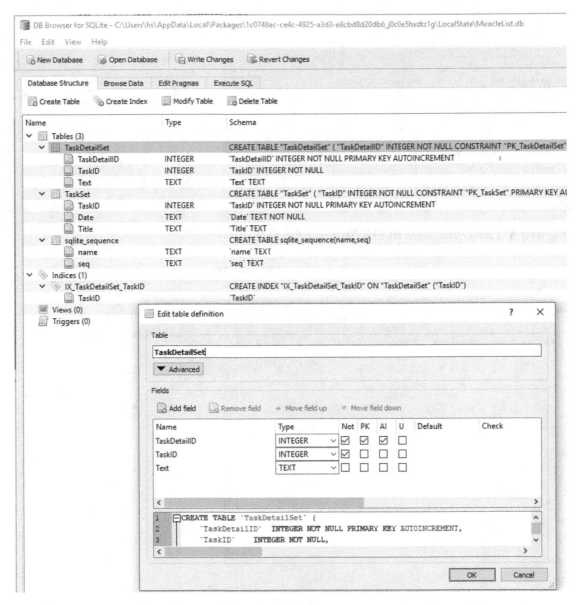

***Figure A-11.*** *Database schema view in DB Browser for SQLite*

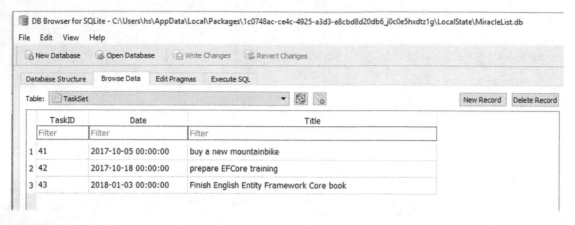

**Figure A-12.** *Data view in DB Browser for SQLite*

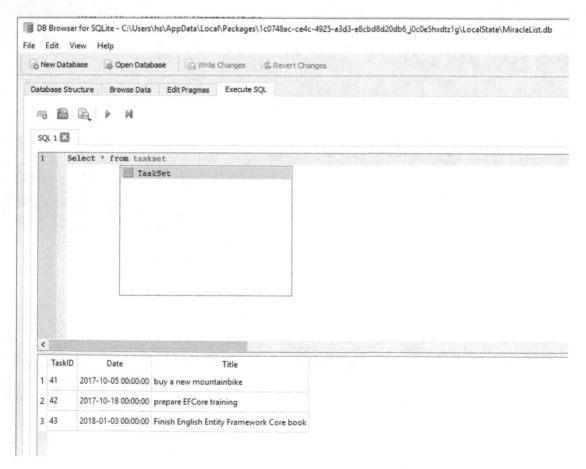

**Figure A-13.** *Executing SQL commands in DB Browser for SQLite*

# Data Access Code

The data access code is not separate from the presentation layer in this simple case study. It also deliberately does not use the Model-View-ViewModel (MVVM) pattern, what will keep the program code in this book manageable.

The data access uses the asynchronous methods of Entity Framework Core to keep the UI responsive.

When creating subtasks and deleting all tasks, two variants are shown (Listing A-18). The commented-out variant is the less efficient one. Thus, deleting all tasks and subtasks without using SQL involves unnecessarily loading all tasks and sending a DELETE command for each task. Explicit deletion of the subtasks is not necessary in both cases since cascading deletion is active in the standard system.

***Listing A-18.*** Data Access Code in the MainPage.xaml.cs File

```
using Microsoft.EntityFrameworkCore;
using System;
using System.Collections.Generic;
using System.Collections.ObjectModel;
using System.ComponentModel;
using System.Linq;
using Windows.Foundation;
using Windows.Storage;
using Windows.UI.Popups;
using Windows.UI.Xaml;
using Windows.UI.Xaml.Controls;
using Windows.UI.Xaml.Input;

namespace EFC_UWP_SQLite
{
 /// <summary>
 /// Main page of the app
 /// </summary>
 public sealed partial class MainPage : Page, INotifyPropertyChanged
 {
  public MainPage()
  {
```

```csharp
  this.DataContext = this;
  this.InitializeComponent();
  Windows.UI.ViewManagement.ApplicationView.PreferredLaunchViewSize = new
  Size(800, 500);
  Windows.UI.ViewManagement.ApplicationView.PreferredLaunchWindowingMode =
  Windows.UI.ViewManagement.ApplicationViewWindowingMode.
  PreferredLaunchViewSize;

  System.Diagnostics.Debug.WriteLine(ApplicationData.Current.LocalFolder.
  Path);
}
public event PropertyChangedEventHandler PropertyChanged;

private ObservableCollection<Task> _Tasks { get; set; }
public ObservableCollection<Task> Tasks
{
 get { return _Tasks; }
 set { _Tasks = value; this.PropertyChanged?.Invoke(this, new Property
 ChangedEventArgs(nameof(Tasks))); }
}

private string _Statustext { get; set; }
public string Statustext
{
 get { return _Statustext; }
 set { _Statustext = value; this.PropertyChanged?.Invoke(this, new Proper
 tyChangedEventArgs(nameof(Statustext))); }
}

private async void Page_Loaded(object sender, RoutedEventArgs e)
{
 var count = await this.LoadTaskSet();
 SetStatus(count + " records loaded!");
}

private void SetStatus(string text)
{
 string dbstatus;
```

```csharp
 using (var db = new EFContext())
 {
  dbstatus = db.TaskSet.Count() + " tasks with " + db.TaskDetailSet.
  Count() + " task details. " + ApplicationData.Current.LocalFolder.Path
  + @"\" + EFContext.FileName;
 }
 Statustext = text + " / Database Status: " + dbstatus + ")";
}

/// <summary>
/// Get all tasks from database
/// </summary>
/// <returns></returns>
private async System.Threading.Tasks.Task<int> LoadTaskSet()
{
 using (var db = new EFContext())
 {
  var list = await db.TaskSet.OrderBy(x => x.Date).ToListAsync();
  Tasks = new ObservableCollection<Task>(list);
  return Tasks.Count;
 }
}

private async void Add(object sender, RoutedEventArgs e)
{
 if (String.IsNullOrEmpty(C_Task.Text)) return;
 if (!C_Date.Date.HasValue) { C_Date.Date = DateTime.Now; }

 // Create new Task
 var t = new Task { Title = C_Task.Text, Date = C_Date.Date.Value.Date };
 var d1 = new TaskDetail() { Text = "Plan" };
 var d2 = new TaskDetail() { Text = "Execute" };
 var d3 = new TaskDetail() { Text = "Run Retrospective" };
 // Alternative 1
 //t.Details.Add(d1);
 //t.Details.Add(d2);
 //t.Details.Add(d3);
```

```csharp
  // Alternative 2
  t.Details.AddRange(new List<TaskDetail>() { d1, d2, d3 });

  using (var db = new EFContext())
  {
   db.TaskSet.Add(t);
   // Save now!
   var count = await db.SaveChangesAsync();
   SetStatus(count + " records saved!");
   await this.LoadTaskSet();
  }
  this.C_Task.Text = "";
  this.C_Task.Focus(FocusState.Pointer);
 }

 private async void SetDone(object sender, RoutedEventArgs e)
 {
  // Get TaskID
  var id = (int)((sender as Button).CommandParameter);
  // Remove record
  using (var db = new EFContext())
  {
   Task t = db.TaskSet.SingleOrDefault(x => x.TaskID == id);
   if (t == null) return; // not found :-(
   db.Remove(t);
   var count = db.SaveChangesAsync();
   SetStatus(count + " records deleted!");
   await this.LoadTaskSet();
  }
 }

 private async void ShowDetails(object sender, RoutedEventArgs e)
 {
  // Get TaskID
  var id = (int)((sender as Button).CommandParameter);
  // Get Details
  using (var db = new EFContext())
```

```
  {
   string s = "";
   Task t = db.TaskSet.Include(x => x.Details).SingleOrDefault(x =>
   x.TaskID == id);
   s += "Task: " + t.Title + "\n\n";
   s += "Due: " + t.Date.Date + "\n\n";
   foreach (var d in t.Details)
   {
    s += "- " + d.Text + "\n";
   }
   SetStatus("Details for task #" + id);
   await new MessageDialog(s, "Details for task #" + id).ShowAsync();
  }
}

private void C_Task_KeyDown(object sender, KeyRoutedEventArgs e)
{
 if (e.Key == Windows.System.VirtualKey.Enter) Add(null, null);
}

private async void RemoveAll(object sender, RoutedEventArgs e)
{
 // Remove all tasks
 using (var db = new EFContext())
 {
  // Alternative 1: unefficient :-(
  //foreach (var b in db.TaskSet.ToList())
  //{
  // db.Remove(b);
  //}
  //db.SaveChanges();

  // Alternative 2: efficient!
  //db.Database.ExecuteSqlCommand("Delete from TaskDetailSet");
  var count = await db.Database.ExecuteSqlCommandAsync("Delete from
  TaskSet");
  SetStatus(count + " records deleted!");
```

```
    Tasks = null;
   }
  }
 }
}
```

# User Interface

Listing A-19 shows the XAML UWP app interface.

***Listing A-19.***  MainPage.xaml

```xml
<Page
    x:Class="EFC_UWP_SQLite.MainPage"
    xmlns="http://schemas.microsoft.com/winfx/2006/xaml/presentation"
    xmlns:x="http://schemas.microsoft.com/winfx/2006/xaml"
    xmlns:local="using:EFC_UWP_SQLite"
    xmlns:d="http://schemas.microsoft.com/expression/blend/2008"
    xmlns:mc="http://schemas.openxmlformats.org/markup-compatibility/2006"
    mc:Ignorable="d"    Loaded="Page_Loaded">

 <Grid Margin="0,0,0,0">
  <Grid.RowDefinitions>
   <RowDefinition Height="*"></RowDefinition>
   <RowDefinition Height="auto"></RowDefinition>
  </Grid.RowDefinitions>
  <Grid.Background>
   <LinearGradientBrush EndPoint="0.5,1" StartPoint="0.5,0">
    <GradientStop Color="#FFA1D7E9" Offset="0.081"/>
    <GradientStop Color="#FF4C94AD" Offset="0.901"/>
   </LinearGradientBrush>
  </Grid.Background>
  <StackPanel Margin="10,10,10,10" Grid.Row="0">
   <!-- ==================== logo -->
   <Image x:Name="Logo" Source="Assets/MiracleListLogo.jpg" Width="130"
   MinHeight="50" HorizontalAlignment="Right"></Image>
```

```xml
<!-- ===================== new Task -->
<TextBlock Text="What do you have to do?" FontSize="20"></TextBlock>
<StackPanel Orientation="horizontal">
 <CalendarDatePicker Name="C_Date" />
 <TextBox Background="White" Name="C_Task" KeyDown="C_Task_KeyDown"
 Width="600"></TextBox>
</StackPanel>
<!-- ===================== actions -->
<StackPanel Orientation="horizontal">
 <Button Click="Add">Add</Button>
 <Button Click="RemoveAll" Margin="10,0,0,0">Remove all</Button>
</StackPanel>
<TextBlock Text="Your task list:" FontSize="20"/>
<!-- ===================== list of tasks -->
<ListView ItemsSource="{Binding Tasks}" ScrollViewer.VerticalScrollBar
Visibility="Visible">
 <ListView.ItemTemplate>
  <DataTemplate>
   <StackPanel Orientation="Horizontal">
    <Button  Background="white"  Content="Done" Name="C_Done"
    CommandParameter="{Binding TaskID}"  Click="SetDone"
    Margin="0,0,10,0" />
    <Button Background="white" FontWeight="Bold" Content="{Binding
    View}" Name="C_Details"  CommandParameter="{Binding TaskID}"
    Click="ShowDetails" />
   </StackPanel>
  </DataTemplate>
 </ListView.ItemTemplate>
</ListView>
</StackPanel>
<!-- ===================== statusbar -->
<StackPanel Background="White" Grid.Row="1">
 <TextBlock Text="{Binding Statustext}" Margin="10,0,0,0"  FontSize="11" />
</StackPanel>
</Grid>
</Page>
```

# Using Entity Framework Core in a Xamarin Cross-Platform App

The MiracleList Light case study in this section is a portrayal of the Universal Windows Platform (UWP) app for Windows 10 that was discussed in the previous section. Here, it's a cross-platform app that is not limited to Windows 10 as the UWP app; it also runs on Android and iOS (Figure A-14 and Figure A-15). For the GUI, Xamarin.Forms is used.

> **Note**    For Entity Framework Core 2.0, you need UWP version 10.0.16299 in the Windows 10 Creators Fall 2017 update.

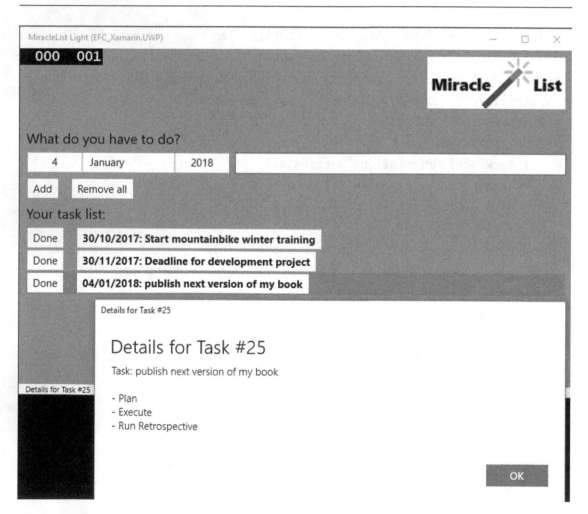

*Figure A-14.*  *The MiracleList Light cross-platform app for Windows 10*

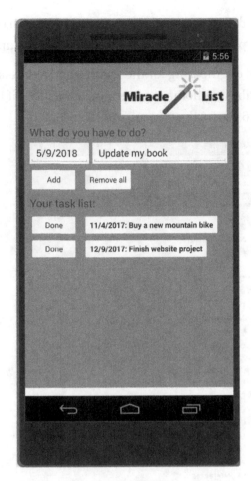

**Figure A-15.** *The MiracleList Light cross-platform app for Android*

# Architecture

Unlike the implementation of this app as a UWP app, this cross-platform version is multilayered.

- The BO project contains the entity classes. The project is a .NET standard library and does not require additional references.

- The DAL project includes the Entity Framework Core context class. The project is a .NET standard library and requires the NuGet packages `Xamarin.Forms` and `Microsoft.EntityFrameworkCore`.

- The UI project contains the UI that uses Xamarin Forms. The project is a .NET standard library and requires the NuGet packages `Xamarin.Forms` and `Microsoft.EntityFrameworkCore`.

- The projects Android, iOS, and UWP contain the platform-specific start code as well as platform-specific declarations for the app.

The app uses forward engineering for the database. If necessary, the database file with the appropriate database schema is generated at runtime if it does not exist when the app starts. Figure A-16 shows the structure of the project.

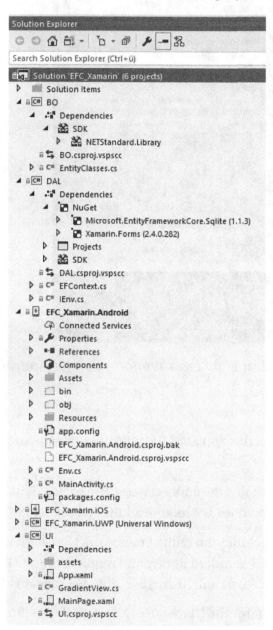

***Figure A-16.*** *Structure of the project*

# Entity

The entity classes for the Xamarin app correspond to the entity classes from the UWP case study.

# Entity Framework Core Context Class

The context class for the Xamarin app differs slightly from the context class of the UWP case study because not every operating system can be specified in the connection string in UseSQLite() simply as a file name without a path. There are platform-specific differences to consider. Therefore, the path for the database file is supplied by dependency injection in an implementation of the self-defined interface IEnv of each of the three frame applications (Listing A-20 and Listing A-21).

---

**Note**  The DAL library requires the NuGet package Xamarin.Forms because dependency injection uses the dependency injection framework built into Xamarin Forms.

---

***Listing A-20.***  Implementation of the Context Class in the File EFContext.cs

```
using Microsoft.EntityFrameworkCore;
using Xamarin.Forms;

namespace EFC_Xamarin
{
 /// <summary>
 /// Entity Framework context
 /// </summary>
 public class EFContext : DbContext
 {

  static public string Path { get; set; }
  public DbSet<Task> TaskSet { get; set; }
  public DbSet<TaskDetail> TaskDetailSet { get; set; }

  protected override void OnConfiguring(DbContextOptionsBuilder
  optionsBuilder)
```

611

```
  {
    EFContext.Path = System.IO.Path.Combine(DependencyService.Get<IEnv>().
    GetDbFolder(), "miraclelist.db");
    // set provider and database file path
    optionsBuilder.UseSqlite($"Filename={ EFContext.Path}");
  }
 }
}
```

***Listing A-21.*** IEnv.cs

```
namespace EFC_Xamarin
{
 /// <summary>
 /// Custom Interface for getting the OS specific folder for the DB file
 /// </summary>
 public interface IEnv
 {
  string GetDbFolder();
 }
}
```

Every operating system has to inject a suitable implementation (Listing A-22, Listing A-23, and Listing A-24).

***Listing A-22.*** Implementation of IEnv on Windows 10 UWP

```
using EFC_Xamarin.UWP;
using Windows.Storage;
using Xamarin.Forms;

[assembly: Dependency(typeof(Env))]
namespace EFC_Xamarin.UWP
{
 public class Env : IEnv
 {
  public string GetDbFolder()
  {
```

```
    return ApplicationData.Current.LocalFolder.Path;
   }
  }
 }
```

**_Listing A-23._** Implementation of IEnv on Android

```
using EFC_Xamarin.Android;
using System;
using Xamarin.Forms;

[assembly: Dependency(typeof(Env))]
namespace EFC_Xamarin.Android
{
 public class Env : IEnv
 {
  public string GetDbFolder()
  {
   return Environment.GetFolderPath(Environment.SpecialFolder.MyDocuments);
  }
 }
}
```

**_Listing A-24._** Implementation of IEnv on iOS

```
using EFC_Xamarin.iOS;
using System;
using System.IO;
using Xamarin.Forms;

[assembly: Dependency(typeof(Env))]
namespace EFC_Xamarin.iOS
{
 public class Env : IEnv
 {
  public string GetDbFolder()
  {
   return Path.Combine(Environment.GetFolderPath(Environment.SpecialFolder.
   MyDocuments),
```

```
        "..", "Library");
  }
 }
}
```

# Start Code

When the application starts, the database file is created in App.xaml.cs by the method Database.EnsureCreated() if the database file does not yet exist (Listing A-25).

*Listing A-25.*   Extract from the App.xaml.cs File in the Project UI

```
using Xamarin.Forms;

//[assembly: XamlCompilation(XamlCompilationOptions.Compile)]
namespace EFC_Xamarin
{
 public partial class App : Application
 {
  public App()
  {
   InitializeComponent();
   // Create Database if it does not exist
   using (var db = new EFContext())
   {
    db.Database.EnsureCreated();
   }
   MainPage = new EFC_Xamarin.MainPage();
  }

  protected override void OnStart()
  {

  }

  protected override void OnSleep()
  {
```

```
  // Handle when your app sleeps
 }

 protected override void OnResume()
 {
  // Handle when your app resumes
 }
 }
}
```

# Generated Database

The generated database for the Xamarin app corresponds to the database from the UWP case study.

# Data Access Code

The data access code is not separate from the presentation layer in this simple case study. It also deliberately does not use the pattern Model-View-ViewModel (MVVM) to keep the program code manageable for printing in this book.

The data access uses the asynchronous methods of Entity Framework Core to keep the UI responsive.

When creating subtasks and deleting all tasks, two variants are shown (Listing A-26). The commented-out variant is the less efficient one. Thus, deleting all tasks and subtasks without using SQL involves unnecessarily loading all tasks and sending a DELETE command for each task. However, explicitly deleting the subtasks is not necessary in both cases since cascading deletion is active in the standard system.

***Listing A-26.***   Data Access Code in the MainPage.xaml.cs File

```
using Microsoft.EntityFrameworkCore;
using System;
using System.Collections.Generic;
using System.Collections.ObjectModel;
using System.Linq;
using Xamarin.Forms;
```

```
namespace EFC_Xamarin
{
 public partial class MainPage : ContentPage
 {
  private ObservableCollection<Task> _Tasks { get; set; }
  public ObservableCollection<Task> Tasks
  {
   get { return _Tasks; }
   set { _Tasks = value; this.OnPropertyChanged(nameof(Tasks)); }
  }

  private string _Statustext { get; set; }
  public string Statustext
  {
   get { return _Statustext; }
   set { _Statustext = value; this.OnPropertyChanged(nameof(Statustext)); }
  }

  public MainPage()
  {
   this.BindingContext = this;
   InitializeComponent();
   var count = this.LoadTaskSet();
   SetStatus(count + " Datensätze geladen!");
  }

  private async System.Threading.Tasks.Task<int> LoadTaskSet()
  {
   using (var db = new EFContext())
   {
    var list = await db.TaskSet.OrderBy(x => x.Date).ToListAsync();
    Tasks = new ObservableCollection<Task>(list);
    return Tasks.Count;
   }
  }
```

```csharp
private void SetStatus(string text)
{
 string dbstatus;
 using (var db = new EFContext())
 {
  dbstatus = db.TaskSet.Count() + " Tasks with " + db.TaskDetailSet.
  Count() + " Task Details";
 }
 Statustext = text + " / Database Status: " + dbstatus + ")";
}

private async void Add(object sender, EventArgs e)
{
 if (String.IsNullOrEmpty(C_Task.Text)) return;
 // Create new Task
 var t = new Task { Title = C_Task.Text, Date = C_Date.Date };
 var d1 = new TaskDetail() { Text = "Plan" };
 var d2 = new TaskDetail() { Text = "Execute" };
 var d3 = new TaskDetail() { Text = "Run Retrospective" };
 // Alternative 1
 //t.Details.Add(d1);
 //t.Details.Add(d2);
 //t.Details.Add(d3);

 // Alternative 2
 t.Details.AddRange(new List<TaskDetail>() { d1, d2, d3 });

 using (var db = new EFContext())
 {
  db.TaskSet.Add(t);
  // Save now!
  var count = db.SaveChangesAsync();

  SetStatus(count + " records saved!");
  await this.LoadTaskSet();
 }
```

```
  this.C_Task.Text = "";
  this.C_Task.Focus();
 }

 private async void SetDone(object sender, EventArgs e)
 {
  // Get TaskID
  var id = (int)((sender as Button).CommandParameter);
  // Remove record
  using (var db = new EFContext())
  {
   Task t = db.TaskSet.Include(x => x.Details).SingleOrDefault(x =>
   x.TaskID == id);
   if (t == null) return; // not found!
   db.Remove(t);
   int count = await db.SaveChangesAsync();
   SetStatus(count + " records deleted!");
   await this.LoadTaskSet();
  }
 }

 private async void ShowDetails(object sender, EventArgs e)
 {
  // Get TaskID
  var id = (int)((sender as Button).CommandParameter);
  // Get Details
  using (var db = new EFContext())
  {
   string s = "";
   Task t = db.TaskSet.Include(x => x.Details).SingleOrDefault(x =>
   x.TaskID == id);
   s += "Task: " + t.Title + "\n\n";
   s += "Due: " + String.Format("{0:dd.MM.yyyy}",t.Date) + "\n\n";
   foreach (var d in t.Details)
```

```
  {
    s += "- " + d.Text + "\n";
  }
  SetStatus("Details for Task #" + id);
  await this.DisplayAlert("Details for Task #" + id, s, "OK");
 }
}

private async void RemoveAll(object sender, EventArgs e)
{
 // Remove all tasks
 using (var db = new EFContext())
 {
  // Alternative 1: unefficient :-(
  //foreach (var b in db.TaskSet.ToList())
  //{
  // db.Remove(b);
  //}
  //db.SaveChanges();

  // Alternative 2: efficient!
  //db.Database.ExecuteSqlCommand("Delete from TaskDetailSet");
  var count = await db.Database.ExecuteSqlCommandAsync("Delete from
  TaskSet");
  SetStatus(count + " records deleted!");
  Tasks = null;
 }
 }
 }
}
```

Table A-1 shows the main differences between the user interface controls of the app in UWP and the app in Xamarin Forms.

***Table A-1.*** *UWP-XAML vs. Xamarin-Forms XAML*

| UWP | Xamarin Forms |
| --- | --- |
| private async void Page_Loaded(object sender, RoutedEventArgs e) | protected async override void OnAppearing() |
| this.DataContext = this; | this.BindingContext = this; |
| await new MessageDialog(s, "Details for Task #" + id).ShowAsync(); | await this.DisplayAlert("Details for Task #" + id,s,"OK"); |
| this.C_Task.Focus(FocusState.Pointer); | this.C_Task.Focus(); |

# User Interface

Listing A-27 shows the Xamarin Forms–based UI of the Xamarin app.

***Listing A-27.*** MainPage.xaml

```
<?xml version="1.0" encoding="utf-8" ?>
<ContentPage  x:Name="MainPage" xmlns="http://xamarin.com/schemas/2014/forms"
             xmlns:x="http://schemas.microsoft.com/winfx/2009/xaml"
             xmlns:local="clr-namespace:EFC_Xamarin"
             x:Class="EFC_Xamarin.MainPage" WidthRequest="800"
             HeightRequest="500" >
<Grid Margin="0,0,0,0" BackgroundColor="CornflowerBlue">
  <Grid.RowDefinitions>
   <RowDefinition Height="*"></RowDefinition>
   <RowDefinition Height="auto"></RowDefinition>
  </Grid.RowDefinitions>
  <StackLayout Margin="10,10,10,10" Grid.Row="0">
   <!-- ==================== logo -->
   <Image x:Name="Logo" Source="miraclelistlogo.jpg" HeightRequest="100"
   WidthRequest="200" HorizontalOptions="End"   ></Image>
   <!-- ==================== new Task -->
   <Label Text="What do you have to do?" FontSize="20"></Label>

   <StackLayout Orientation="Horizontal">
     <ContentView BackgroundColor="White">  <DatePicker x:Name="C_Date" />
     </ContentView>
```

```
      <Entry BackgroundColor="White" x:Name="C_Task" HorizontalOptions=
      "FillAndExpand" Completed="Add"></Entry>
    </StackLayout>
    <!-- ==================== actions -->
    <StackLayout Orientation="Horizontal">
     <Button Clicked="Add" BackgroundColor="White" Text="Add"></Button>
      <Button Clicked="RemoveAll"  BackgroundColor="White" Text="Remove all"
      Margin="5,0,0,0"></Button>
    </StackLayout>
    <Label Text="Your task list:" FontSize="20"/>
    <!-- ==================== list of tasks -->
    <ListView x:Name="C_Tasks" ItemsSource="{Binding Tasks}">
     <ListView.ItemTemplate>
      <DataTemplate>
       <ViewCell>
        <StackLayout Orientation="Horizontal" >
         <Button BackgroundColor="White" Text="Done"
         x:Name="C_Done"  Clicked="SetDone" Margin="0,0,5,0"
         CommandParameter="{Binding TaskID}" />
         <Button BackgroundColor="White"  CommandParameter="{Binding
         TaskID}" FontAttributes="Bold" Text="{Binding View}" x:Name=
         "C_Details"  Clicked="ShowDetails" />
        </StackLayout>
       </ViewCell>
      </DataTemplate>
     </ListView.ItemTemplate>
    </ListView>
   </StackLayout>
   <!-- ==================== statusbar -->
   <StackLayout BackgroundColor="White" Grid.Row="1">
    <Label Margin="10,0,0,0"  x:Name="C_StatusBar" FontSize="11"
    Text="{Binding StatusText}" />
   </StackLayout>
  </Grid>
</ContentPage>
```

Table A-2 shows the main differences between the app in UWP XAML and the app in Xamarin Forms XAML. You can see that there are many differences that make the migration expensive.

***Table A-2.***  *UWP XAML vs. Xamarin Forms XAML*

| UWP XAML | Xamarin Forms XAML |
|---|---|
| `<Page>` | `<ContentPage>` |
| `<TextBlock>` | `<Label>` |
| `<CalendarDatePicker>` | `<DatePicker>` |
| `<StackPanel>` | `<StackLayout>` |
| `<TextBox>` | `<Entry>` |
| `Name="abc"` | `x:Name="abc"` |
| `Orientation="horizontal" oder Orientation="Horizontal"` | `Orientation="Horizontal"` |
| `Background="white"` | `BackgroundColor="White"` |
| `Click="Add"` | `Clicked="Add"` |
| `<Button Content="Add"> oder <Button>Add</Button>` | `<Button Text="Add">` |
| `FontWeight="Bold"` | `FontAttributes="Bold"` |
| `<TextBox Background="White" Name="C_Task" KeyDown="C_Task_KeyDown" Width="600">` | `<Entry BackgroundColor="White" x:Name="C_Task" Completed="Add" WidthRequest="600" >` |
| `<ListView> <ListView.ItemTemplate> <DataTemplate>` | `<ListView> <ListView.ItemTemplate> <DataTemplate> <ViewCell>` |
| `<Image HorizontalAlignment="Right">` | `<Image HorizontalOptions="End">` |

# Many-to-Many Relationship to Oneself

Geography poses a seemingly simple task: Say a country has borders with any number of other countries. The question is, how do you express this in an object model so that Entity Framework Core makes it an N:M relationship of a table on itself?

The first naive approach would look like this: The `Country` class has a list of "borders" that lead back to `Country` objects. You would establish a relationship between Denmark and Germany like this: `dk.Borders.Add(de)`.

But that does not satisfy the requirements.

- In Entity Framework Core, there is no N:M mapping. Therefore, there must be an explicit `Border` class in the object model, just as in the data model. This can be encapsulated with a helper method; see `AddBorderToCounty(Country c)`.

- A single relationship between two countries is not enough because by creating a relationship such as `dk.Borders.Add(de)`, Denmark now knows that it borders Germany, but Germany knows nothing of its luck to have the Danes as neighbors.

The problem is that a navigation property describes a unidirectional relationship, but you need a bidirectional relationship so that both countries know about the relationship. Therefore, you have to create a navigation property in the class `Country` for both directions of a neighborhood relationship; see `IncomingBorders` and `OutgoingBorders` in the `Country` class as well as `IncomingCountry` and `OutgoingCountry` in the `Border` class. In `OnModelCreating()` in the context class, `WorldContext IncomingCountry` is connected via the Fluent API with `IncomingBorders`, and `OutgoingCountry` is connected with `OutgoingBorders`.

Entity Framework Core is then smart enough to build the counter-relationship when a relationship is established. This is ensured by the internal feature relationship fixup, which is always running when you call `DetectChanges()` or a method that triggers this automatically (for example, `SaveChanges()`).

If you apply the relationship from Denmark to Germany via `OutgoingBorders` from Denmark, Denmark will also appear in `IncomingBorders` from Germany after the next relationship fixup.

Listing A-28 shows the entity classes and the context class. Figure A-17 shows the resulting database in Microsoft SQL Server.

***Listing A-28.*** Solution with Entity Framework Core

```
using Microsoft.EntityFrameworkCore;
using Microsoft.EntityFrameworkCore.Metadata;
using System.Collections.Generic;
```

```csharp
using System.Linq;

class Border
{
 // foreign key for IncomingCountry
 public int Country_Id { get; set; }
 // foreign key for OutgoingCountry
 public int Country_Id1 { get; set; }

 public virtual Country IncomingCountry { get; set; }
 public virtual Country OutgoingCountry { get; set; }
}

class Country
{
 public int Id { get; set; }
 public string Name { get; set; }

 // N-M relationship via Borders
 public virtual ICollection<Border> IncomingBorders { get; set; } = new
List<Border>();
 public virtual ICollection<Border> OutgoingBorders { get; set; } = new
List<Border>();

 public void AddBorderToCounty(Country c)
 {
  var b = new Border() {Country_Id = this.Id, Country_Id1 = c.Id};
  this.OutgoingBorders.Add(b);
 }
}

class WorldContext : DbContext
{
 public DbSet<Country> Countries { get; set; }
 public DbSet<Country> Borders { get; set; }
 protected override void OnConfiguring(DbContextOptionsBuilder optionsBuilder)
 {
```

```csharp
  optionsBuilder.UseSqlServer(@"Server=.;Database=EFC_NMSelf;Trusted_
  Connection=True;MultipleActiveResultSets=True");
}

protected override void OnModelCreating(ModelBuilder modelBuilder)
{
  // Configure primary key
  modelBuilder.Entity<Border>().HasKey(x => new {x.Country_Id, x.Country_Id1});
  // Configure relationships and foreign keys
  modelBuilder.Entity<Border>().HasOne<Country>(x => x.IncomingCountry).
  WithMany(x => x.IncomingBorders).HasForeignKey(x=>x.Country_Id1).
  OnDelete(DeleteBehavior.Restrict);
  modelBuilder.Entity<Border>().HasOne<Country>(x => x.OutgoingCountry).
  WithMany(x => x.OutgoingBorders).HasForeignKey(x => x.Country_Id).
  OnDelete(DeleteBehavior.Restrict); ;
}

/// <summary>
/// Get all neighbors by the union of the two sets
/// </summary>
/// <param name="countryId"></param>
/// <returns></returns>
public IEnumerable<Country> GetNeigbours(int countryId)
{
  var borders1 = this.Countries.Where(x => x.IncomingBorders.Any(y =>
  y.Country_Id == countryId)).ToList();
  var borders2 = this.Countries.Where(x => x.OutgoingBorders.Any(y =>
  y.Country_Id1 == countryId)).ToList();
  var allborders = borders1.Union(borders2).OrderBy(x=>x.Name);
  return allborders;
 }
}
```

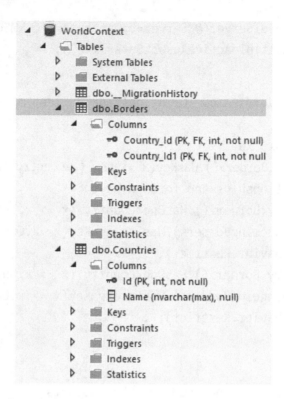

*Figure A-17.* *N:M relationship between country objects via Borders in the data model*

Although the database can now store all necessary relationships, you cannot get just a list of all border relations since these are distributed on two lists (OutgoingBorders). So, you would have to form the union with the LINQ operator Union() and write each time.

```
var borders1 = country.OutgoingBorders;
var borders2 = ctx.Countries.Where(x => x.OutgoingBorders.Any(y => y.Id ==
country.Id)).ToList();
var borders = borders1.Union(borders2).OrderBy(x=>x.Name);;
```

However, the user of the class should not care if the relationship Germany to Denmark or the relationship Denmark to Germany was created.

In the classic Entity Framework, you could elegantly encapsulate this union in the Country class and use lazy loading to load the connected objects. Since there is no lazy loading (so far) in Entity Framework Core and, as you know, the entity class instance does not know the context instance, the encapsulation works only in the context class (see GetNeigbours() in Listing A-29). Figure A-18 shows the output.

***Listing A-29.*** Using the Implementation of Listing A-28

```
using System;
using Microsoft.EntityFrameworkCore;
using System.Linq;

namespace EF_CodeFirst_NMSelf
{
 class Program
 {
  static void Main(string[] args)
  {
   var ctx = new WorldContext();
   ctx.Database.EnsureCreated();

   // Reset example
   ctx.Database.ExecuteSqlCommand("delete from Border");
   ctx.Database.ExecuteSqlCommand("delete from country");

   // Create countries
   var de = new Country();
   de.Name = "Germany";
   ctx.Countries.Add(de);
   ctx.SaveChanges();

   var nl = new Country();
   nl.Name = "Netherlands";
   ctx.Countries.Add(nl);
   nl.AddBorderToCounty(de);
   ctx.SaveChanges();

   var dk = new Country();
   dk.Name = "Denmark";
   ctx.Countries.Add(dk);
   dk.AddBorderToCounty(de);
   ctx.SaveChanges();

   var be = new Country();
   be.Name = "Belgium";
```

```
ctx.Countries.Add(be);
be.AddBorderToCounty(de);
be.AddBorderToCounty(nl);
ctx.SaveChanges();

var fr = new Country();
fr.Name = "France";
ctx.Countries.Add(fr);
fr.AddBorderToCounty(de);
ctx.SaveChanges();

var cz = new Country();
cz.Name = "Czech Republic";
ctx.Countries.Add(cz);
cz.AddBorderToCounty(de);
ctx.SaveChanges();

var lu = new Country();
lu.Name = "Luxembourg";
ctx.Countries.Add(lu);
lu.AddBorderToCounty(de);
lu.AddBorderToCounty(fr);
lu.AddBorderToCounty(be);
ctx.SaveChanges();

var pl = new Country();
pl.Name = "Poland";
ctx.Countries.Add(pl);
pl.AddBorderToCounty(de);
pl.AddBorderToCounty(cz);
ctx.SaveChanges();

var at = new Country();
at.Name = "Austria";
ctx.Countries.Add(at);
at.AddBorderToCounty(de);
at.AddBorderToCounty(cz);
ctx.SaveChanges();
```

```
var ch = new Country();
ch.Name = "Switzerland";
ctx.Countries.Add(ch);
ch.AddBorderToCounty(de);
ch.AddBorderToCounty(fr);
ch.AddBorderToCounty(at);
ctx.SaveChanges();

Console.WriteLine("All countries with their borders");
foreach (var country in ctx.Countries)
{
 Console.WriteLine("--------- " + country.Name);

 // now explicitly load the neighboring countries, as Lazy Loading in
 EFC does not work
 //var borders1 = ctx.Countries.Where(x => x.IncomingBorders.Any(y =>
 y.Country_Id == country.Id)).ToList();
 //var borders2 = ctx.Countries.Where(x => x.OutgoingBorders.Any(y =>
 y.Country_Id1 == country.Id)).ToList();
 //var allborders = borders1.Union(borders2);

 // better: encapsulated in the context class:
 var allborders = ctx.GetNeigbours(country.Id);

 foreach (var neighbour in allborders)
 {
  Console.WriteLine(neighbour.Name);
 }
}

 Console.WriteLine("=== DONE!");
 Console.ReadLine();
 }
 }
}
```

```
H:\TFS\Demos\EF\EFC_CodeFirst_NMSelf\EF_CodeFirst_NMSelf\bin\Debug\EF_CodeFirst_NMSelf.exe

All countries with their borders
--------- Germany
Austria
Belgium
Czech Republic
Denmark
France
Luxembourg
Netherlands
Poland
Switzerland
--------- Netherlands
Belgium
Germany
--------- Denmark
Germany
--------- Belgium
Germany
Luxembourg
Netherlands
--------- France
Germany
Luxembourg
Switzerland
--------- Czech Republic
Austria
Germany
Poland
--------- Luxembourg
Belgium
France
Germany
--------- Poland
Czech Republic
Germany
--------- Austria
Czech Republic
Germany
Switzerland
--------- Switzerland
Austria
France
Germany
=== DONE!
```

***Figure A-18.***  *Output of Listing A-29*

# APPENDIX B

# Internet Resources

You can find the source code for Entity Framework Core on GitHub at `https://github.com/aspnet/EntityFramework`.

You can download the examples for this book from GitHub. Please go to `www.EFCore.net` for details.

The official but not very detailed documentation on Entity Framework Core can be found on the website: `https://docs.microsoft.com/en-us/ef/core/`.

The Entity Framework Core development team used to run a blog at `https://blogs.msdn.microsoft.com/adonet`, but the current development team uses Microsoft's main .NET blog: `https://blogs.msdn.microsoft.com/dotnet/tag/entity-framework`.

On Twitter, the development team announces innovations with the account @efmagicunicorns. The unicorn is the mascot of the development team; see `https://twitter.com/efmagicunicorns` (Figure B-1).

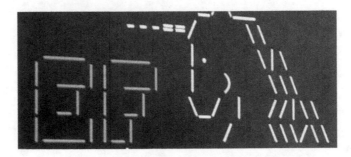

***Figure B-1.*** *The unicorn mascot as ASCII graphic in the tool ef.exe*

631

# New Features in Entity Framework Core 2.1

You will find this document on the GitHub page, as release 2.1 was in early preview while writing this book. Please go to `www.EFCore.net` for additional details.

© Holger Schwichtenberg 2018
H. Schwichtenberg, *Modern Data Access with Entity Framework Core*,
https://doi.org/10.1007/978-1-4842-3552-2_23

# Index

## A

Aggregate functions, 172

Alternative key
HasAlternateKey(), 348
Microsoft SQL Server Management
Studio, 348, 350
Project EFC_MappingTest, 350
unique index *vs.*, 347

Analysis functions, 493–494

ASP.NET Core application, 14
assemblies, 39
dependency injection
ApplicationDbContext, 589
ApplicationUser class, 588
database, local users, 588
DbContext pooling, 592
FlightManager class, 590–591
Individual User Accounts, 587
Startup.cs, 589
WWWingsController class, 591
MiracleList (*see* MiracleList)

Asynchronous extension methods
ForEachAsync(), 291
SaveChangesAsync(), 289–291
ToListAsync(), 288–289

Attach() method, 396–400

AutoMapper
BeforeMap() and AfterMap(), 539
class-based type converter, 527

conventions, 514–516
DTOs, 501, 503–504
ForMember() method, 522
generic class
lists, 534–535
parameters, 537–538
for partnership, 535
Ignore() method, 522
inheritance
class hierarchy for, 529
DTO class, 529
Include() methods, 531
manual mapping
configuration, 531
Initialize(), 511
Map() method, 512–513
method-based type
converter, 526–527
.NET variants, 507
nonstatic API, 513
NuGet package, 507
profile class, 521
with reflection and explicit
mapping, 541–542
ReverseMap() method, 512
SeparatorCharacter, 519–520
SplittingExpression, 519
string, 525
value resolver class, 525

635

© Holger Schwichtenberg 2018
H. Schwichtenberg, *Modern Data Access with Entity Framework Core*,
https://doi.org/10.1007/978-1-4842-3552-2

# G, H

# I, J, K

# L

# M